Glass Science

WILEY SERIES ON THE SCIENCE AND TECHNOLOGY OF MATERIALS

Advisory Editors: **J. E. Burke, B. Chalmers, James A. Krumhansl**

THERMODYNAMICS OF SOLIDS, SECOND EDITION
Richard A. Swalin
GLASS SCIENCE
Robert H. Doremus
THE SUPERALLOYS
Chester T. Sims and William C. Hagel, editors
X-RAY DIFFRACTION METHODS IN POLYMER SCIENCE
L. E. Alexander
PHYSICAL PROPERTIES OF MOLECULAR CRYSTALS, LIQUIDS, AND GLASSES
A. Bondi
FRACTURE OF STRUCTURAL MATERIALS
A. S. Tetelman and A. J. McEvily, Jr.
ORGANIC SEMICONDUCTORS
F. Gutmann and L. E. Lyons
INTERMETALLIC COMPOUNDS
J. H. Westbrook, editor
THE PHYSICAL PRINCIPLES OF MAGNETISM
Allan H. Morrish
HANDBOOK OF ELECTRON BEAM WELDING
R. Bakish and S. S. White
PHYSICS OF MAGNETISM
Sōshin Chikazumi
PHYSICS OF III-V COMPOUNDS
Otfried Madelung (translation by D. Meyerhofer)
PRINCIPLES OF SOLIDIFICATION
Bruce Chalmers
THE MECHANICAL PROPERTIES OF MATTER
A. H. Cottrell
THE ART AND SCIENCE OF GROWING CRYSTALS
J. J. Gilman, editor
SELECTED VALUES OF THERMODYNAMIC PROPERTIES OF METALS AND ALLOYS
Ralph Hultgren, Raymond L. Orr, Philip D. Anderson and Kenneth K. Kelly
PROCESSES OF CREEP AND FATIGUE IN METALS
A. J. Kennedy
COLUMBIUM AND TANTALUM
Frank T. Sisco and Edward Epremian, editors
TRANSMISSION ELECTRON MICROSCOPY OF METALS
Gareth Thomas
PLASTICITY AND CREEP OF METALS
J. D. Lubahn and R. P. Felgar
INTRODUCTION TO CERAMICS
W. D. Kingery
PHYSICAL METALLURGY
Bruce Chalmers
ZONE MELTING, SECOND EDITION
William G. Pfann

Glass Science

ROBERT H. DOREMUS

Rensselaer Polytechnic Institute

A WILEY-INTERSCIENCE PUBLICATION

JOHN WILEY & SONS, New York · London · Sydney · Toronto

Library of Congress Cataloging in Publication Data

Doremus, R. H.
Glass science.

(Wiley series on the science and technology of materials)
"A Wiley-Interscience publication."
Includes bibliographical references.
1. Glass. 2. Glass manufacture. I. Title.

TP857.D67 666'.1 73-2730
ISBN 0-471-21900-2

Printed in the United States of America
10 9 8 7 6 5 4 3 2 1

Preface

This book was written for three different audiences. First, it was designed as a text and reference source for advanced undergraduate and graduate courses in glass science. In an increasing number of universities a course in the science and technology of glasses is offered to students of materials science and engineering and related disciplines, but there is no suitable text for these courses. To satisfy this audience various subject areas of glass science are introduced at an elementary level, with a minimum of mathematical complications. An introductory course in glass science at Rensselaer for seniors and graduate students covers about half the material in this book (not the first half, but a selection of material). Thus the whole book could serve as a text for a two-semester course meeting three times a week.

A second audience includes those persons who are involved in manufacture and use of glass in industry. These persons have diverse backgrounds and often need specific information about glass properties and the mechanism behind them. Many references are included so that those who desire fuller treatments can find them in more detailed reviews and original papers.

In addition I have written this book as an aid and stimulus for my colleagues studying the science of glass. In several instances I have explained phenomena in new ways, ignoring the traditional approaches. In other cases I have offered tentative explanations for experimental results even though the explanations are not completely established by the data. I hope these speculations will lead to critical work and more satisfying insights into the science of glass.

When citing references I have tried to include the most important papers, with emphasis on recent work; I offer my apologies to those whose work has not been given just recognition. Students using the book as a text may be annoyed at the brief mention of much work without full explanation of it, and with the many reference citations in the text. I hope they will be indulgent for the other audiences for whom the collection of these references in a single volume will, I hope, be of assistance.

I wish to acknowledge my debt to the institutions in which I have worked during the writing of this book, the General Electric Research Laboratory and Rensselaer Polytechnic Institute, for providing a congenial atmosphere and many helpful services. The stimulation, criticism, and encouragement of Dr. J. E. Burke of General Electric has been especially valuable to me in building up the background for and actually carrying out the task of writing this book. Mrs. Gene Marsh and Miss Lorraine Drago of General Electric helped with the preparation of the early chapters, and Mrs. Mary Ishkanian of Rensselaer has shown great patience and skill in preparing the final manuscript in spite of many changes and obscurities.

I welcome criticisms, suggestions, and corrections from those who wish to send them to me, and hope that this unified treatment of glass science, in spite of omissions and inadequacies, will be of value to the many different groups using and trying to understand glass.

ROBERT H. DOREMUS

Troy, New York
January 1973

Contents

1	Introduction	1
Part One	Formation and Structure of Glasses	
2	Glass Formation	11
3	Molecular Structure	23
4	Phase Separation	44
5	Crystallization	74
Part Two	Transport	
6	Viscosity	101
7	Glass Transition	115
8	Molecular Solution and Diffusion in Glass	121
9	Electrical Conductivity and Ionic Diffusion	146
10	Electronic Conduction	177
11	Dielectric and Mechanical Loss	190
Part Three	Chemical and Surface Properties	
12	Surface Properties	213
13	Chemical Reactions	229
14	Ion Exchange and Potentials of Glass Electrodes	253

Part Four Strength

15 Fracture of Glass 281

16 Static Fatigue 296

17 Strengthening of Glass 310

Part Five Optical Properties

18 Optical Absorption in Glasses 319

Author Index 333

Subject Index 345

Glass Science

1

Introduction

Glass is an amorphous solid. A material is amorphous when it has no long-range order, that is, when there is no regularity in the arrangement of its molecular constituents on a scale larger than a few times the size of these groups. For example, the average distance between silicon atoms in vitreous silica (SiO_2) is about 3.6 Å, and there is no order between these atoms at distances above about 10 Å. A solid is a rigid material; it does not flow when it is subjected to moderate forces. More quantitatively, a solid can be defined as a material with a viscosity of more than about 10^{15} P (poises).

Many glass technologists object to the above definition of a glass. These workers prepare a glass by cooling a liquid in such a way that it does not crystallize, and feel that this process is an essential characteristic of a glass. Many earlier writers insist on this criterion: "A glass ... is a material, formed by cooling from the normal liquid state, which has shown no discontinuous change ... at any temperature, but has become more or less rigid through a progressive increase in its viscosity," according to Jones (Ref. 1, p. 1), or, more succinctly, "glass is an inorganic product of fusion which has been cooled to a rigid condition without crystallization" as taken from the *ASTM Standards for Glass.* The difficulty with this view is that glasses can be prepared without cooling from the liquid state. Glass coatings are deposited from the vapor or liquid solution, sometimes with chemical reactions. Thus sodium silicate glass can be made by evaporating an aqueous solution of sodium silicate (water glass) and baking the deposit to remove water. The product of this process is indistinguishable from sodium silicate glass of the same composition made by cooling from the liquid. It seems wise to use the same name for materials with the same molecular structure and properties no matter how they are made; consequently, the broader definition given here is preferred. Hutchins and Harrington[2] and Prins[2a] use a similar definition.

Traditional glasses have been made of inorganic materials such as silica sand, sodium and calcium carbonates, feldspars, borates, and phosphates that react to form metallic oxides in the final glass. In recent years "glassy" polymers have been widely used, and they are often called glasses in the technical literature, but there is little use of this term in everyday language. The ASTM definition of glass given above limits it to inorganic constituents, whereas Jones' definition does not. The present definition includes newer and more exotic materials such as "splat-cooled" metallic glasses, the electronically conducting glasses of the arsenic, sulfur, selenium, and tellurium halide families, and glasses of ionic salts and aqueous solutions, as well as oxide glasses and glassy organic polymers.

The word "glass" is derived from an Indo-European root meaning "shiny," which has also given us the words glare, glow, and glaze. The word "vitreous" comes from the Latin word for glass. No distinction is made here between the words glassy and vitreous.

The structure and properties of silicate glasses are emphasized in this book because these glasses are the most important commercially and have been studied more extensively than other types. In Chapters 2 and 3 an attempt is made to include information on formation and structure of all types of glasses known at the time of writing. In the succeeding chapters data on nonoxide glasses are included when this information is available and helps in understanding, or is of particular interest, such as the electrical properties of the semiconducting chalcogenide glasses.

HISTORY OF GLASS SCIENCE

The history of glass is summarized in articles in the *Encyclopedia Britannica* and by Morey (Ref. 3, pp. 1 ff.). Natural glasses have been used by man from the earliest times of which there is archeological evidence. Glass and glazes were manufactured far back in human history. According to Sir W. M. Flinders Petrie, as quoted by Morey (Ref. 3, p. 4), the earliest known glaze dates from about 12,000 B.C. and the earliest pure glass from about 7000 B.C.; both were found in Egypt, and were probably brought there from Asia. At first glass was used only for decorative objects, but later it was molded or pressed into vessels. The invention of glass blowing in about the first century B.C. greatly increased the use of glass for practical purposes in Roman times, mainly for vessels but later for windows. In the West glass manufacturing was dispersed to isolated sites after the fall of the Roman Empire, but was continued in Byzantium and later in the Middle East by the Arabs. Venice became the center of a resurgent glass industry in the West from about 1300 on. The art of glass making was summarized in 1612 by Neri in *L'Arte Vetraria*. Progress in techniques of glass manufacture and in the application of glass was subsequently rapid, in parallel with many other areas of technology. Until the twentieth century most of these advances were made empirically, using common sense to guide experimentation. The application of basic scientific understanding to the improvement of manufacture and to new applications of glass has occurred only in the last few decades.

Among the first to study glass in a more basic way was Michael Faraday. He described glass "rather as a solution of different substances one in another than as a strong chemical compound,"[4] which can still stand today as a characterization of a multicomponent glass. Faraday studied the electrolysis and conductivity of melts of various glasses,[5] and found that some "decomposed" under the effect of a field and others did not. Apparently the glasses with higher conductivity or with easily reduced components were the ones that decomposed. Faraday was also the first to conclude correctly that the red color of a gold ruby glass was caused by very small gold particles in the glass,[6] as described in Chapter 18. Also in the midnineteenth century Buff studied the conductivity and electrolysis of glass,[7] and later Warburg and Tegetmeier[8,9] showed that the electrolysis of glass followed Faraday's law. About this same time Tammann initiated work on the viscosity of glass, the glass transition, the relationship of crystallization rate and viscosity, and the reasons for glass formation.[10] Schulz[11] made a detailed study of exchange and diffusion of silver ions for sodium ions in a commercial glass that was unsurpassed until the last few years.

The understanding of glass strength received its greatest impetus from the theory of Griffith,[12] in which the fracture of a brittle solid results from the propagation of a surface flaw. At first this theory was considered by some to be a mathematical curiosity without application to real materials, but more recently it has provided the basis for much work and improved understanding of the strength of glass, as discussed in Chapter 15.

The Department of Glass Technology at Sheffield, England, under Professor W. E. S. Turner, was very active during the 1920s in measuring such properties as density, electric conductivity, chemical durability, viscosity, and thermal expansion of a wide variety of commercial and laboratory glasses.

In the 1930s the understanding of the reasons why certain molecules are glass formers, and of the structure of glass, was enlarged by the papers of Zachariasen[13] and Warren.[14] Since World War II the activity in glass science has grown sharply, along with all science and technology. The 1950s might be characterized as the "golden age" of metallurgical science. At that time physics and chemistry were intensively applied to the understanding of metallic behavior that had been found empirically and was previously understood only in terms of macroscopic phenomena. In a similar way the

1960s could perhaps be described as a "golden age" of glass science because of the profitable application during this period of the basic sciences to understanding glass in terms of its structure and composition. It is curious that glass, which was in common use long before many other metals, polymers, colloids, solutions, solvents, and so on, should be one of the last to yield some of its mysteries to scientific exploration. Perhaps the large variation in compositions and properties in glasses, coupled with the lack of long-range structure in a rigid material, has deterred systematic and broad characterization of glasses.

Some of the areas of glass science that have been most active in the last few years are phase separation, uniform crystallization, ionic and gaseous diffusion, electronic conductivity, internal friction, surface reactions and structure, ion exchange, glass electrodes, and static fatigue.

This brief survey of the history of glass science is not intended to be exhaustive, and many deserving names and studies have undoubtedly been omitted. It is to be hoped that the next decades will see a continued expansion of knowledge about glass and its useful application.

USES OF GLASS AND IMPORTANT COMMERCIAL COMPOSITIONS

The original use of glass was for decorative purposes. Next glass was used for containers, and this use is still the most important today. The production of flat glass, chiefly for windows in buildings and vehicles, is now the second largest item of glass manufacture. Lamp envelopes are another major area of use. There are many special applications of glass, some in small quantity, most of which have been developed over the past few decades. Some of these special applications are glass ceramics and surface-strengthened glass for higher strength, fiber glass, mirror blanks of fused silica, glass electrodes for alkali ions, and glass lasers. A good summary of uses of glass is given by Hutchins and Harrington.[2]

The most important commercial glasses are based on sodium calcium silicates. These glasses are cheap and weather well; they are easily melted and formed. Minor additions to the base compositions are made to improve certain properties: alumina for improved weathering and less devitrification (crystallization), borates for easier working and lower thermal expansion, and arsenic or antimony oxide for fining (removal of bubbles). Another important class of compositions is the borosilicates. They have lower thermal expansion and thus better thermal shock resistance, as well as improved chemical durability, for such applications as automobile headlamps, cookingware, and laboratory apparatus. Aluminosilicate glasses are used for chemical durability, resistance to devitrification, higher temperatures, and greater strength in cookingware, glass ceramics, fiber glass, and seals. Lead glasses are used for their high refractive index, easier working, and greater density as lamp envelopes, seals, optical glass, and "crystal" glass for art and tableware. Fused silica is especially valuable for its high-temperature stability, low thermal expansion and consequent high thermal-shock resistance, excellent chemical durability, purity, and good optical transmission in the ultraviolet. It is used for lamp envelopes, crucibles for melting silicon and germanium, optical components, and many special applications requiring its unique properties.

A vast number of other compositions, including nonsilicate glasses, has been made for all sorts of applications. The infinite variability of glass compositions leads to a great variety of possible properties and consequent uses. Only a beginning has been made in exploiting this advantage of glass.

A table giving the compositions and properties of some of the most important commercial glasses is included in Chapter 6 on viscosity. A table of impurity concentrations and methods of manufacture of fused silicas is given in Chapter 18 on optical absorption.

BOOKS AND REVIEWS

One of the most useful books on glass is *Properties of Glass* by Morey.[3] This book has chapters on all the important properties of glass. Each chapter contains background information on the property as well as an extensive numerical graphical compilation of the property for glasses of different compositions. Volf's book[15] includes discussions of compositions and properties of the most important commercial glasses. Jones' monograph[1] emphasizes the glass transition. Shand's book[16] contains sections on manufacture and uses of glass, and on fiber glass. The *Handbook of Glass Manufacture* describes equipment, materials, and methods of glass making.[17] *The Physical Chemistry of the Silicates* by Eitel[18] is a comprehensive summary of the literature in this field up to 1952. The five volumes of *Silicate Science*, by Eitel, cover the same area from 1952 to 1962, Volume II being devoted to glasses, enamels, and slags.[19] Two interpretive books on glass properties are *The Constitution of Glasses*, in three volumes, by Weyl and Marboe,[20] and *The Structure and Mechanical Properties of Inorganic Glasses* by Bartenev.[21] Rawson's book gives information on the formation and structure of a wide variety of glasses,[22] and *Glas* by Scholze summarizes glass properties in German.[23] The properties of vitreous silica are discussed in Sosman's book,[24] which is still valuable although it was published in 1927, and two review articles on silica have been published.[25,26] Other more specialized books are referred to in the relevant chapters.

The *Encyclopedia of Chemical Technology* contains a useful review article on glass.[2] The series of volumes entitled *Modern Aspects of the Vitreous*

State, edited by J. D. Mackenzie,[27] includes review articles on many different aspects of glass science. The summaries of recent research by N. J. Kreidl in *Glass Industry* are very valuable for keeping up with new developments.[28] A volume entitled *Amorphous Materials* contains the papers presented at the Third Conference on the Physics of Noncrystalline Solids in 1970.[29] Other reviews and articles are discussed in the appropriate chapters.

OUTLINE OF THE BOOK

The following chapters describe the state of glass science in 1972. They are grouped into sections on structure, transport, chemical properties, strength, and optical properties, each of which includes introductory remarks. These sections are a general way of separating the chapters, and there is much overlapping. For example, properties are related to structure and can often be used as tools to deduce structural information. Areas of work that have been active recently are emphasized, whereas those that have not seen much progress in the last two decades or so are more briefly considered. In the chapters on properties, such as viscosity, electrical conductivity, chemical durability, and dielectric loss, it is assumed that Morey's book[3] is available, and therefore no attempt is made to reproduce the extensive tabulations of properties given in it. In each chapter the commercial applications that have resulted from or stimulated the investigations described are discussed. A brief description of experimental methods is included whenever they are not standard or cannot be referred to easily.

REFERENCES

1. G. O. Jones, *Glass*, Methuen, London, 1956.
2. J. R. Hutchins and R. V. Harrington, in *Encyclopedia of Chemical Technology*, Vol. 1, 2nd ed., Wiley, New York, 1966, p. 533.

2a. J. A. Prins, in *Proc. Int. Conf. on Physics of Noncrystalline Solids*, J. A. Prins, Ed., North Holland, Amsterdam, 1965, p. 1.

3. G. W. Morey, *The Properties of Glass*, 2nd ed., Reinhold, New York, 1954.
4. M. Faraday, *Phil. Trans. Roy. Soc.*, 49 (1830).
5. M. Faraday, *Experimental Researches in Electricity*, reprinted by J. M. Dent, London, 1914, pp. 38 and 115.
6. M. Faraday, *Phil. Mag.*, **14**, 401, 512 (1857).
7. H. Buff, *Liepzig Ann.*, **90**, 257 (1854).
8. E. Warburg, *Ann. Phys.*, **21**, 622 (1884).
9. E. Warburg and F. Tegetmeier, *Ann. Phys.*, **35**, 455 (1888).
10. G. Tammann, *Der Glaszustand*, Voss, Leipzig, 1933.
11. G. Schulze, *Ann. Phys. Liepzig*, **40**, 335 (1913).
12. A. A. Griffith, *Phil. Trans. Roy. Soc.*, **221A**, 163 (1921).

13. W. H. Zachariasen, *J. Am. Chem. Soc.*, **54**, 3841 (1932).
14. B. E. Warren and co-workers, *J. Am. Ceram. Soc.*, **17**, 249 (1934); **18**, 239 (1935); **19**, 202 (1936); **21**, 287 (1938).
15. W. B. Volf, *Technical Glasses*, Pitman, London, 1961.
16. E. B. Shand, *Glass Engineering Handbook*, McGraw-Hill, New York, 1958.
17. F. V. Tooley, Ed., *Handbook of Glass Manufacture*, Vols. I and II, Ogden, New York, 1961.
18. W. Eitel, *The Physical Chemistry of the Silicates*, University of Chicago Press, Chicago, 1954.
19. W. Eitel, *Silicate Science*, Academic, New York, 1969.
20. W. A. Weyl and E. C. Marboe, *The Constitution of Glasses*, Vols. I and II, Wiley, New York, 1964–1967.
21. G. M. Bartenev, *The Structure and Mechanical Properties of Inorganic Glasses*, Wolters-Neordheff, Groningen, The Netherlands, 1970.
22. H. Rawson, *Inorganic Glass-Forming Systems*, Academic, London, 1967.
23. H. Scholze, *Glas*, Vieweg, Braunschweig, Germany, 1965.
24. R. B. Sosman, *The Properties of Silica*, Chemical Catalog, New York, 1927.
25. W. H. Dumbaugh and P. C. Schultz, see Ref. 2, p. 73.
26. R. Bruckner, *J. Noncryst. Solids*, **5**, 123 (1970).
27. J. D. Mackenzie, Ed., *Modern Aspects of the Vitreous State*, Vols. I to III, Butterworths, London, 1960–1964.
28. N. J. Kriedl, in *Glass Ind.* (1969–1971).
29. R. W. Douglas and B. Ellis, *Amorphous Materials*, Wiley, London, 1972.

Part One

FORMATION AND STRUCTURE OF GLASSES

In Chapter 2 the properties of materials that determine whether or not they can be made into glasses, and the conditions for glass formation, are discussed. This chapter includes lists of materials that have been made into glasses.

Molecular arrangements in glasses are discussed in Chapter 3. An attempt is made to include as many different types of glasses listed in Chapter 2 as possible.

Structural regularities on a larger scale than molecular, or microstructure, are considered in Chapters 4 and 5 on the formation of two liquid phases and crystallization in glasses. The microstructure of glasses has been studied intensively in the last 15 years because it strongly affects properties and also because new materials with valuable characteristics have been developed by crystallizing glasses uniformly.

2

Glass Formation

What materials can be glasses? In principle any substance can be made into a glass by cooling it from the liquid state fast enough to prevent crystallization. The final temperature must be so low that the molecules move too slowly to rearrange to the more stable crystalline form. Alternatively one can imagine building up the material onto a substrate, either by deposition or chemical reaction from the vapor, the substrate being so cold that rearrangement cannot take place. In actual practice glass formation has been achieved with a relatively limited number of substances. After substances that have been made into glasses are listed, the question of why so few substances are amenable to glass formation is considered in subsequent sections.

Materials that have been made into glasses are listed in Tables 1 and 2. Undoubtedly these lists will be extended as new methods of more rapidly cooling melts and of deposition from vapor or solution are devised. Thus the lists show those materials most easily formed into glasses and are representative rather than exhaustive. Some of the glasses in the tables may contain fine crystals of size 100 Å or less. It is often difficult to be certain that small crystalline regions are absent, even with x-ray or electron diffraction measurements. A material containing crystallites of size from 20 to 100 Å can be mistaken for a glass even though it does not strictly fit the definition of a glass given in Chapter 1. If the existence of amorphous water[27,28] is confirmed, condensation from the vapor should be a method of glass formation for many common liquids.

The materials in Table 1 can be placed in several categories. The oxides are by far the most important commercially, as mentioned in the last chapter. Multicomponent oxide glass results from the mixing of other oxides with the main "glass formers" SiO_2, B_2O_3, GeO_2, and P_2O_5. The "conditional" glass-forming oxides in Table 1 do not form glasses alone, but can do so in binary or multicomponent mixtures with other oxides.

Table 1 Glasses Formed by Cooling from the Liquid

	Refs.
Elements	
S, Se	
Te(?)	1,2
P	1,3
Oxides	
B_2O_3, SiO_2, GeO_2, P_2O_5, As_2O_3, Sb_2O_3	
In_2O_3, Tl_2O_3, SnO_2, PbO_2, SeO_2	4
"Conditional" TeO_2, SeO_2, MoO_3, WO_3, Bi_2O_3, Al_2O_3, Ba_2O_3, V_2O_5, SO_3	1,5
Sulfides	
As_2S_3, Sb_2S_3	6
various compounds of B, Ga, In, Te, Ge, Sn, N, P, Bi	4,7
CS_2	8,9
Selenides	
various compounds of Tl, Sn, Pb, As, Sb, Bi, Si, P	4,10,11
Tellurides	
various compounds of Tl, Sn, Pb, As, Sb, Bi, Ge	1,4
Halides	
BeF_2, AlF_3, $ZnCl_2$, Ag(Cl, Br, I), $Pb(Cl_2, Br_2, I_2)$, and multicomponent mixtures	1,4,10
Nitrates	
KNO_3—$Ca(NO_3)_2$ and many other binary mixtures containing alkali and alkaline earths nitrates	14,15
Sulfates	
$KHSO_4$ and other binary and ternary mixtures	16
Carbonates	
K_2CO_3—$MgCO_3$	17,18
Simple organic compounds	
O-Terphenyl, toluene, 3-methyl hexane, 2,3-dimethyl ketone, diethyl ether, isobutyl bromide, ethylene glycol, methyl alcohol, ethyl alcohol, glycerol, glucose	8,9,13
As droplets only: m-xylene, cyclopentane, *n*-heptane, methylene chloride	8,9
Polymeric organic compounds	
Example—polyethylene $(—CH_2—)_n$, and many others	19
Aqueous solutions	
Acids, bases, chlorides, nitrates, and others	20,21
Metallic alloys by "Splat Cooling"	
Au_4Si, Pd_4Si	21a
Te_x—Cu_{25}—Au_5	22

Table 2 Glasses Formed by Deposition or Reaction from the Vapor, and not included in Table 2.1

	Refs.
Elements	
Boron	24
Silicon, Germanium	23
Bismuth, Gallium	32
Oxides	
Aluminum(?)	26
Tantalum	25
Niobium	25
Water(?)	27,28
Other compounds	
Silicon carbide	23
Indium antimonide	29
Various combinations of silicon, or germanium, with oxygen, sulfur, selenium, and tellurium	29
Magnesium with antimony or bismuth	30
Nickel, cobalt, or iron with phosphorous or sulfur	31

A second category is the chalcogenide glasses based on sulfur, selenium, or tellurium. In chemistry the term chalcogenide refers to compounds containing oxygen, sulfur, selenium, or tellurium; thus its use to describe glasses of these elements, excluding oxides, is unfortunate. Many of these glasses of commercial interest include halide elements. The root chalco- comes from a Greek word for copper, and the chalcogenide elements have traditionally been those that form strong compounds with copper. These glasses are primarily interesting because some are electronic conductors (semiconductors). They also soften at relatively low temperatures and transmit infrared radiation well.

Another category is the ionic glasses—halides, nitrates, sulfates, and carbonates. The latter three are quite similar because they form glasses only as binary or multicomponent mixtures. Angell has reviewed work on these glass formers.[33]

Other categories are the simple organic compounds, the organic polymers, the aqueous solutions, and metallic alloys.

CRYSTALLIZATION AND GLASS FORMATION

Turnbull[34] has discussed the question of whether a glass is ever the most stable state of a solid below its melting point. He concludes that there is no rigorous proof that the most stable state of a substance at low temperature is crystalline rather than glassy, but that with the exception of helium it is found experimentally that the most stable forms of pure substances are crystalline. Furthermore the viscosities of liquids at their freezing points are invariably much lower than the 10^{15} P characterizing a rigid glass, the highest being silica with a viscosity of 10^{7} P at the melting point of cristobalite.

Therefore, if a glass is formed from a liquid, it must be cooled below its melting point so fast that the supercooled liquid does not crystallize. Thus the rate of crystallization of a material may control whether it can form a glass. As an example, the velocity of crystallization of cristobalite from fused silica is shown as a function of temperature in Fig. 1, from the work of

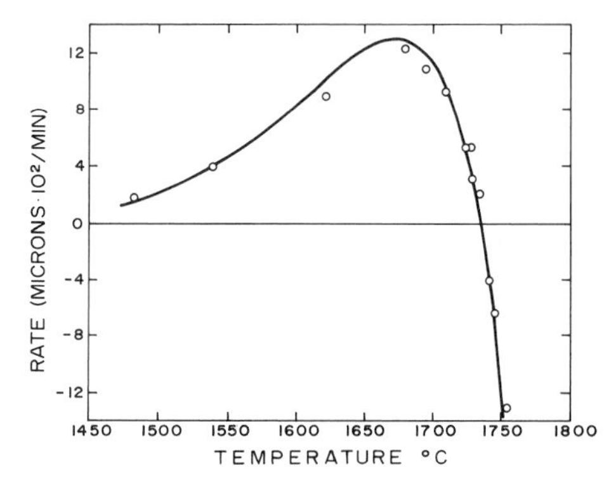

Fig. 1 The rate of crystallization of cristobalite from fused silica as a function of temperature.[35]

Wagstaff.[35] As the silica is cooled below the melting point of cristobalite, the rate of crystallization rises to a maximum value and then decreases. At much lower temperatures (below about 900°C for pure vitreous silica) the rate of crystallization is so slow that even for long holding times no appreciable amount of crystalline material is formed. Thus, if the silica is cooled rapidly enough from above the melting point to below 900°C, no appreciable amount of crystalline cristobalite is formed, and the silica becomes a stable glass.

Maximum crystallization velocities for a number of glass-forming materials are given in Table 3. The rates of crystallization are lowest for substances easily cooled to glasses, such as silica and germania, whereas they are higher for alkali and alkaline earth compounds, which must be cooled more rapidly to avoid crystallization. From the table one might guess that an upper limit for crystallization rate of a material easily formed as a glass by cooling the liquid would be about 10^{-4} cm/sec. With this rate it is necessary to cool the

Table 3 Crystallization Velocities and Viscosities of Glass-Forming Liquids

Material	Melting Point (°C)	Max cryst. velocity (cm/sec)	Refs. for velocity	Temp. of max. v (°C)	Log viscosity at m.p. (P)	Refs. for viscosity
Vitreous silica, SiO_2	1734	$2.2(10)^{-7}$	35	1674	7.36	36
Vitreous germania, GeO_2	1116	$4.2(10)^{-6}$	37	1020	5.5	38
Phosphorous pentoxide, P_2O_5	580	$1.5(10)^{-7}$	39	561	6.7	40
Sodium disilicate, $Na_2O \cdot 2SiO_2$	878	$1.5(10)^{-4}$	41,41a,42	762	3.8	42
Potassium disilicate, $K_2O \cdot 2SiO_2$	1040	$3.6(10)^{-4}$	41	930		
Barium diborate, $BaO \cdot 2B_2O_3$	910	$4.3(10)^{-3}$	43	849	1.7	44
Lead diborate, $PbO \cdot 2B_2O_5$	774	$1.9(10)^{-4}$	45	705	1.0	45
1,2-Diphenylbenzene	55.5	$2.5(10)^{-3}$	46	38	−0.46	46
1,3,5-Tri-α-naphthylbenzene	197	$9.3(10)^{-5}$	47	175	−0.34	47
Glycerol	18.3	$1.8(10)^{-4}$	48	−6.7	1.0	
Salol	43	$6.4(10)^{-3}$	49	20		
Polyethylene adipate	55	$9.5(10)^{-6}$	51	26		
Toluene	−95	>0.5	58			
Ethyl ether	−116	>0.13	58			
Methyl alcohol	−98	>0.09	58			

glass through the temperature range of maximum crystallization rate in less than 10^{-2} sec to prevent growth of crystals larger than 100 Å in diameter. This time seems too short for most practical conditions, so there must be another reason for the inhibition of crystallization for many materials, as described in the next section. For glasses such as silica the rate of crystallization alone is slow enough to allow glass formation under most conditions.

The relationship between the rates of crystallization and other properties of the liquid is now considered. A simple theoretical equation for the velocity v of crystallization of a liquid at a temperature T is

$$v = \frac{L(T_m - T)}{3\pi a^2 \eta T_m} \tag{1}$$

where L is the heat of fusion at the melting temperature T_m, η is the viscosity of the liquid, and a is a distance of the order of a lattice spacing. This equation is derived and discussed in Chapter 5 on crystallization. Of the factors in the equation the viscosity varies much more than L/T_m or a_o from one material to another. Thus a liquid with a high viscosity near its melting point has a low crystallization velocity and can be more easily formed into a glass. This relationship is borne out by the viscosities listed in Table 3. Furthermore the high viscosity of a glass-forming material helps to prevent the rearrangement of its molecules to the crystalline state at temperatures below that of maximum crystallization rate.

NUCLEATION AND GLASS FORMATION

The cooling rates necessary to prevent crystallization in many of the materials in Table 3 are impractically large, yet all the substances listed in the table are easily cooled to the glassy state. Therefore another barrier to bulk crystallization must exist in many glass-forming materials. This barrier can be the rate of nucleation of crystals; it is discussed in this section. The equations for nucleation are derived and discussed in more detail in Chapter 5.

The rate of homogeneous nucleation of a crystal from a liquid is

$$I = K \exp\left(\frac{W^*}{RT}\right) \tag{2}$$

where W^* is the work involved in forming the critical nucleus, and K is a coefficient whose dependence on temperature is neglected here. The work done in forming a spherical, isotropic, crystalline critical nucleus with isotropic surface energy γ from a liquid of the same composition is approximately (see Chapter 5)

$$W^* = -\frac{16\pi\gamma^3 V^2 T_m^2}{3L^2(\Delta T)^2} \tag{3}$$

where V is the molar volume of the liquid, L is its heat of fusion, and ΔT is the difference between the actual temperature and the melting temperature T_m.

Equations 2 and 3 are valid for homogeneous nucleation. In actual practice nucleation is rarely homogeneous, but takes place on impurity particles, vessel walls, defects, or some other heterogeneity. These heterogenieties effectively lower the interfacial tension γ.

In condensed systems the nucleation rate can be limited by the rate of molecular rearrangement of the liquid, and this transport limitation is particularly important in viscous systems such as the glasses being considered here. One way to take account of this transport limitation is to multiply Eq. 2 for the nucleation rate by a factor that is inversely proportional to the viscosity of the liquid.

In order to make quantitative calculations of the nucleation rate from Eqs. 2 and 3 it is necessary to estimate the interfacial tension γ. From experiments on nucleation in small droplets, Turnbull and Cohen deduced that γ should be proportional to the heat of fusion L. The small droplet experiments were carried out chiefly on organic liquids, and it is uncertain whether or not this relation applies to other liquids such as oxides. From this relation, Eq. 2 (with a transport term), and Eq. 3 Turnbull and Cohen[52] calculated that the nucleation rate should be high when the rate of crystallization is above 10^{-7} cm/sec. Any heterogeneous nucleation would make this rate even higher. Turnbull and Cohen concluded that a low nucleation rate could not explain glass formation in the materials listed in Table 3 with fairly high crystallization rates. The simple organic liquids are an exception. They are discussed in more detail below. However, this calculation is based on many uncertain assumptions. For example, the transport limitation may be much more severe than is accounted for by simply multiplying Eq. 2 by the viscosity. This is true because in complicated liquids rearrangements to the crystalline state can involve mechanisms additional to viscous flow or diffusion in the liquid. In a later paper Turnbull[34] described broader conditions for glass formation by suppression of nucleation and concluded that this mechanism may be important, especially when the viscosity of the material is low at its melting point. Thus it seems likely that in many complex liquids nucleation is slow enough to aid glass formation, although direct evidence to support this contention is lacking.

Uhlmann discussed glass formation in terms of the minimum cooling rate needed to suppress nucleation and growth of crystals in a liquid.[54] He treated only homogeneous nucleation since he was unable to make accurate estimates of nucleation rates. Uhlmann deduced that the most important factors determining the tendency to glass formation are the viscosity at the melting point and the rate of decrease of viscosity with temperature below the melting point. This deduction is in agreement with the conclusions of the last section and the next one.

FORMATION OF GLASSES FROM SIMPLE ORGANIC LIQUIDS

Several organic liquids listed in Table 1 can form glasses, even though they show rapid crystallization rates near their melting points and are quite fluid there. Different explanations have been proposed to account for this anomaly, but none is entirely satisfactory. This problem is considered in the following paragraphs.

The organic liquids that form glasses have relatively low melting points, and most have asymmetric molecules. Turnbull and Cohen[52] have pointed out that the ratio of the boiling temperature to the melting temperature (T_b/T_m) for these liquids is greater than 1.8, whereas it is less than this value for organic liquids that do not form glasses. They deduced that the temperature of glass formation is related to the cohesive energy of the liquid, which is proportional to the boiling point. Thus the closer the melting temperature is to the temperature of glass formation (the larger is T_b/T_m), the greater is the tendency to glass formation. Another qualitative way to view glass formation in these liquids is that they have lower melting temperatures because of the greater difficulty in rearranging their asymmetric molecules, and this difficulty, enhanced by the lower crystallization temperature, leads to reduced crystalline nucleation and growth and therefore to glass formation. These considerations explain the relative glass-forming tendency of the organic liquids with asymmetric molecules, but it is still difficult to account for glass formation in view of their relatively low viscosity and high crystallization rate near the melting points.

The viscosity of these liquids increases sharply below their melting points,[46,47,55] and consequently the crystallization rate decreases. Stavely et al.[8,9] found that they had to cool droplets of organic liquids to temperatures from 0.7 to 0.8 of the melting temperature before they crystallized. At this fraction of the melting temperature the viscosity of the liquids with asymmetric molecules is so high that the rate of crystallization is low, and as droplets they form glasses. However, this reasoning does not explain why these liquids form glasses in the bulk, since many nucleation catalysts are present, in contrast to the droplets. It may be that these catalysts are ineffective because of the complex shapes of the organic molecules. Further work is needed to establish with certainty the reason for glass formation in these liquids.

GLASS FORMATION IN OXIDES

The most important commercial glasses are based on oxides, as described in the preceding chapter and listed in Table 1. Zachariasen[56] considered the relative glass-forming ability of oxides and concluded that the ultimate condition for glass formation is that a substance can form extended three-dimensional networks lacking periodicity with an energy content comparable with that of the corresponding crystal network. From this condition he derived four rules for oxide structure that allow one to choose those oxides that tend to form glasses. These rules were remarkably successful in predicting new glass-forming oxides as well as including such oxides known at the time of their formulation. The rules are the following:

1. An oxygen atom is linked to not more than two glass-forming atoms.
2. The coordination number of the glass-forming atoms is small.
3. The oxygen polyhedra share corners with each other, not edges or faces.
4. The polyhedra are linked in a three-dimensional network.

Oxides A_2O or AO, where A is a metal atom, do not satisfy the rules. Oxides A_2O_3 satisfy rules 1, 3, and 4 if the oxygen atoms form triangles around each A atom, and AO_2 or A_2O_5 satisfy these rules if the oxygen atoms form tetrahedra around each A. Higher coordination is apparently excluded by rule 2. From these considerations Zachariasen concluded that the following oxides should be glass formers: B_2O_3, SiO_2, GeO_2, P_2O_5, As_2O_5, P_2O_3, As_2O_3, Sb_2O_3, V_2O_5, Sb_2O_5, Nb_2O_5, and Ta_2O_5. At the time of Zachariasen's research, only B_2O_3, SiO_2, GeO_2, P_2O_5, As_2O_5, and As_2O_3 had been made into glasses, but since then glasses of several other oxides in his list have been prepared, as shown in Table 1. Oxides not listed either have not been made into glasses, or in a few cases are made as glasses only with difficulty.

Thus Zachariasen's rules are quite accurate in predicting glass formation, and it is interesting to question their relationship to the criteria for glass formation involving crystallization rates. The requirement that the oxide form a three-dimensional network means that viscous flow is relatively difficult because it requires breaking of primary chemical bonds. In fact, Sun[57] has shown that the glass-forming tendency of an oxide is directly related to the strength of the bonds between its oxygen and metal atoms.

Glass formers have bond strengths above about 80 kcal/mole, and modifying ions that are not part of the oxide network have oxygen-metal bond strengths below this value. The requirement that the energy of the glass and crystal be close means that the heat of fusion for a glass former is less than that for other chemically similar materials. A lower heat of fusion in Eqs. 1 and 2 leads to lower rates of nucleation and crystallization; therefore for similar materials this factor may be of some significance in predicting the tendency to glass formation. However, it is much less important than factors influencing the viscosity of the liquid.

Other correlations of glass-forming tendency of oxides have been made. Stanworth[59] has suggested the following criteria:

1. The cation valence must be three or greater.
2. The tendency to glass formation increases with decreasing cation size.
3. The electronegativity should be between about 1.5 and 2.1 on Pauling's scale.

Using these criteria Stanworth finds four groups of oxides: the strong glass formers, Si, Ge, As, P, and B; "intermediate" glass formers which form glasses only with "splat" cooling,[60] Sb, V, W, Mo, and Te; other oxides that do not form glasses on rapid cooling, but do form them on oxidized surfaces of their metals or in binary combination with nonglass-forming oxides, Al, Ga, Ti, Ta, Nb, and Bi; and other oxides that do not form glasses. The second category includes most of the "conditional" glass-forming oxides in Table 1. Some oxides do not fit into this scheme, such as tin and chromium.

Rawson[1,61] has modified Sun's comparison of bond strength and glass-forming tendency of oxides mentioned above to a comparison of the ratios of bond strength to melting temperature. Rawson claims this ratio accounts for the thermal energy available to break the bonds as well as their strength. This criterion separates glass-forming tendency somewhat more accurately than the criterion of bond strength alone, and also helps in understanding the greater glass-forming tendency of binary eutectics, particularly when neither of the components forms an oxide by itself. This effect of greater glass-forming tendency for lower melting temperature is reminiscent of the ease of glass formation for lower melting organic liquids mentioned previously.

CONCLUSIONS

The tendency to form a glass is related to the rates of growth and nucleation of crystals in a liquid. In glass-forming material with rapid crystallization rates a low rate of crystalline nucleation probably aids glass formation, although direct evidence for this conclusion is still needed in many different systems.

In a particular class of materials, such as oxides, some structural criteria can be established for glass formation, but these criteria seem to be rather specific for the particular class. Thus Zachariasen's rules for oxides cannot be applied to glass formation in organic liquids or molten ionic salts, and the finding of asymmetric molecules in glass-forming organic liquids is irrelevant to glass formation in oxides. Therefore one should expect correlation between glass-forming ability and structure only with classes of glass-forming materials. The more general criteria of low rates of nucleation or growth of crystals must be examined to establish relative glass-forming ability of different classes or unclassified materials.

REFERENCES

1. H. Rawson, *Inorganic Glass-Forming Systems*, Academic, London, 1967.
2. R. Frerichs, *J. Opt. Soc. Am.*, **43**, 1153 (1953).
3. R. C. Ellis, *Inorg. Chem.*, **2**, 22 (1963).
4. A. Winter, *J. Am. Ceram. Soc.*, **40**, 54 (1957).
5. M. Gerding, *Naturwiss.*, **25**, 251 (1937).
6. G. Tammann, *The States of Aggregation*, Van Nostrand, New York, 1925 p. 233.
7. A. Stock and K. Thiel, *Research*, **38**, 2719 (1905).
8. D. G. Thomas and L. A. K. Stavely, *J. Chem. Soc.*, 4569 (1952).
9. H. J. deNordwall and L. A. K. Stavely, *J. Chem. Soc.*, 224 (1954).
10. A. Weiss and A. Weiss, *Z. Naturforsch.*, **8b**, 104 (1953).
11. P. L. Robinson and W. E. Scott, *Z. Anorg. Allgem. Chem.*, **210**, 57 (1933).
12. K. H. Sun, *Glass Ind.*, **27**, 552, 580 (1946).
13. D. Turnbull and M. H. Cohen, in *Modern Aspects of the Vitreous State*, Vol. I, J. D. Mackenzie, Ed., Butterworths, London, 1960, p. 54.
14. Ref. 1, pp. 213 ff.
15. E. Thile, C. Wiecker, and W. Wiecker, *Silikattech.*, **15**, 109 (1964).
16. Ref. 1, pp. 209 ff.
17. Ref. 1, p. 201.
18. R. K. Datta, D. M. Roy, S. P. Faile, and O. F. Tuttle, *J. Am. Ceram. Soc.*, **47**, 153 (1964).
19. R. F. Boyer, *Rubber Chem. Tech.*, **36**, 1303 (1963).
20. G. E. Vuilland, *Ann. Chim.*, **2**, 233 (1957).
21. C. A. Angell and E. J. Sayre, *J. Chem. Phys.*, **52**, 1058 (1970).

21a. P. Duwez, *Trans. ASM*, **60**, 605 (1970).

22. P. Duwez and C. C. Tsuei, *J. Noncryst. Solids*, **4**, 345 (1970).
23. M. H. Brodsky, *J. Vac. Sci. Tech.*, **8**, 125 (1971).
24. C. Feldman and K. Moorjani, *J. Noncryst. Solids*, **2**, 82 (1970).
25. L. Young, *Anodic Oxide Films*, Academic, New York, 1961, p. 171.
26. *Ibid.*, pp. 212 ff.
27. J. A. Prade and G. O. Jones, *Nature*, **170**, 685 (1952).
28. J. Yannas, *Science*, **160**, 298 (1968).
29. R. S. Allgaier, *J. Vac. Sci. Tech.*, **8**, 113 (1971).
30. R. P. Ferrier and D. J. Herrell, *J. Noncryst. Solids*, **4**, 338 (1970).
31. A. S. Nowich and S. R. Mader, *IBM J Res. Dev.*, **9**, 358 (1965).
32. R. Hilsch, in *Noncrystalline Solids*, V. D. Frechetts, Ed., Wiley, New York, 1960, p. 348.

32a. W. B. Hillig and D. Turnbull, *J. Chem. Phys.*, **24**, 914 (1956).

33. C. A. Angell, *J. Phys. Chem.*, **70**, 2793 (1966).

34. D. Turnbull, *Contemp. Phys.*, **10**, 473 (1969).
35. F. E. Wagstaff, *J. Am. Ceram. Soc.*, **52**, 650 (1969).
36. G. Hofmaier and G. Urbain, in *Science of Ceramics*, Vol. 4, British Ceramic Soc., 1968, p. 25.
37. P. J. Vergano and D. R. Uhlmann, *Phys. Chem. Glasses*, **11**, 30 (1970).
38. E. J. Fontana and W. A. Plummer, *Phys. Chem. Glasses*, **7**, 139 (1966).
39. R. L. Cormia, J. D. Mackenzie, and D. Turnbull, *J. Appl. Phys.*, **34**, 2239 (1963).
40. *Ibid.*, p. 2245.
41. A. Leontewa, *Acta Physicochem., USSR*, **16**, 97 (1942).
41a. W. D. Scott and J. A. Pask, *J. Am. Ceram. Soc.*, **44**, 181 (1961).
42. G. S. Meiling and D. R. Uhlmann, *Phys. Chem. Glasses*, **8**, 62 (1967).
43. J. A. Laird and C. G. Bergeron, *J. Am. Ceram. Soc.*, **53**, 482 (1970).
44. P. Li, A. C. Shore, and G. Su, *J. Am. Ceram. Soc.*, **45**, 86 (1962).
45. J. P. DeLuca, R. J. Eagan, and C. G. Bergeron, *J. Am. Ceram. Soc.*, **52**, 322 (1969).
46. R. J. Greet, *J. Crystal Growth*, **1**, 195 (1967).
47. J. D. Magill and D. J. Plazek, *J. Chem. Phys.*, **46**, 3757 (1967); **45**, 3038 (1966).
48. M. Volmer and A. Maider, *Z. Phys. Chem.*, **154A**, 97 (1931).
49. K. Neumann and G. Micus, *Z. Phys. Chem.*, **2**, 25 (1954).
50. O. Jantsch, *Z. Krist*, **108**, 185 (1956).
51. M. Takayangi, *Mem. Fac. Eng., Kyushu Univ.*, **16**, 111 (1957).
52. D. Turnbull and M. H. Cohen, *J. Chem. Phys.*, **29**, 1049 (1958).
53. D. Turnbull, *J. Appl. Phys.*, **21**, 1022 (1950).
54. D. R. Uhlmann, *J. Noncryst. Solids*, **7**, 337 (1972).
55. D. J. Denney, *J. Chem. Phys.*, **30**, 159 (1959).
56. W. H. Zachariasen, *J. Am. Chem. Soc.*, **54**, 3841 (1932).
57. K. H. Sun, *J. Am. Ceram. Soc.*, **30**, 277 (1947).
58. A. Van Hook, *Crystallization*, Reinhold, New York, 1961, p. 168.
59. J. E. Stanworth, *J. Am. Ceram. Soc.*, **54**, 61 (1971).
60. P. T. Sarjeant and R. Roy, *J. Am. Ceram. Soc.*, **50**, 500, 503 (1967).
61. H. Rawson, in *Proc. IV Inst. Conf. on Glass*, Imprimerie Chaix, Paris, 1956, p. 62.

3

Molecular Structure

Orderly persons such as physicists and crystallographers who are used to the regular arrangement of atoms in crystals are inclined to consider the molecular structure of glassy solids to be chaotic as well as somewhat mysterious. In fact a good deal is known about the molecular structure of certain glasses, even if it is not yet possible to describe this structure in the elegant detail reserved for crystals. One key to the understanding of glassy structure is the recognition of different classes of glass-forming materials, described in the preceding chapter. The molecular structure of materials in these different classes is quite different; therefore it is important to consider them separately.

Table 1 Classification of Glass-Forming Materials by the Type of Bonding

Bond type	Examples
Covalent	Oxides (silicates, borates, phosphates, germanates, etc.) Chalcogenides Organic polymers
Ionic	Halides, nitrates, carbonates, sulfates
Hydrated ionic	Aqueous solutions of salts
Molecular	Organic liquids
Metallic	Splat-cooled alloys

A classification of glass-forming substances based on the type of bonding is useful in grouping structural types, and is given in Table 1. The glassy structure can be quite different in each class. By far the most studied is the covalently bonded materials. The covalent bonds give rise to molecular chains and networks, which are favorable to glass formation. This chapter concentrates on results in the first class, especially the silicates, because of their overwhelming commercial importance and because they have been studied more extensively than other glasses.

The coordination number of an atom in a glass, or the number of other atoms of a specific kind that are bonded to it, is particularly important in understanding the structure of glassy solids. Since a glass is rigid the coordination number of atoms is relatively fixed as a function of time, in contrast to the rapid motion of molecules in a liquid. Thus the concept of a spacial average of coordination numbers is useful in glasses. The coordination number of an atom is important in determining its role in the glass as a part of a three-dimensional network, a chain, as a disruptor of networks and chains, or as a source of incipient crystallization. Therefore considerable attention must be paid to determining the coordination numbers of atoms in glasses.

VITREOUS SILICA

The simplest of the silicate glasses is vitreous silica (SiO_2), and the study of its molecular structure has been fundamental to understanding the structure of other silicates. This understanding began with Goldschmidt's recognition of the importance of the tetrahedral arrangement of oxygen ions around silicon ions in silicates,[1] and Zachariasen's postulate of a three-dimensional network without periodicity, formed by the union of these tetrahedra at their corners.[2]

In all crystalline silicates the coordination number of silicon-to-oxygen ions is four; therefore the silicon-oxygen tetrahedron is the basic building block for silicate structures. These tetrahedra can be attached to none, one, two, three, or four other tetrahedra by silicon-oxygen bonds at their corners, depending on the concentration of other oxides present. In vitreous silica each tetrahedron is attached to four others, giving a three-dimensional network. Zachariasen[2] proposed that this network lacked any symmetry or periodicity, and this random-network model is generally accepted as the best description of the structure of fused silica.

The initial x-ray diffraction study of Warren et al.[3] was consistent with the random-network model, and later work of Mozzi and Warren[4] has confirmed this agreement. There were theoretical and experimental limitations in the early studies that have been removed in the later work. The strong Compton scattering at large values of $\sin \theta/\lambda$, where θ is the scattering angle and λ the wavelength of the x-radiation, limited the accuracy of the diffraction results at these values. The method of fluorescence excitation has largely eliminated this background scattering, and a new treatment in terms of "pair-distribution" functions allows the interatomic distances to be calculated directly from the peak positions on a distribution curve, and the areas under the peaks give the number of neighboring atoms.

The "pair-function distribution" curve measured by Mozzi and Warren is shown in Fig. 1. The first peak corresponds to the silicon-oxygen distance of

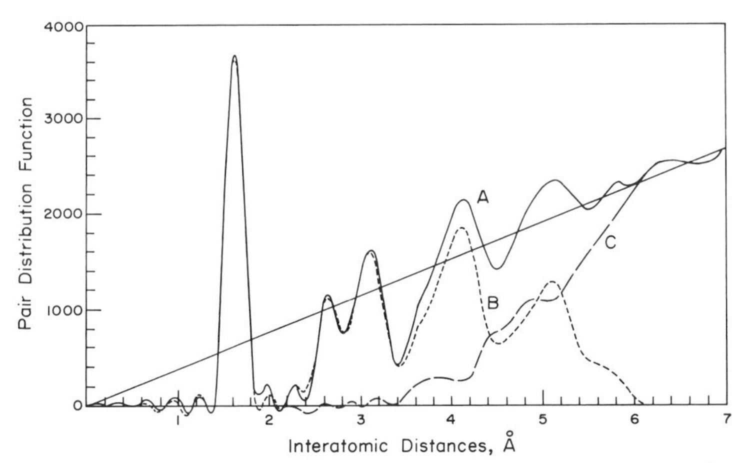

Fig. 1 The pair distribution function for fused silica, measured by Mozzi and Warren.[4] *A* is the measured curve. *B* is the sum of the calculated curves for the first six contributions: Si—O, O—O, Si—Si, Si—2nd O, O—2nd O, and Si—2nd Si. *C* is the difference between *A* and *B*.

1.62Å, and the second peak corresponds to an oxygen-oxygen distance of 2.65 Å. These distances are close to those found in crystalline silicates. The widths of these peaks are consistent with a tetrahedral distribution of oxygen atoms around each silicon atom, as shown by the calculated peaks in the figure. The third peak corresponds to the silicon-silicon distance of 3.12 Å, and is much broader than the other two peaks because of the distribution in silicon-oxygen-silicon bond angle. The fourth peak, at 4.15 Å, is the silicon-second oxygen contribution, and the fifth peak, at about 5.1 Å, is a combination of the oxygen-second oxygen and silicon-second silicon peaks. A sixth peak, at about 6.4 Å, could result from the silicon-third oxygen contribution. No more peaks would be expected from a random network, and none were found experimentally.

The shapes of the third, fourth, and fifth peaks were fitted by assuming a distribution in the silicon-oxygen-silicon bond angle, as shown in Fig. 2. The maximum of the distribution curve is at 144°, with a maximum angle of 180° and a minimum angle of 120°. Almost all the angles are within ±10% of the maximum. Although this distribution of bond angles is wide enough to distinguish the structure clearly from a crystalline arrangement, it is rather

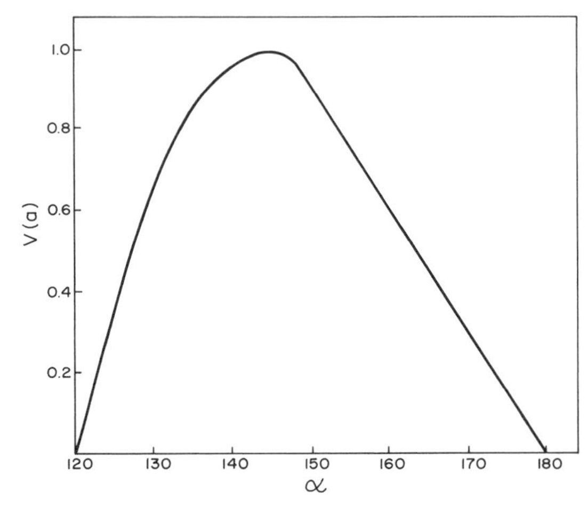

Fig. 2 The distribution of silicon-oxygen-silicon angles in fused silica assumed by Mozzi and Warren.[4] The function $V(\alpha)$ is the fraction of bonds with angles α normalized to the most probable angle.

narrow compared to a completely random distribution of bond angles from 90 to 180°. Thus the structure of vitreous silica is quite uniform at a short range, although there is no order beyond several layers of tetrahedra.

The silica structure is illustrated schematically in Fig. 3. The size of the silicon atoms is increased somewhat and only three oxygen atoms are attached to each silicon. The oxygen-silicon-oxygen bond angle is varied instead of the Si—O—Si angle. As in Fig. 2, most of the bond angles fall between $\pm 10\%$ of the mean angle, which is 120° in two dimensions. This distribution gives quite a regular structure on the short range, with a gradual distortion over a distance of three or four rings, corresponding to a distance of 20 to 30 Å. The illustration should be a more accurate schematic representation of the structure of vitreous silica than the often-copied drawing of Zachariasen and Warren, in which the variation in bond angles is much greater. It is similar to the schematic diagram of Oberlies and Dietzel.[14] Such a diagram does not show ordered regions and, therefore, is consistent with the random-network model.

Histograms calculated from physical models of the structure of vitreous silica, based on a random network and constructed in various ways,[5–7] agree well with the experimental results of Warren et al. and Mozzi and Warren.[4] Various properties of vitreous silica calculated from the random-network model are also in accord with experimental results, which will be described in following chapters. For example, the spectrum of atomic vibrations as measured[8,9] by infrared absorption, raman emission, and inelastic neutron

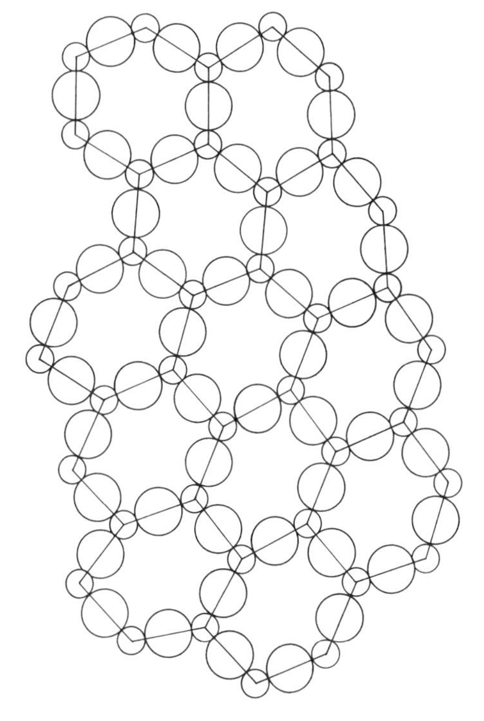

Fig. 3. Schematic diagram of a two-dimensional structure of fused silica.

scattering is consistent with the random-network model[10] and shows that, at most, only a small amount of crystalline material could be present in completely melted vitreous silica.

Some alternate models for the structure of vitreous silica have been advocated in which there are regions of order alternating with connecting regions of disorder. These crystallite models,[11,12] where very small crystalline regions are connected by disordered material, were examined by Warren and Briscoe.[13] They showed that the breadth of the first diffraction peak in vitreous silica required a crystallite size of no larger than 8 Å. At this size the term crystal loses any meaning. Furthermore, vitreous silica showed little low-angle scattering of x-rays; any appreciable amount of fine nonuniformities would lead to such scattering.

More recently other models for the structure of vitreous silica based on regions with structures close to that of quartz[14a] or trydimite[14b] have been advocated. These models ignore the distribution of Si—O—Si angles deduced from the breadth of the third diffraction peak, and also the scattering results of Warren and Briscoe.

Models built up of pentagonal dodecahedra[15,16] require silicon-oxygen-silicon bond angles of 180°, which are inconsistent with the experimental results of Mozzi and Warren for the distribution of these angles. There is also disagreement of this model with entropy considerations.[17,18]

Another objection to the random-network model for vitreous silica is that its configurational entropy value is too high,[17,19] and the small entropy of fusion of cristobalite indicates the presence of a considerable amount of order in vitreous silica. However, Bell and Dean[18] have shown that the configurational entropy of a random-network model of vitreous silica is quite low, so that this model is consistent with the experimentally measured entropy.

There have been reports of some structure in fused silica on the scale of 100 to 200 Å as seen in the electron microscope.[20,21] However, a careful study of thin slivers of different vitreous silicas by transmission in the electron microscope reveals no structure down to a resolution of about 20 Å.[22] In dark-field electron microscopy coherently scattering regions 10–20 Å in size were revealed in thin films of silica, silicon, germanium, and germanium-tellurium alloys.[22a] The authors concluded that their results were in better agreement with the random network model than with a microcrystalline model.

The conclusion of this discussion is, therefore, in support of the random-network model of Zachariasen and Warren, with the distribution of silicon-oxygen-silicon bond angles as shown in Fig. 2.

Vukcevich has suggested a model of the structure of fused silica in which two different silicon-oxygen-silicon bond angles close to 144° are preferred.[22b] Then there would be two maxima in the distribution curve of Fig. 2 rather than one, as shown. This distribution is still consistent with the data of Mozzi and Warren. With this model Vukcevich explained anomalous properties of fused silica, such as acoustic loss, compressibility, thermal expansion and specific heat.

The distribution of impurity ions in vitreous silica is not necessarily uniform. Optical nonuniformities are sometimes visible along the boundaries between regions that were crystalline grains.[23,24] Because of the high viscosity of silica at its melting temperature these regions are not mixed, even though fusion is complete. Impurities on the surface of the quartz starting material lead to a different refractive index. There is also evidence for a nonuniform distribution of sodium ions in synthetic vitreous silica,[25] and in the chapter on electrical conductivity some possible evidence for chains of aluminum ions in vitreous silica is discussed.

MULTICOMPONENT SILICATE GLASSES

Some ions can substitute for silicon in the silicon-oxygen network, either with charge-compensating ions (aluminum and phosphorous) or without them (germanium). The substitution of boron into the silicon-oxygen network with low alkali concentrations is difficult or impossible because of the three-coordinated planar structure of boron-oxygen (described below). Some evidence for this difficulty is given by the finding of Hair and Chapman that borate groups are not found in the silica phase of a phase-separated sodium borosilicate glass.[26]

When an alkali or alkaline earth oxide reacts with silica to form a glass, the silicon-oxygen network is broken up by the alkali or alkaline earth ions, as evidenced by the much lower viscosity of these glasses compared to fused silica. As long as the number of A_2O or AO units, A being the metallic ion, is in less than a one-to-one ratio to the number of SiO_2 units, the silicon-oxygen network is preserved because each silicon-oxygen tetrahedron is linked to at least three other tetrahedra, and the glass-forming tendency of the mixture is retained. The backbone network of the glass is preserved, and probably is similar to the random network of fused silica. Thus the results of Warren and Briscoe in an x-ray diffraction study of five different sodium silicate glasses were consistent with a random network of silicon-oxygen tetrahedra in which some oxygen ions were bonded to one silicon ion and some to two. They also were able to rule out the existence of discrete molecules such as SiO_2, Na_2O, $Na_2O \cdot 2SiO_2$, and $Na_2O \cdot SiO_2$.

However, the distribution of alkali or alkaline earth ions in these glasses is uncertain. There is convincing evidence from x-ray diffraction studies that these ions are not uniformly distributed throughout the glass, but that their average separation is considerably less than for a uniform distribution.[28–31] Whereas there is a possibility of phase separation in some of these glasses, the nonuniform distribution seems to exist even when phase separation is not present.[31] This "clustering" of ions is perhaps not so surprising in view of the structure of the alkali disilicates, for example, lithium disilicate as shown in Fig. 4, after Liebau.[32] In these crystals there are layers of silicon-oxygen tetrahedra bonded together by the alkali ions, and the alkali ion distances in the layers are considerably less than if the ions had been uniformly distributed throughout the crystal. Thus in the alkali silicate glasses one might expect a tendency for sheets of Si—O—A layers to form with a shorter than uniform A—A ion distance. Milberg and Peters[31] proposed a somewhat similar model of clusters of Si—O—Tl groups to explain their results on thallium silicate glasses, and Domenici and Pozza[32a] suggested that lithium silicate glasses have short range order similar to that in crystalline lithium disilicate. These models still retain the random-network structure at longer distances.

The coordination number of alkali and alkaline earth ions in silicate glasses is uncertain. In the alkali disilicates the coordination number of the alkali ions with oxygen is four,[32] although the alkali ion is closely associated with one negatively charged oxygen ion. It may be most profitable to

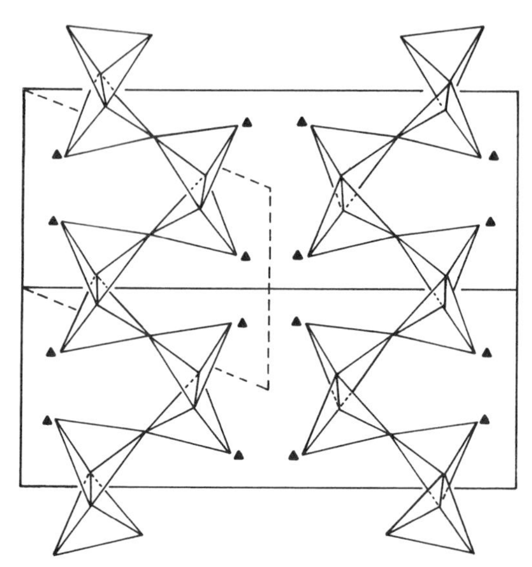

Fig. 4 Structure of crystalline lithium disilicate.[32] Small black triangles are lithium ions, and tetrahedra represent silicon-oxygen groups. The projections of two unit cells are shown.

consider the alkali ion coordinated with the one oxygen ion, particularly when examining models for ionic transport and exchange, as discussed in Chapters 9 and 14. Furthermore, the infrared spectra of silicate glasses containing hydroxyl groups show the typical —OH absorption band, indicating that the hydrogen ions are closely bound to one oxygen ion.

As more alkali or alkaline earth oxide is added beyond the one-to-two ratio the network becomes more and more disrupted as some tetrahedra are bonded to only two other tetrahedra. Glass formation becomes progressively more difficult because the rates of nucleation and crystallization in the glass become much more rapid as alkali or alkaline earth oxide is added. In the composition range between disilicate ($A_2O \cdot 2SiO_2$ or $AO \cdot 2SiO_2$) and metasilicate ($A_2O \cdot SiO_2$ or $AO \cdot SiO_2$) chainlike structures should exist, although direct evidence for them is lacking. At alkali concentrations above the metasilicate, isolated islands and rings, as well as chains, of silicon-oxygen tetrahedra should be the major structural features, although the evidence for them is indirect, coming from studies of the melt.[34,35]

Optical absorption spectra are valuable in determining the coordination number and configuration of higher valent ions such as those of the transition metals, which will be discussed in detail in Chapter 18. In general these studies show that the immediate surroundings of these ions in a silicate glass are much like the surroundings in a corresponding crystal, or even in a solution with complexed ions. This finding is consistent with the picture of glass structure in which the short-range environment is quite regular, and the random characteristic of the glassy state occurs only over dimensions of several structural units.

Mossbauer spectroscopy is another tool that is useful in answering questions about certain ions in glass, such as whether or not an ion is molecularly dispersed throughout a glass or is clustered in colloidal particles, and what is the coordination number. This technique and its application to glasses have been reviewed by Kurkjian.[33] So far iron, tin, and thulium ions in glass have been examined. These results are tentative and need confirmation with other techniques before conclusive structural information can be deduced from them.

Nuclear magnetic resonance is another technique that can give information about the immediate surroundings of ions in glass, as will be shown in more detail in the section on borate glasses. From "chemical shifts" in nuclear magnetic resonance of lead 207, Leventhal and Bray[33a] have shown that lead in binary $PbO—SiO_2$ glasses is in the same environment as in the crystalline compound $PbO \cdot SiO_2$, and does not show the character of Pb^{2+} ions, even at low lead concentrations. The structure of lead silicate glasses was examined by x-ray diffraction by Mydlar et al.[33b]

BORATE GLASSES

Borate glasses are of little commercial importance because they react with atmospheric water and degrade. These glasses are of interest, however, because their structure and properties are quite different from the silicates, and boric oxide is used extensively in borosilicate glasses.

In crystalline oxides boron occurs either in triangular coordination with three oxygen atoms or in tetrahedral coordination. Early x-ray results indicated a triangular coordination in B_2O_3 glass,[34] and this result was confirmed by nuclear magnetic resonance studies.[36] Krogh-Moe deduced from nuclear magnetic resonance, infrared and raman spectra, and other physical properties that the boroxyl group is an important element in the structure of vitreous B_2O_3. This group is a planar ring containing six members, alternatively boron and oxygen atoms. These rings are linked together in a three-dimensional network by boron-oxygen-boron bonds, as shown in Fig. 5.

The importance of the boroxol groups was confirmed by Mozzi and Warren in an x-ray diffraction study, using the same fluorescence excitation method as for vitreous silica, as described previously.[37] In this study, the experimental "pair-distribution" curve showed peaks that could be assigned to different atom-atom distances. The first indicated a boron-oxygen distance of 1.37 Å, which is the same as for triangular coordination in

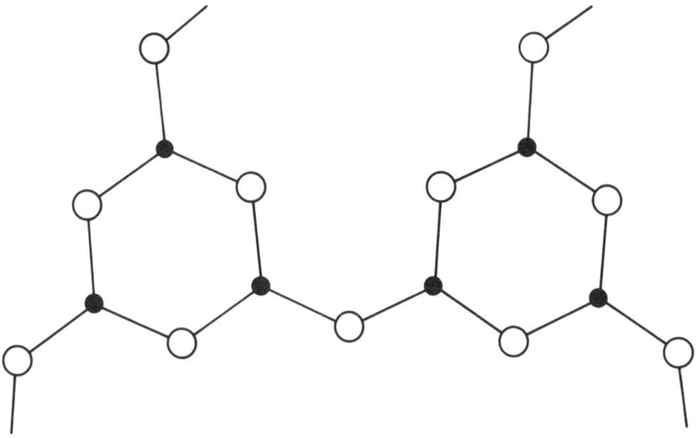

Fig. 5 Two boroxyl groups linked by a shared oxygen. ●, boron; ○, oxygen.

crystalline borates and less than the value of 1.48 Å for tetrahedral coordination. The second peak gave an oxygen-oxygen distance of 2.40 Å, close to the expected value of 2.37 Å. A model of randomly oriented BO_3 triangular groups was inconsistent with the remaining peaks. The presence of definite peaks out to a distance of about 6 Å required structural units larger than the BO_3 triangles. A model built up of boroxyl groups, as shown in Fig. 5, was capable of explaining the data. Within the six-membered rings the boron-oxygen-boron bond angle was 120°. The boron-oxygen-boron bond between rings was assumed to have an angle of 130°, with a random orientation of the linked rings about this bond. This model of linked rings gave good agreement with the experimental x-ray curve; somewhat better agreement was possible by assuming that a small part of the BO_3 groups was linked randomly to the boroxyl groups and was not in rings.

This structure of linked rings is very different from the random network of silicon-oxygen tetrahedra in vitreous silica. The space around any boroxyl group is not completely filled by linked neighboring groups, and there is a low probability that two atoms near one another, for example, at 10-Å separation, are linked by a direct and unbroken path of B—O and O—B bonds. Mozzi and Warren suggest that this structural feature may be the reason for the low viscosity of glassy B_2O_3 compared to other glass-forming oxides, such as silica and germania. This structure is also very different from that of any crystalline form of B_2O_3, which may be the reason for the difficulty in preparing crystalline B_2O_3 from the glass.

Many multicomponent borates show wide ranges of glass formation, as reviewed by Rawson.[38] The changes in the properties of alkali borates as more alkali is added is quite different from the changes in corresponding alkali silicates; this difference in borate properties is called the "boron oxide anomaly." The viscosity at certain temperatures can increase as alkali is added in some composition ranges, and the activation energy for viscous flow increases as alkali is added. In sodium borate the coefficient of thermal expansion decreases as alkali is added to about 16% Na_2O. In contrast the viscosity of the silicates increases and the coefficient of thermal expansion decreases as alkali is added, as one would expect from the simple picture of additional alkali breaking up the silicate network.

It seems likely that these anomalies are at least partially related to the change of coordination number of the boron with oxygen as alkali is added. This coordination number can be studied by nuclear magnetic resonance.[39,39a] The B^{11} nucleus shows strong quadruple coupling with a broad resonance line when it is in triangular coordination, whereas in tetrahedral coordination the quadruple coupling is weak, with a narrow resonance. These differences allowed Bray and O'Keefe to determine the fraction of three- and four-coordinated boron in these alkali borates, as shown in Fig. 6.

Several other intriguing and tantalizing pieces of structural information about borate glasses have resulted from nuclear magnetic resonance studies.[42,43] In most cases these results are rather preliminary and need substantiation by other techniques and properties. In lead borate glasses the lead appears to be present as lead ions at low concentrations,[33a] in contrast to the results on lead silicate glasses mentioned above. As the lead concentration increases the lead becomes more covalently bonded, as in lead

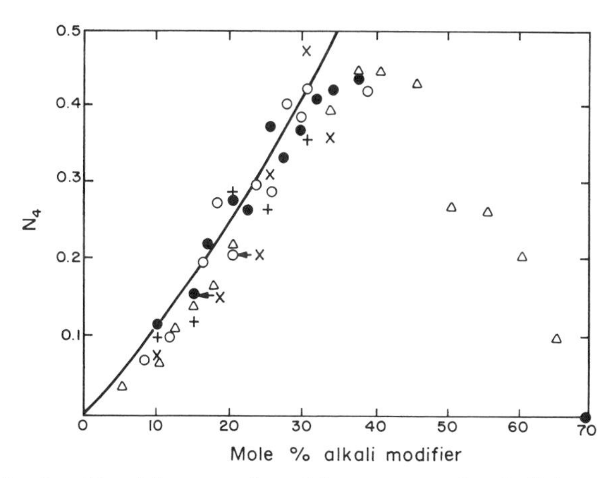

Fig. 6 The fraction N_4 of four coordinated boron atoms in alkali borate glasses.[39] ●, Na_2O; ○, K_2O; △, Li_2O; +, Rb_2O; ×, Cs_2O.

oxide and crystalline lead borates. There is evidence[42,44] for a change in the structure of cesium borate glasses at about 25% Cs_2O, although the curve of boron coordination in Fig. 6 shows no break at this composition. In zinc borate glasses the number of four-coordinated boron atoms is found to be much lower than in crystalline compounds of the same composition.[42,45] In calcium aluminoborate glasses (Cabal glasses) boron and aluminum compete for the oxygen introduced as CaO, since there is some four-coordinated boron in compositions with not all of the aluminum in four coordination.[42,44a] Germania-borate binary glasses can be formed over the entire composition range from pure GeO_2 to pure B_2O_3. There is no evidence for phase separation in these glasses,[42,46] yet all the boron atoms are in three coordination. It is difficult to imagine a network of tetrahedral germania and triangular borons, so it is possible that germanium has a different coordination number from four in these glasses. There is some evidence of a change from four to six coordination of germanium in alkali germanate glasses (Ref. 38, pp. 182 ff.).

PHOSPHATE GLASSES

As with the borate glasses, the phosphate glasses are of little commercial importance because of their reaction with atmospheric water. However, aqueous solutions of these glasses are important because of the complexing ability of the long phosphate chains, and P_2O_5 is an important constituent in a limited number of silicate glasses.

The basic building block in phosphate glasses and crystals is the phosphorous-oxygen tetrahedron. However, in contrast to the four-valent glass formers, the phosphorous has a double bond to one of its surrounding oxygen atoms. Thus it seems likely that the structure of glassy P_2O_5 is a three-dimensional network of these phosphorous-oxygen tetrahedra, each tetrahedron being bonded to three rather than four other tetrahedra as in vitreous silica. Apparently there has not been an x-ray study of P_2O_5 glass. It is also possible that sheets of PO_4 tetrahedra exist in some phosphate glass, although there is no direct evidence for them.

As other oxides that do not form part of the network, for example, alkali or alkaline earth oxides, are added to P_2O_5, the phosphate network is presumably broken up in the same way as the silicate, although these glasses have apparently not been studied in much detail in compositions up to the metaphosphate. An x-ray study of two calcium phosphate glasses of composition 42% CaO and 49% CaO is consistent with this picture.[47] This study showed that one-fourth of the phosphorous-oxygen bonds are π-bonds (directed double bonds), and that the phosphorous-oxygen-phosphorous bond angle is about 140°. The glass with less calcium had more P—O—P links.

In metaphosphate glasses with a one-to-one molar ratio of alkali or alkaline earth oxide to P_2O_5 each phosphorous-oxygen tetrahedron is linked to two other tetrahedra, and one oxygen ion per tetrahedron is associated with a metal ion. This structure leads to long chains of tetrahedra. An x-ray study confirmed the presence of such chains.[50] As the amount of metal oxide is increased, the average chain length decreases, and eventually some rings of phosphate tetrahedra appear. The distribution of chain lengths in alkali and "acid" (H_2O) phosphate glasses has been determined by dissolving the glass in water and analyzing the dissolved molecules chromatographically.[48,49] There is good evidence that the molecules are not degraded during solution, so that those dissolved are an accurate representation of the molecules present in the glass. The long-chain phosphate molecules in the glass are entwined in much the same way as long thin organic molecules are in polymers.

OTHER OXIDE GLASSES

Germania glass is built up of germanium-oxygen tetrahedra in a random network very much like that of silica glass, according to x-ray diffraction studies.[51,51a] This result is expected because of the similarity between the structure of many crystalline germanates and silicates. However, in crystals germanium can be six coordinated to oxygen as well as four coordinated. As mentioned before, there is some evidence that the coordination number of germanium changes from four to six as alkali oxide is added to GeO_2.[52,53] More structural work is needed to confirm this change.

Rawson gives some information on coordination numbers in tellurite, vanadate, aluminate, titanate, and molybdate glasses.[38] It seems likely that at least some of these glasses are more like the ionic glasses considered below than the covalently bonded glasses, since the viscosities of their melts tend to be low; thus it is probable that they do not form network structures.

The viscosities of melts of As_2O_3 and Sb_2O_3 are higher, and therefore it is possible that glasses of these oxides have chain or partial network structure. X-ray results on As_2O_3 glass were interpreted in terms of a layer structure.[38,54]

CHALCOGENIDE GLASSES

Glasses containing Group V elements, particularly arsenic and antimony, combined with Group VI elements such as sulfur, selenium, and tellurium, and sometimes with additions of halogens and other elements, have excited much interest because of their peculiar semiconducting and electrical switching properties. Most studies of these glasses have naturally involved electrical properties, but some structural work is beginning and will undoubtedly increase in the coming years. Therefore this discussion is preliminary and will need revision as more work is done.

The presently accepted definition of chalcogenide glasses, as stated in the preceding chapter, includes those glasses with any substantial amount of sulfur, selenium, or tellurium. The structures of elemental glasses of sulfur and selenium are discussed first, followed by discussions of binaries and multicomponent glasses.

The many forms of solid and liquid sulfur have been the subject of much study.[55,56] From the melting point of about 114° to 160°C liquid sulfur is made up of S_8 ring molecules. Above 160°C the rings break up and form long chains of average length 10^5 to 10^6 sulfur atoms.[57] This structural change leads to a sharp rise in the viscosity by a factor of about 10^5. At higher temperatures the chains break down, causing a decrease in the viscosity. Glasses can be formed by quenching the liquid sulfur from above 160°C to below about −30°C. These glasses contain the entwined chains of the liquid, with chain lengths depending on the temperature from which they were quenched. Some S_8 rings also are present in the liquid and glass. Thus the structure of glassy sulfur is much like that of the metaphosphate glasses and the organic polymers, except for the presence of the rings.

Selenium forms a glass by cooling from above its melting point of 217°C to room temperature.[56,58] Again the structure consists of long chains of selenium atoms. Near the melting point the average chain length in the liquid is about 10^4 atoms, and decreases as the temperature decreases. Thus the structure is very similar to that of glassy sulfur, the chain length being determined by the quenching temperature. Some Se_8 rings are also present in the glass and liquid, but less than in sulfur. Sulfur-selenium glasses are also known, with chains of randomly arranged sulfur or selenium atoms, similar to organic copolymers, mixed with S_8 and Se_8 rings.[59]

Several x-ray[60,61] and electron diffraction studies[62] of arsenic trisulfide glass have been made. The arsenic atoms are bonded to three sulfur atoms, and there is evidence for a distorted layer structure in the glass, similar to what is found in the crystal. Within the layers there are cross-linked chains or rings of arsenic-sulfur-arsenic links. The structure of arsenic triselenide is very similar to that of arsenic trisulfide.[61] Mixtures of the arsenic compounds have similar structures, with some evidence for more ionic character of the bonding in equimolar mixtures.

When a halide such as iodine is added to arsenic trisulfide, some arsenic-sulfur bonds are broken with formation of arsenic-iodine and sulfur-sulfur bonds. This bond breaking disrupts the cross-linked structure in the layers, forming individual chains and leading to a marked decrease in the melt viscosity.[60]

Crystalline germanium sulfide consists of GeS_4 tetrahedra (Ref. 38, p.

282), so the glass should form a random-network structure of linked tetrahedra, similar to those of glassy silica and germania. In crystalline silicon sulfide the silicon tetrahedra form chains by sharing edges, and the glass is probably also made up of such chains.

An x-ray diffraction study of amorphous arsenious telluride As_2Te_3 indicated that the coordination numbers in the glass were different from those in crystalline As_2Te_3.[63] The authors proposed a model in which the coordination of arsenic was three and that of tellurium two, the glass being a random network of covalently bonded $AsTe_{3/2}$ groupings.

An x-ray diffraction study of several germanium-tellurium mixtures[64] showed that the germanium was four coordinated and the tellurium was two coordinated in them, in contrast to the six coordination of both atoms in crystalline GeTe. The radial distribution functions were strikingly similar for mixtures varying from 11 to 72% germanium. These results suggest the possibility of a random-network structure of germanium-tellurium tetrahedra, but the positions of excess atoms in the mixtures different from $GeTe_2$ are uncertain. More recently Betts et al.[64a] proposed a model in which all germanium and some tellurium atoms have three nearest neighbors with some tellurium being twofold coordinated. Further work is needed to decide between this model and the "random covalent" one.

OTHER COVALENTLY BONDED GLASSES

An x-ray diffraction study of evaporated silicon films indicated an absence of long-range ordering for distances greater than 10 to 15 Å, whereas the short-range arrangement of the diamond lattice is preserved.[65] Electron spin resonance indicated a large amount of crystalline surface area, with an average distance of roughly 17 Å between surface-like defects. Thus "amorphous" silicon made by evaporation may be microcrystalline, but this question is still being debated.[22a] Evaporated germanium and silicon carbide are similar.

A curious series of glasses based on cadmium arsenide, $CdAs_2$, have been made.[66] The structure of two compositions containing germanium ($CdGe_XAs_2$, with $X = 0.1$ and 1.1) were examined by x-ray diffraction. The basic building block is a cadmium-arsenic tetrahedron, with four arsenic atoms surrounding each cadmium atom. In crystalline $CdAs_2$ the tetrahedra are bound together by cadmium-arsenic-cadmium bonds, forming three interpenetrating networks held together loosely by bonds between arsenic atoms. The authors speculate that the introduction of another element such as germanium, which is essential for glass formation, causes tetrahedral bonding between the germanium and arsenic, disrupting the $CdAs_2$ lattice and leading to a distorted network in the glass.

ORGANIC POLYMERS

A great variety of organic polymers are known, based on carbon-carbon bonding. The simplest polymers are linear chains, such as polyethylene:

$$\left(\begin{array}{c} H \\ | \\ -C- \\ | \\ H \end{array} \right)_n$$

The length of the carbon-carbon chain depends on conditions of preparation. Many different side groups and branches can be attached to the backbone chain. Other atoms than carbon, such as oxygen or nitrogen, can also be linked in the chain, and copolymers with blocks of different molecules in the chain are possible.

In the melt the chains are so entangled with one another that translational motion of the entire molecule is impossible, and thermal energy is manifested as Brownian motion of free segments of the chain. Different chains can be held together by Van der Waals forces at points or limited regions. If the melt is cooled rapidly the chains will not be able to orient with respect to one another even partially, and a rigid glass results. However, with slower cooling small crystalline regions will form, consisting of several chains precisely oriented. Thus most solid polymers consist of a mixture of very small crystalline regions in a matrix of entangled chains.

The chains can be tied together into a three-dimensional network if substituents on the side chains can react and form covalent bonds. This cross-linking leads to substantial changes in the properties of the polymer, such as increasing the viscosity and softening temperature. A simple chainlike polymer is called thermoplastic because it softens as it is heated. A thermosetting polymer becomes more rigid as it is heated because of the formation of cross-links resulting from chemical reactions at the higher temperature.

Rubbers are cross-linked polymers that can be deformed elastically by large amounts. When a rubber is cooled sufficiently it becomes a glass.

For those readers who wish more information on organic polymers than is contained in this brief sketch, a short bibliography is provided as a starting point.[67–70] There are also many glassy polymers composed of combinations of carbon atoms with inorganic atoms.[71]

GLASSES OF IONIC SALTS

Different types of ionically bonded compounds can form glasses. The halides beryllium fluoride and zinc chloride have special structures that lead to glass formation. Binary mixtures of nitrates, carbonates, and sulfates form glasses when one metallic ion in the mixture is univalent and the other divalent. Aqueous solutions of certain acids, bases, and salts can be cooled to glasses.

The structures of crystalline modifications of beryllium fluoride, BeF_2, are particularly curious in that they closely resemble those of silica. The basic structural unit of these crystals is a beryllium-fluorine tetrahedron with each beryllium ion coordinated to four fluorine atoms and each fluorine bonded to two beryllium ions. There are two crystalline forms of BeF_2 : a quartz form and a cristobalite form. The ionic sizes of the beryllium and fluoride ions are almost the same as those of the silicon and oxygen ions in silica, and the strength of the beryllium-fluorine bond is high. The four coordination of the beryllium ion and the bonding of the fluorides to two beryllium ions are quite unusual, in view of the strongly ionic character of these ions.

The structure of glassy beryllium fluoride as derived from x-ray diffraction studies is built up of beryllium-fluorine tetrahedra.[72,73,73a] Presumably the tetrahedra are linked at their corners in the same random-network structure as vitreous silica.[72]

There is a series of binary glasses of alkali or alkaline earth fluorides mixed with BeF_2 (Ref. 38, pp. 235 ff.), and by analogy with binary silicates these glasses are expected to be based on a beryllium-fluoride network, broken up by the modifying metal cations. The viscosity of BeF_2 above its melting point is quite high and is reduced by addition of the alkali and alkaline earth fluorides, in analogy with the silicates.

Zinc chloride also forms a glass, although it is much less stable than the beryllium chloride glass. Apparently the zinc ions are surrounded by four chloride ions in tetrahedral coordination, giving a network structure like that of beryllium fluoride and the silicates,[74] although the zinc-chlorine bond is not as strong as that of beryllium and fluorine. The reason for this structure in zinc chloride is not clear.

Glasses can be formed from binary mixtures of nitrates, sulfates, and carbonates when one cation in the mixture is divalent and the other monovalent. Several different structural explanations for glass formation in the salts have been put forward,[75,76] but none seems to be definitive. Stevels proposed that the distortion of the nitrate ion should be greater as the difference in field strength of the cations becomes greater; however, there is no evidence that such distortion occurs.[77,78] Urnes[77] and Thilo[79] suggested that the divalent cations should have monovalent cations near them when the difference in the field strength of the ions is great. This "clustering" of ions inhibits crystallization and leads to glass formation. Kleppa[79a] found that barium and strontium nitrate undergo a disordering process during fusion consisting of a rotation of the nitrate ions.

Nitrate ions in calcium nitrate do not show this rotation. Kleppa suggests that this difference results from the higher field strength of the calcium ion, and the absence of rotation leads to a network structure in the glass. These explanations for the structure of glassy salts are speculative and need to be tested by other experiments, preferably x-ray diffraction.

Potassium acid sulfate $KHSO_4$ is probably not strictly a member of this class of glassy salts since it contains only monovalent cations. In the crystal of this salt there are sulfate chains held together by hydrogen ions,[82] and it is possible that these chains exist in the glass also.

A number of aqueous solutions of acid, bases, and salts form glasses.[80,81] Their structures undoubtedly involve hydration of the ions and breaking up of the hydrogen-bonded structure of water and ice, but the details of these arrangements are still uncertain. Multivalent cations in concentrated aqueous solution can have more than a single hydration layer, and the anions then can pack around them octahedrally. Whether this structure can be extrapolated to glasses is unclear. More structural work on the glasses is needed to clarify these questions.

ORGANIC LIQUIDS

A number of simple organic liquids form glasses, as listed in Chapter 2. Most of these liquids have asymmetric molecules. The bonding between these molecules should be weak, so the structure of the glasses probably consists of a random arrangement of these molecules, essentially a frozen liquid without much interaction between them. It is possible that some of the molecules have preferred orientations with respect to one another, giving rise to small crystallike regions separated by disordered molecules. These conclusions are speculative because there has apparently been no direct study of the structure of these glasses.

METALLIC ALLOYS

A variety of metallic alloys have been prepared as glasses by deposition techniques,[83,84] or rapid cooling of the liquid.[85] Cargill examined the x-ray scattering from electrodeposited nickel-phosphorous in the composition range Ni_3P to Ni_4P. He found that the results were inconsistent with the structures of crystalline Ni_3P. The radial distribution functions indicated more short-range order than in liquid noble metals. The results did agree well with the radial distribution function calculated for Bernal's dense random packing of hard spheres[86] by Finney,[87] and Cargill[88] proposed this packing as a model for the structure of glassy alloys. Polk[89] extended this model by proposing that the nonmetallic atoms fill holes in the dense random packing of metallic atoms. The structure is stabilized by the nonmetallic atoms, which interact strongly with the metal atoms.

Amorphous films of copper, nickel, gold and silver have been formed by vapor deposition on a substrate cooled to liquid nitrogen temperature.[90] The authors concluded that these films also had the structure of dense random packing of hard spheres, stabilized by gaseous impurities.

REFERENCES

1. V. M. Goldschmidt, *Skrifter Norske Videnskaps Akad. (Oslo), Mat.-natur.*, **1**(8), 7 (1926).
2. W. H. Zachariasen, *J. Am. Chem. Soc.*, **54**, 3841 (1932).
3. B. E. Warren, H. Krutter, and O. Morningstar, *J. Am. Ceram. Soc.*, **19**, 202 (1936).
4. R. L. Mozzi and B. E. Warren, *J. Appl. Cryst.*, **2**, 164 (1969).
5. F. Ordway, *Science*, **143**, 800 (1964).
6. D. L. Evans and S. V. King, *Nature*, **212**, 1353 (1966).
7. R. J. Bell and P. Dean, *Nature*, **212**, 1354 (1966).
8. M. Hass, *J. Phys. Chem. Solids*, **31**, 415 (1970).
9. A. J. Leadbetter, *J. Chem. Phys.*, **51**, 779 (1969).
10. R. J. Bell and P. Dean, *Disc. Far. Soc.*, **50**, 55 (1970).
11. J. F. Randall, H. R. Rooksby, and B. S. Cooper, *J. Soc. Glass Tech.*, **14**, 219 (1930).
12. N. Valenkov and E. Porai-Koshitz, *Z. Krist.*, **45**, 195 (1936).
13. B. E. Warren and J. Biscoe, *J. Am. Ceram. Soc.*, **21**, 49 (1938).
14. F. Oberlies and A. Dietzel, *Glastech. Ber.*, **30**, 37 (1957).

14a. A. H. Narten, *J. Chem. Phys.*, **56**, 1905 (1972).

14b. J. H. Konnert and J. Karle, *Nature Phys. Sci.*, **236**, 92 (1972).

15. L. W. Tilton, *J. Res. Nat. Bur. Stand.*, **59**, 139 (1957).
16. H. A. Robinson, *J. Phys. Chem. Solids*, **26**, 209 (1965).
17. J. F. G. Hicks, *Science*, **155**, 459 (1967); *Phys. Chem. Glasses*, **10**, 164 (1969).
18. R. J. Bell and P. Dean, *Phys. Chem. Glasses*, **9**, 125 (1968); **10**, 164 (1969).
19. S. Urnes, *Trans. Brit. Ceram. Soc.*, **60**, 88 (1961).
20. A. F. Prebus and J. W. Michener, *Ind. Eng. Chem.*, **46**, 147 (1953).
21. J. Zarzycki and R. Mezard, *Phys. Chem. Glasses*, **3**, 163 (1962).
22. A. M. Turkalo, General Electric Research Laboratory, unpublished results.

22a. P. Chaudhari, J. F. Graczyk and L. R. Herd, Phys. Stat. Sol. **51b**, 801 (1972); Phys. Rev. Letters **29**, 425 (1972).

22b. M. R. Vukcevich, J. Noncryst. Solids **11**, 25 (1972).

23. G. Hetherington and K. H. Jack, *Phys. Chem. Glasses*, **3**, 129 (1962).
24. R. Bruckner, *J. Noncryst. Solids*, **5**, 123 (1970).
25. R. H. Doremus, *Phys. Chem. Glasses*, **10**, 28 (1969).
26. M. L. Hair and J. D. Chapman, *J. Am. Ceram. Soc.*, **49**, 651 (1966).
27. B. E. Warren and J. Biscoe, *J. Am. Ceram. Soc.*, **21**, 259 (1938).
28. C. Brosset, *Trans. Soc. Glass Tech.*, **42**, 125 (1958); *Phys. Chem. Glasses*, **4**, 99 (1963).
29. S. M. Ohlberg and J. M. Parsons, in *Physics of Noncrystalline Solids*, North Holland, Amsterdam, 1965, p. 31.
30. S. Urnes, *Phys. Chem. Glasses*, **10**, 69 (1969).
31. M. E. Milberg and C. R. Peters, *Phys. Chem. Glasses*, **10**, 46 (1969).
32. F. Liebau, *Acta Cryst.*, **14**, 389, 395 (1961).

32a. M. Domenici and F. Pozza, *J. Mat. Sci.*, **5**, 746 (1970).

33. C. R. Kurkjian, *J. Noncryst. Solids*, **3**, 157 (1970).

33a. M. Levanthal and P. J. Bray, *Phys. Chem. Glasses*, **6**, 113 (1965).

33b. M. F. Mydlar, N. J. Kreidl, J. K. Hendren, and G. T. Clayton, *Phys. Chem. Glasses*, **11**, 196 (1970).
34. J. D. Mackenzie, in *Modern Aspects of the Vitreous State*, Vol. I, J. D. Mackenzie, Ed., Butterworths, London, 1960, p. 188.
35. H. L. Trap and J. M. Stevels, *Glastech. Ber.*, **32**, 51 (1959); *Phys. Chem. Glasses*, **1**, 107, 181 (1960).
36. A. H. Silver and P. J. Bray, *J. Chem. Phys.*, **29**, 984 (1958).
37. B. L. Mozzi and B. E. Warren, *J. Appl. Cryst.*, **3**, 251 (1970).
38. H. Rawson, *Inorganic Glass-Forming Systems*, Academic, London, 1967, pp. 97 ff.
39. P. J. Bray and J. G. O'Keefe, *Phys. Chem. Glasses*, **4**, 37 (1963).
39a. H. M. Kriz, M. J. Park, and P. J. Bray, *Phys. Chem. Glasses*, **12**, 45 (1971).
40. J. Krogh-Moe, *Phys. Chem. Glasses*, **3**, 1 (1962).
41. S. E. Svanson, E. Forslind, and J. Krogh-Moe, *J. Phys. Chem.*, **66**, 174 (1962).
41a. J. Wong and C. A. Angell, in *Applied Spectroscopy Reviews*, Vol. 4, Marcel Dekker, New York, 1970, p. 103.
42. P. J. Bray, in *Interaction of Radiation with Solids*, A. Bishay, Ed., Plenum, New York, 1967, p. 25.
43. P. J. Bray, in *Magnetic Resonance*, Plenum, New York, 1970.
44. M. E. Milberg, K. Otto, and T. Kuchida, *Phys. Chem. Glasses*, **7**, 14 (1966).
44a. S. G. Bishop and P. J. Bray, *Phys. Chem. Glasses*, **7**, 73 (1966).
45. J. Krogh-Moe, *Z. Krist.*, **117**, 166 (1962).
46. M. K. Murthy and B. Scroggie, *Phys. Chem. Glasses*, **7**, 68 (1966).
47. J. Biscoe, A. G. Pincus, C. A. Smith, and B. E. Warren, *J. Am. Ceram. Soc.*, **24**, 116 (1941).
48. A. E. R. Westman, *Modern Aspects of the Vitreous State*, J. D. Mackenzie, Ed., Butterworths, London, 1960, pp. 63 ff.
49. J. R. Van Wazer, *Phosphorous and Its Compounds*, Vol. I, Interscience, N.Y., 1958. pp. 717 ff.
50. G. W. Brady, *J. Chem. Phys.*, **28**, 48 (1958).
51. J. Zarzycki, *Verres et Refract.*, **11**, 3 (1957).
51a. A. J. Leadbetter and A. C. Wright, *J. Noncryst. Solids*, **7**, 37 (1972).
52. A. O. Ivanov and K. L. Estropiev, *Dokl. Akad. Naak, USSR.*, **145**, 747 (1962).
53. E. F. Riebling, *J. Chem. Phys.*, **39**, 3022 (1963).
54. K. Plieth, E. Reuber, and J. N. Stranski, *Z. Anorg. Allgem. Chem.*, **280**, 205 (1955).
55. B. Meyer, *Chem. Rev.*, **64**, 429 (1964).
56. Ref. 38, pp. 254 ff.
57. G. Gee, *Trans. Far. Soc.*, **48**, 515 (1952).
58. H. Krebs, *Angew. Chem.*, **65**, 293 (1953).
59. A. V. Tobolsky, G. D. T. Owen, and A. Eisenberg, *J. Colloid. Sci.*, **17**, 717 (1962).
60. T. E. Hopkins, R. A. Pasternak, E. S. Gould, and J. R. Herndon, *J. Phys. Chem.*, **66**, 733 (1962).
61. A. A. Vaipolin and E. A. Porai-Koshitz, *Sov. Phys. Solid State*, **5**, 497 (1963).
62. P. A. Young and W. G. Thege, *Thin Solid Films*, **7**, 41 (1971).
63. J. R. Fitzpatrick and C. Maghrabi, *Phys. Chem. Glasses*, **12**, 105 (1971).
64. F. Betts, A. Bienenstock, and L. R. Ovshinsky, *J. Noncryst. Solids*, **4**, 554 (1970).
64a. F. Betts, A. Bienenstock, D. T. Keating and J. P. deNeufville, *J. Noncryst. Solids*, **7**, 417 (1972).
65. M. H. Brodsky, R. S. Title, K. Weiser, and G. D. Pettit, *J. Noncryst. Solids*, **4**, 328 (1970).
66. L. Cervinka et al., *J. Noncryst. Solids*, **4**, 258 (1970).
67. P. J. Flory, *Principles of Polymer Chemistry*, Cornell University Press, Ithaca, New York, 1953.

68. C. C. Winding and G. D. Hiatt, *Polymeric Materials*, McGraw Hill, New York, 1961.
69. C. Tanford, *Physical Chemistry of Macromolecules*, Wiley, New York, 1961, Chapters 2 and 3, pp. 15 ff.
70. P. Mears, *Polymers: Structure and Bulk Properties*, Van Nostrand, New York, 1965, Chapter 2, pp. 17 ff.
71. F. G. A. Stone and W. A. G. Graham, *Inorganic Polymers*, Academic, New York, 1962.
72. B. E. Warren and C. F. Hill, *Z. Krist*, **A89**, 481 (1941).
73. J. Zarzycki, *Phys. Chem. Glasses*, **12**, 97 (1971).
73a. A. J. Leadbetter and A. C. Wright, *J. Noncryst. Solids*, **7**, 156 (1972).
74. B. Brehler, *Naturwiss*, **46**, 554 (1959).
75. Ref. 38, p. 217.
76. C. A. Angell, *J. Phys. Chem.*, **70**, 2793 (1966).
76a. J. M. Stevels, *Philips Tech. Rev.*, **13**, 293 (1952).
77. S. Urnes, *Glastech. Ber.*, **31**, 337 (1958); **34**, 213 (1961).
78. C. A. Angell, *J. Phys. Chem.*, **68**, 218, 1917 (1964).
79. E. Thilo, C. Wiecker, and W. Wiecker, *Silikattech.*, **15**, 109 (1964).
79a. O. J. Kleppa, *J. Phys. Chem. Solids*, **23**, 819 (1962).
80. Ref. 38, pp. 228 ff.
81. C. A. Angell and E. J. Sare, *J. Chem. Phys.*, **52**, 1058 (1970).
82. L. H. Loopstra and C. H. MacGillavry, *Acta Cryst.*, **11**, 349 (1958).
83. A. S. Nowick and S. Mader, *IBM J. Res. Dev.*, **9**, 358 (1965).
84. B. G. Bagley and D. Turnbull, *J. Appl. Phys.*, **39**, 5681 (1968); *Acta Met.*, **18**, 857 (1970).
85. P. Duwez, *Trans. ASM.*, **60**, 605 (1970).
86. J. D. Bernal, *Proc. Roy. Soc.*, **280A**, 299 (1964).
87. J. L. Finney, *Proc. Roy. Soc.*, **319A**, 479, 495 (1970).
88. G. S. Cargill, *J. Appl. Phys.* **41**, 2248 (1970).
89. D. E. Polk, *J. Noncryst. Solids*, **5**, 365 (1971).
90. L. B. Davies and P. J. Grundy, *Phys. Stat. Sol.*, **8a**, 189 (1971); *J. Noncryst. Solids*, **11**, 179 (1972).

4

Phase Separation

The separation of borosilicate glasses into two liquid phases has long been recognized,[1] and is the basis of the Vycor process for the manufacture of 96% silica.[1a] In this process a sodium borosilicate glass is heat treated to separate the two liquid phases. Then the borate-rich phase is leached out with acid, leaving a honeycomb matrix of 96% silica. The channels in this matrix, which is sometimes called "thirsty glass," are a few hundred Å or so in diameter, as shown in the electron micrograph in Fig. 1. This matrix sinters to a clear glass when it is heated to about 1000°C. Thus a glass similar to fused silica is produced without going to the high temperatures (above 1800°C) required to prepare fused silica.

It is only within the last decade or two that glass scientists have recognized the importance of phase separation in other commercially important glasses. For example, certain soda-lime silicate glasses can separate into two liquid phases after proper heat treatment,[2] as shown in Fig. 2, and Pyrex borosilicate glass is phase separated on a very fine scale,[3] as shown in Fig. 3. Phase separation can influence properties of glasses. For example, Pyrex borosilicate glass owes its chemical durability to phase separation in the glass. In this glass the matrix phase is rich in silica, and the borosilicate phase is separated at such a fine scale that it cannot be leached out, so that the glass shows a chemical durability approaching that of vitreous silica.

It is therefore important to a glass technologist to be aware of the possibility of phase separation. He may wish to cause phase separation to enhance the properties of his glasses, or to avoid it to change the properties in another direction. Furthermore, phase separation in a glass is sometimes important in bringing about uniform crystallization in glass ceramics, as described in the next chapter.

Variations in glass composition often affect properties because of enhancement or elimination of phase separation. The addition of alumina to

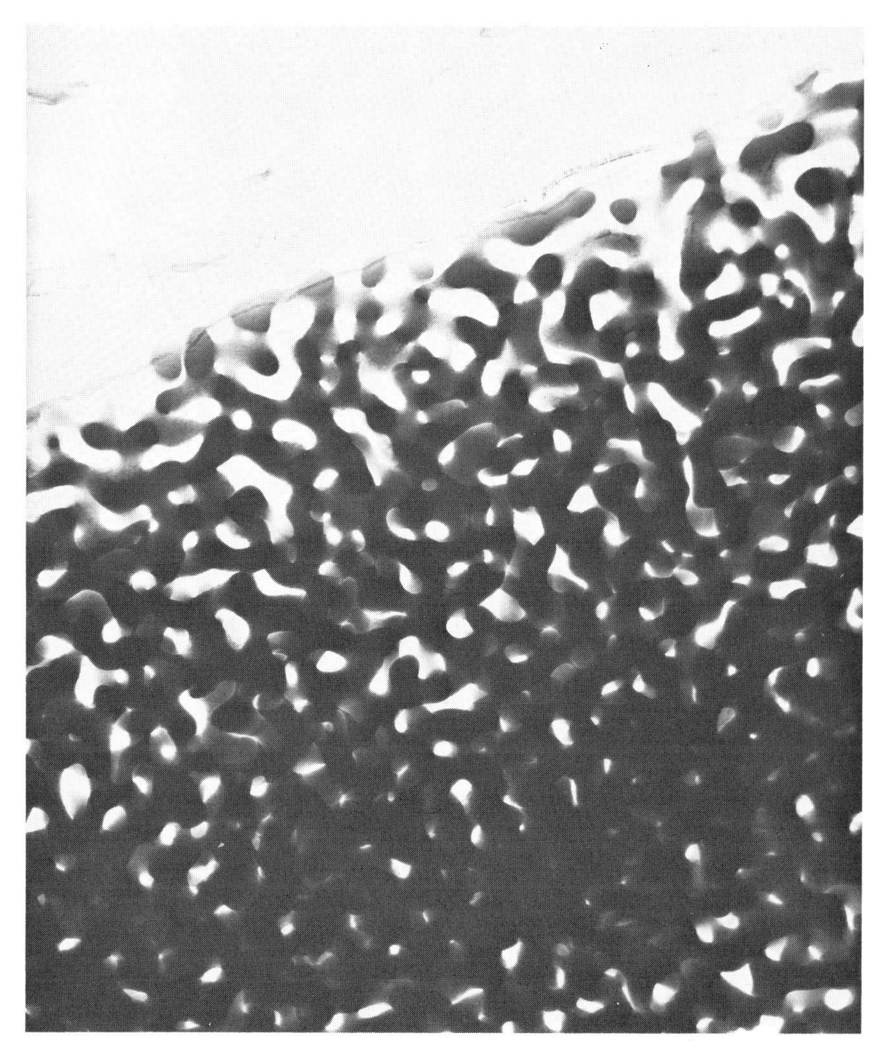

Fig. 1 Phase separation in a sodium borosilicate glass. The sodium borate phase was etched out with HCl. Transmission electron micrograph. Courtesy of Miss A. Turkalo. Magnification 26,000 ×.

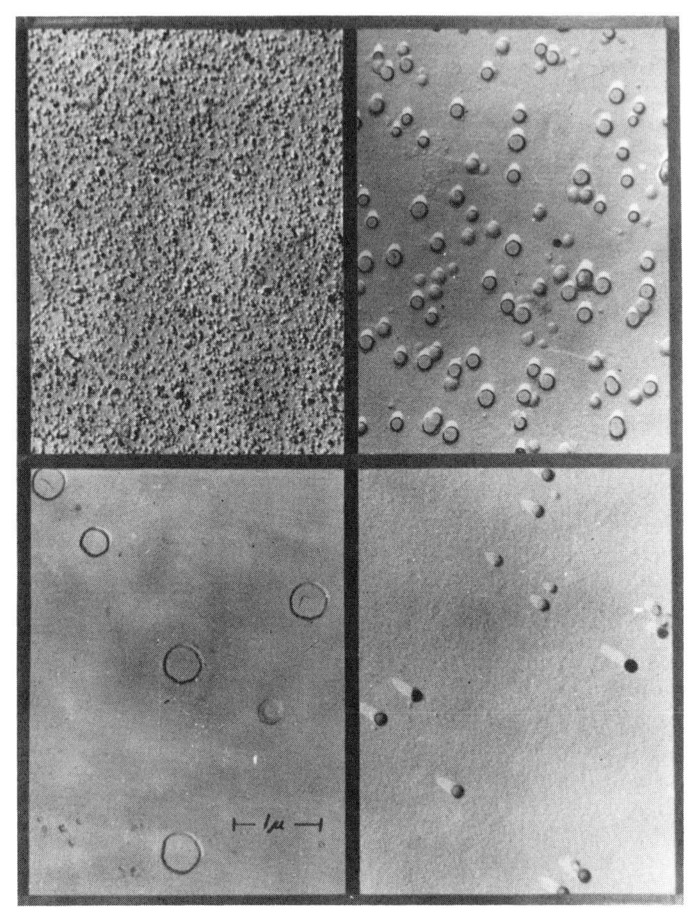

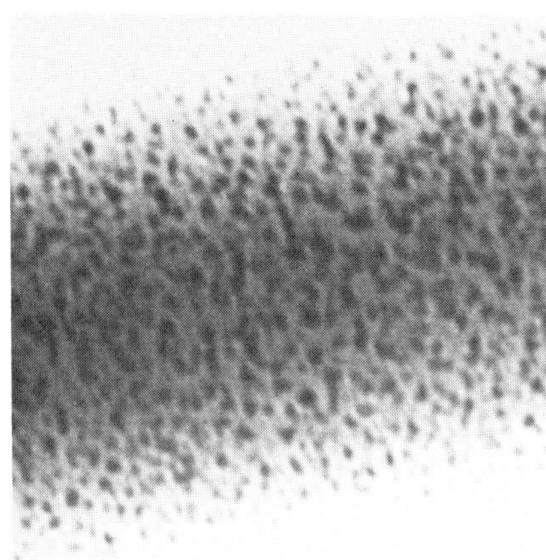

Fig. 2 Fracture surfaces of phase-separated glass (13% Na_2O, 11% CaO, and 76% SiO_2) after heating for different times at different temperatures.[2]

Fig. 3 Electron transmission photograph of Pyrex borosilicate glass in which silver ions have been exchanged for sodium ions to enhance contrast between the phases.[3] Magnification 240,000 ×.

commercial soda-lime glass reduces the tendency to phase separation, thus reducing in turn weathering and devitrification (gross crystallization and consequent lower strength) of these glasses caused by a matrix phase high in soda.

Phase separation has been studied most intensively in the alkali silicates and borosilicates, and therefore this discussion deals mostly with these glasses. Phase separation is also important in other glasses, such as the chalcogenides and beryllium fluorides.

In this chapter phase diagrams for phase separation in glasses are described first, and some theoretical proposals to explain these diagrams are also discussed. Then classical nucleation theory is briefly reviewed, since it will be useful in this chapter and the next one on crystallization. A different mode of phase separation called spinodal decomposition is then described. This type of decomposition leads to different transformation rates and morphologies in phase-separated glasses. There is a vigorous controversy over the importance and breadth of applicability of spinodal decomposition in glasses; both sides of this question are described, and some judgements made about their validity.

PHASE DIAGRAMS

Immiscibility in liquid solution is common. Many single-phase mixtures of organic liquids separate into two liquid phases as they are cooled. Certain liquid silicate mixtures also separate into two phases; examples of phase separation in alkaline earth silicates are shown in Fig. 4. The figure shows

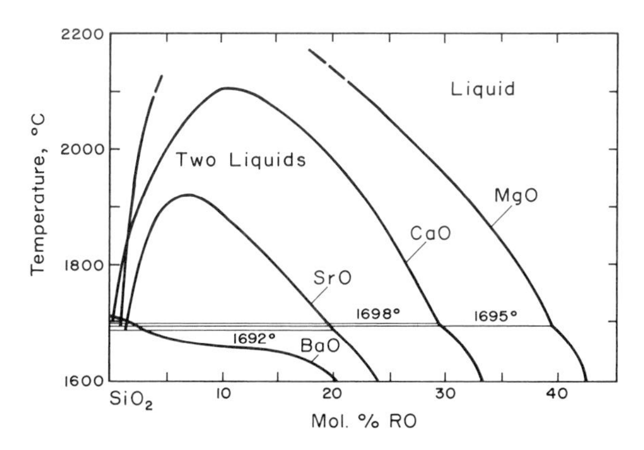

Fig. 4 Phase diagrams for binary alkaline-earth oxide-silica systems showing the miscibility gaps.[4]

that the tendency to separate into two phases becomes greater as the cation becomes smaller; barium silicate mixtures do not show stable immiscibility. This correlation with ionic size has been the basis of certain explanations of phase separation in silicates and is discussed in the next section. In these explanations the ratio of ionic charge Z to ionic radius r, called the "ionic potential," is used as a measure of the tendency to phase separation of silicates.

The ratio Z/r is plotted as a function of the width of the equilibrium immiscibility gap at its base for various binary silicate melts in Fig. 5. The

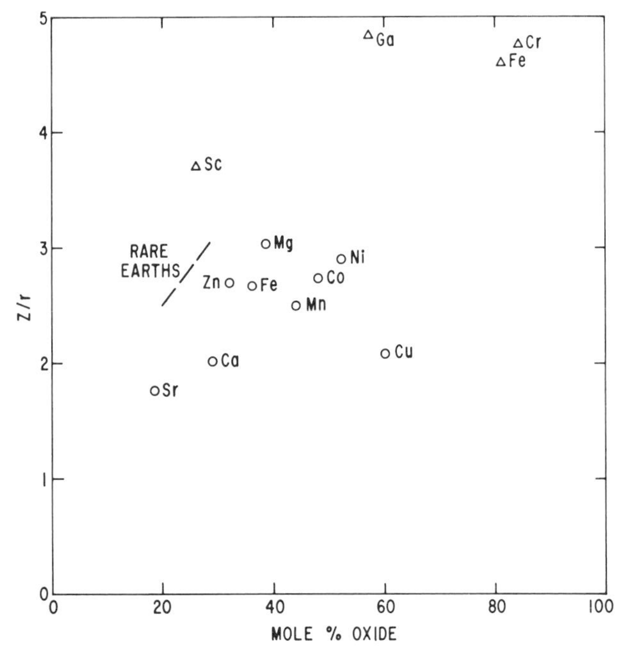

Fig. 5 Width of miscibility gap at its base for various binary silicate mixtures as a function of the charge over ionic radius (Z/r). ○, two-valent ions; △, three valent ions; data on phase diagrams from Ref. 4.

temperature at the base is close to 1700°C for all the melts plotted, as shown in Fig. 4, so the base width should be an approximate measure of the tendency to phase separation, as suggested by Glasser et al.[5] The base widths were taken from phase diagrams in Ref. 4. If the tendency towards phase separation were determined by the Z/r ratio, all the points in Fig. 4 would fall on the same straight line. One could draw a line through the data for the divalent ions of Ca, Sr, Mn, Co and Ni, and a line from the Sr point to those for trivalent Fe and Cr does not deviate too much from the points for divalent Ca, Mn, Fe, Co and Ni. However, the data for divalent ions Zn, Mg, and Cu do not fit with those already mentioned, and the trivalent rare earths together with scandium and gallium appear to form a separate group. Beryllium and lead silicates apparently show no stable separation, possibly because these ions are tetrahedrally coordinated and act as glass formers at least in some concentration ranges. Several silicates of four-valent ions show stable two-liquid regions, but the extent of separation is less than would be expected from the Z/r ratio, especially for thorium, uranium, and zirconium silicates. Germanium and phosphorous silicates show no stable phase separation because they substitute for silicon in the silicon-oxygen network. Vanadium silicate with a high Z/r ratio shows no phase separation, although the similar niobium silicate separates into two liquids at 1695°C over most of the binary composition range. Therefore other factors than the charge and ion size can be quite important in determining the tendency of a binary silicate to separate into two liquid phases.

The presence of stable immiscibility is important in glass manufacture. Two mixed glassy phases often have quite different properties than a single phase of the same average composition. For example, weathering and devitrification are faster the lower the silica content of a phase. Thus a glass separated into a silica-rich phase and another containing less silica might weather or devitrify more rapidly than a homogeneous glass. The resultant properties depend on the scale of the separation, since in some borosilicate glasses, as mentioned before, the very fine scale of phase separation leads to a more chemically resistant glass.

In binary alkali silicates there is no stable separation into two liquid phases. However, these silicates are easily cooled to glasses without crystallization, and some of them show liquid-liquid phase separation or immiscibility at temperatures below the liquidus temperature of crystallization. This type of phase separation is often called metastable, because crystalline phases are more stable than liquid ones at the temperatures of phase separation. The immiscibility curves for lithium and sodium silicates are shown in Fig. 6. There is some indication of phase separation in binary potassium silicates,[6] but the immiscibility boundary has apparently not been determined. There is as yet no evidence for phase separation in binary rubidium and cesium silicates. The metastable immiscibility boundary for binary barium silicates is shown in Fig. 4.

Most of the data for these curves were found from the "clearing" temperature of the glasses, or the temperature above which the opalescent phase-separated glass becomes optically clear. This method is not entirely reliable, because the amount of light scattering depends on the size of the

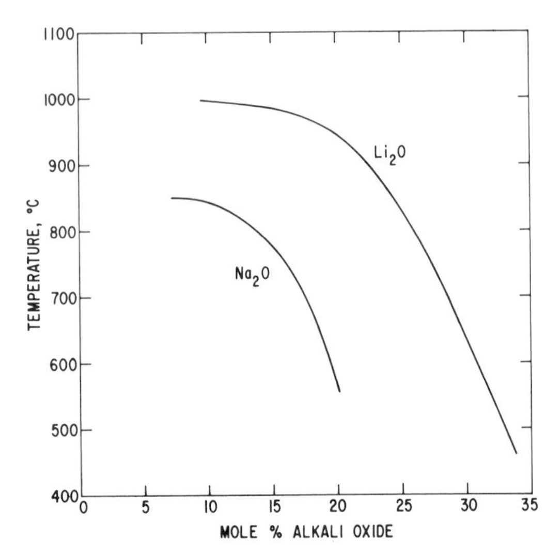

Fig. 6 Phase-separation diagrams for lithium[7,12,57] and sodium[55,57] silicates.

particles and the refractive index difference between the two phases. The light scattering per unit volume of particles is a maximum at an intermediate particle size. For particles very much larger or smaller than the wave length of the incident light the scattering is less. Thus if the size of the particles changes, or the refractive index difference changes as the composition difference between the particles and matrix changes, the clearing temperature may not always be at the phase boundary. The phase boundary can be determined most reliably from electron micrographs of quenched samples, if the quenching is fast enough to preserve the structure at the quenching temperature. If the composition of one phase is known, the volume fractions of the separated phases as measured from the micrographs[7] can also be used to calculate the phase boundary.

The metastable immiscibility boundaries of some alkali borates[8] and borate-rich borate-silicates[9] have been determined. The curves for the borates of lithium, sodium, potassium, rubidium, and cesium are remarkably similar, in contrast to the large differences for the corresponding silicates. The phase diagrams for the borates are no more alike than those for the silicates.

Burnett and Douglas have made a detailed study of phase separation in the commercially important ternary system of sodium calcium silicates,[10] which form the soda-lime glasses. The addition of small amounts of soda to binary calcium silicates drops the immiscibility region below the liquidus temperature. No phase separation is observed below about 580°C because diffusional processes are too slow below this temperature. The maximum temperature of phase separation (consolute temperature) occurs at about 90 mole % SiO_2 for all ratios of the other components; in fact the consolute temperature in all alkaline earth and alkali silicates seems to occur at about 90% SiO_2. The composition of the silica-rich phase in these glasses becomes almost 100% silica at temperatures not much below the consolute temperature.[11]

In both the soda-lime-silica[10] and the soda-lithia-silica[12] systems Douglas and his co-workers found that the reciprocal of the absolute immiscibility temperature was linear with changing composition ratio of the nonsilicate ions, at constant silica composition. This relationship is probably valid for most multicomponent silicates. It provides a convenient means for finding immiscibility temperatures in complicated systems in which these temperatures for binary or ternary mixtures are known.

Burnett and Douglas also found that the composition c of the immiscibility curves for several binary and pseudobinary (ternaries with constant silica) silicate mixtures was related to the temperature T by the relation

$$|c_c - c| \propto (T_c - T)^{1/2} \tag{1}$$

where c_c and T_c are the composition and temperature at the critical consolute point. This relation is of theoretical importance because it can be derived from a simple treatment of the free energy of the glass as a function of temperature of the type described in the next section.

Burnett and Douglas have also measured the immiscibility curves for the soda-baria-silica system[13]; they are similar to the soda-lime-silica system except for the lower consolute temperature of the binary baria-silica system.

Another important commercial ternary system is that of sodium borosilicates. Results of several studies in this system have been summarized by Haller et al.[14,14a] and are shown in Fig. 7. There is some question about results in the silica-rich corner, and the positions of tielines[14b]. As mentioned above, there is good evidence that silica-rich phases are nearly 100% silica in alkali and alkaline earth glasses. The results of Hair and Chapman[15] indicate that much of the boron that remains after leaching phase-separated borosilicate glass is on the internal surface of the leached glass and not in solution in the silica. Thus it seems likely that there is little solubility of borate in silicate, as would be expected from the different structures of these amorphous oxides described in the last chapter. Therefore, the silica-rich phases in sodium borosilicate glasses are also probably close to 100% silica.

The effect of small additions of "nucleating agents," such as TiO_2 and P_2O_5, on phase separation in sodium and lithium silicates has been studied because these agents enhance nucleation of crystals in these glasses. James

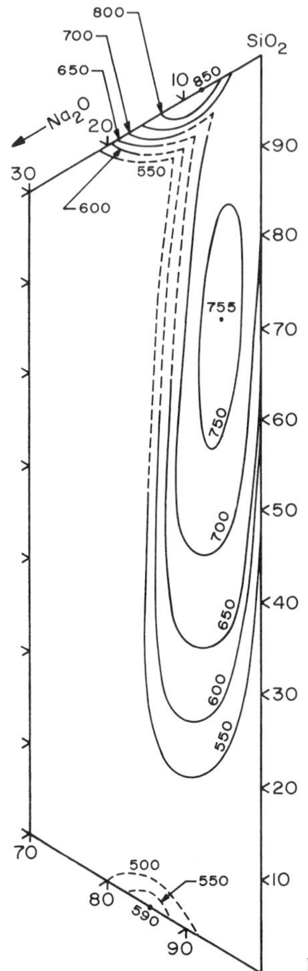

Fig. 7 Phase separation in the sodium borosilicate system.[14] Dashed lines are uncertain.

and McMillan concluded that P_2O_5 had only a small effect on the solubility of silica in the lithium-rich phase in lithium silicate glasses. Tomozawa[16] found that the "clearing temperature" of phase-separated sodium silicate glasses was increased by the addition of P_2O_5, decreased slightly by the addition of TiO_2, and remained unchanged when $AlPO_4$ was added. He explained the greater tendency to phase separation of P_2O_5 over TiO_2 as a result of the higher ionic potential (Z/r) of phosphorous than titanium ions, and the lack of an effect with $AlPO_4$ by the substitution of $AlPO_4$ for SiO_2 in the glass network.

Aluminum ion is known to have a strong tendency to reduce phase separation in silicates. For example, addition of 1.8 mole % Al_2O_3 to a sodium borosilicate glass of the consolute composition (in mole %, 70.5 SiO_2, 27.7 B_2O_3, 6.8 Na_2O) lowered the consolute temperature 111°C from 754°C.[16a] The reason for this effect is uncertain.

The immiscibility range in alkaline earth silicates and borates is sharply reduced by the addition of alkali oxides. Shartiss, Shermer, and Bestal[17] made a detailed study of this reduction in borates. They showed that the effectiveness of alkali oxides in reducing phase separation was in the order Cs > K > Na > Li, which is the same order of decreased tendency to phase separation in the binary alkali silicates, and corresponds to a reduction in ionic size and increase in ionic field strength.

Phase separation has been found in several other of the glass-forming systems discussed in the last two chapters. Various alkali and alkaline earth fluoride-beryllium fluoride glasses show phase separation.[18] A variety of chalcogenide glasses have regions of liquid-liquid immiscibility.[19]

THEORIES OF IMMISCIBILITY

Various microscopic theories have been put forward to explain oxide miscibilities for different systems and compositions. In order to understand these theories it is first useful to describe some thermodynamic parameters involved in the development of phase separation. This discussion is also important for an understanding of the proposed mechanism of spinodal decomposition for phase separation in glasses described in a later section.

The phase boundary between two liquid phases is related to the Gibbs free energies of the various possible phases. The free energy of two miscible components A and B as a function of composition is shown in Fig. 8*a*. The free energy of a single-phase mixture is always lower than the sum of the free energies of any two other phases. As the temperature is lowered the single phase separates into two phases, and the free energy diagram has two branches, as shown in Fig. 8*b*. The compositions of the two phases in equilibrium correspond to the tangent points of the common tangent line. The shape of the free energy curve between the compositions of the equilibrium phases is uncertain because it is difficult to conduct experiments on the individual phases in the unstable region; they very quickly transform to the stable state. Two possible shapes of the free energy curves in the unstable region are shown in Figs. 8*b* and *c*.

Various theories can be used to derive free energy curves of the types shown in Fig. 8. The simplest is often called the regular solution or nearest neighbor model.[20,21] In this model each component is assumed to be randomly distributed and coordinated with a certain fixed number Z of

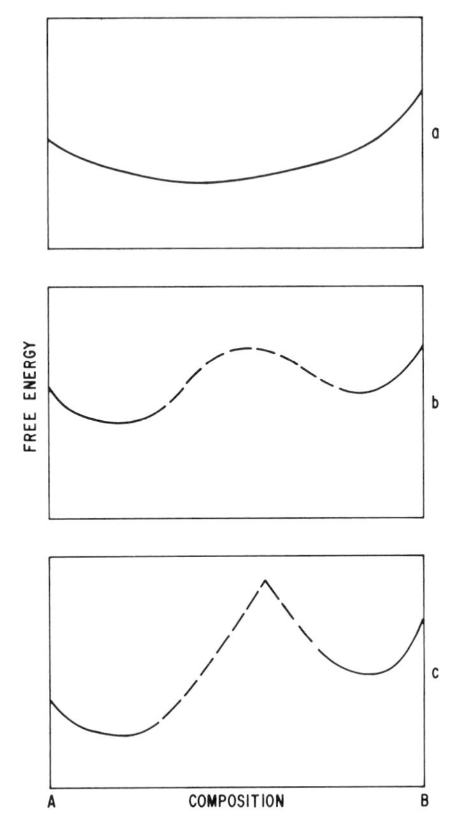

Fig. 8 Schematic diagrams of Gibbs free energy as a function of composition. (*a*) Complete miscibility; (*b*) separation into two phases; (*c*) separation into two phases, another possible functional dependence.

neighbors. In metallic alloys or mixtures of simple molecules such as organic liquids the elementary mixing units are clearly the metal atoms or small molecules. However, in a network compound such as a silicate it is often difficult to identify the mixing units that should be used for theoretical calculations, and this uncertainty leads to weaknesses in theoretical descriptions of miscibility in silicates, as described below.

The molar free energy of mixing, ΔG_m, is the free energy change when pure components are mixed to the desired composition, per mole of final mixture. In terms of the enthalpy and entropies of mixing

$$\Delta G_m = \Delta H_m - T\Delta S_m \tag{2}$$

In this model it is assumed that the entropy change is the ideal entropy of mixing, or

$$\Delta S_m = -R(X_A \ln X_A + X_B \ln X_B) \tag{3}$$

where X_A is the mole fraction of component A and X_B is the mole fraction of component B.

The enthalpy change ΔH_m is assumed to equal the energy change ΔE_m. The total energy of the mixture is calculated from bond energies of nearest neighbors, there are on the average ZX_A molecules of A and ZX_B molecules of It is also assumed that the bond energies are not a function of composition or temperature. In a mixture with A and B molecules, each with Z nearest neighbors, there are on the average ZX_A molecules of A and ZX_B molecules of B around any given molecule in the mixture. Then if the total concentration of molecules in the mixture is c, the number of A-A pairs per unit volume is $ZX_A^2c/2$, the number of B-B pairs is $ZX_B^2c/2$, and the number of A-B pairs is ZX_AX_Bc. If the energy of A-A bonds is $-E_A$, B-B bonds is $-E_B$, and A-B bonds is $-E_{AB}$, where the Es are positive quantities, then the total energy E per unit volume of the mixtures is

$$E = -\left(\frac{ZX_A^2cE_A}{2} + \frac{ZX_B^2cE_B}{2} + ZX_AX_BcE_{AB}\right)$$

or (4)

$$E = -\frac{Zc}{2}\left[X_AE_A + X_BE_B + 2X_AX_B\left(E_{AB} - \frac{E_A + E_B}{2}\right)\right]$$

The first two terms in the brackets give the energies of pure A and B, so the energy of mixing is

$$\Delta E_m = -ZcX_AX_B\left(E_{AB} - \frac{E_A + E_B}{2}\right) \tag{5}$$

If the energy of A-B bonds is greater (more negative) than the energy of A-A and B-B bonds, then $E_{AB} > (E_A + E_B)/2$, and ΔE_m is negative at all compositions. Under these conditions the free energy function is as shown in Fig. 8*a* for all temperatures, and the two components are completely miscible at the temperatures where the equations apply. If, however, $E_{AB} < (E_A + E_B)/2$, then there is a tendency for like atoms to segregate, and below a certain temperature the mixture separates into two different phases. The free energy as a function of composition is then given by Fig. 8*b*.

In applying these simple considerations to immiscibility in silicates there is uncertainty about the basic units for entropy considerations and about coordination numbers and bond strengths for energies. Warren and Pincus[22] made the following argument to explain the greater tendency of binary silicates to show immiscibility as the field strength (Z/r) of the cation increases. There are two tendencies for bonding in a binary silicate melt: one for the silicon ions to coordinate four oxygen ions tetrahedrally and the other for the other cation to surround itself with nonbridging (negatively charged) oxygen ions of the network. If the field strength of the cation is low, its coordination is incomplete, and a single phase dominated by the silicon-oxygen network results. On the other hand, if the field strength of a cation is high, it is better able to coordinate nonbridging oxygens, and these groups then form a separate phase, the other phase being predominantly silica. Similar arguments were made by Levin and Block in their extension[23] of the work of Warren and Pincus. In terms of the nearest-neighbor theory this point of view allows the coordination numbers of the different bond types to be different. Thus as the coordination number of the ion becomes larger in the mixture, Z_{AB} can be considered to be larger, and the energy of mixing becomes positive, leading to phase separation. One can also consider the larger field strength of the modifier ion as leading to a larger bond strength E_{AB}, again favoring phase separation, as suggested by Rawson.[24] Thus the structural arguments of these authors involve the energetic contribution to the free energy of mixing. Understanding of the lack of phase separation for ions that substitute for the network ions, and more detailed calculations of energies are hampered by the uncertainty about just what units are mixing and bonding.

Charles[25] rejected the Warren-Pincus explanation of silicate immiscibility, at least in alkali and alkaline earth silicates, and substituted a description based on the deformation of the silica chains in the network. He argues that a more positive value of the energy of mixing ΔE_m as the cation concentration increases, required for immiscibility, results because of the distortion of the bridging silicon-oxygen bonds caused by the coulombic forces of the metal cations and negatively charged oxygen ions. The maximum energy occurs at some intermediate composition of cation determined by its field strength. Support for this view depends critically on the choice of the expression for the entropy of mixing. Charles asserts that the simple use of R_2O or RO and SiO_2 groups as the mixing entities is structurally unrealistic. He considered that the important mixing units are bridging and nonbridging oxygen ions, following a suggestion of Forland. The expression derived for ΔS_m from this model is

$$\Delta S_m = -R\left[X_A \ln \frac{X_A}{2X_B} + (2X_B - X_A) \ln \frac{2X_B - X_A}{2X_B}\right] \tag{6}$$

Whereas this equation seems more reasonable than Eq. 3 for a binary silicate, there is no direct experimental support for it, so that one must reserve judgement about Charles' approach until more experimental data on thermodynamic properties of silicate mixtures are available.

Charles also calculated the metastable immiscibility boundaries of the lithium and sodium silicate systems, shown in Fig. 5, from a subregular solution model.[26] In this model a more complicated equation for ΔE_m is used than Eq. 5, and the consolute temperature does not necessarily occur at the 50% composition (Ref. 21, pp. 196 ff.). Charles derived the activities of the components from the stable liquidus phase boundary. The consolute temperature of the immiscibility curve for sodium silicates was accurately predicted by Charles' treatment, but the predicted width was too small. The consolute temperature predicted for the lithium silicate system was 200°C too high, and the width was again too narrow. It seems likely that these discrepancies derive from the many simplifications implicit in the subregular solution model and also the uncertainties of deriving thermodynamic data at much lower temperatures from the liquidus phase boundary. To make more reliable predictions for other systems, more thermodynamic data, and perhaps an improved model, are needed.

NUCLEATION

The rates of phase separation are important to the development of the morphologies of the phases and thus in some properties of phase-separated glasses such as electrical conductivity and chemical durability. In this and the next two sections various aspects of the kinetics of phase separation in glasses are considered and related to the morphology of the phase-separated structures. Also considered are the fundamentals of "classical" nucleation and growth for understanding kinetics of both phase separation and crystallization. In the following section an alternative mechanism for phase separation, that of spinodal decomposition, is treated. In this discussion nucleation of a crystal from a pure liquid is first considered and then extended to nucleation of a second liquid phase or a crystal from a two-component liquid.

When a liquid is cooled below its melting point small crystalline regions form by molecular fluctuations. However, the formation of these embryos is retarded because of the interfacial energy between them and the liquid. As the size of an embryo increases its chemical potential decreases until it is less than that of the liquid, as shown in Fig. 9; then the embryo becomes a growing crystalline nucleus. The size at which the chemical potential of the embryo equals that of the liquid is called the critical size. The probability p of formation of an embryo of radius r is proportional to the exponential of the reversible work W_r done on the surroundings by the fluctuation to form the embryo:

$$p \propto \exp\left(\frac{W_r}{kT}\right)$$

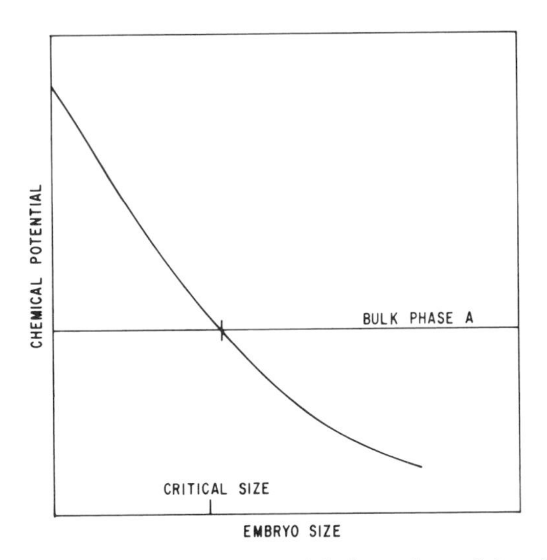

Fig. 9 Schematic drawing of the chemical potential of an embryo of phase *B* forming in phase *A* as a function of embryo size.

where k is Boltzmann's constant and T the absolute temperature. The rate of formation I of growing crystalline nuclei is proportional to the probability of forming a critical nucleus, or

$$I = K \exp\left(\frac{W^*}{kT}\right) \tag{7}$$

where W^* is the work done in forming the critical nucleus. A variety of methods of calculating the coefficient K have been suggested;[27–29] the exponential term is usually predominant, especially in condensed systems. In this discussion the dependence of K on composition and temperature is assumed to be negligible.

The work done in forming a spherical isotropic critical nucleus with isotropic surface energy γ is[29a]

$$W^* = -4\pi r^{*2}\gamma + \frac{4\pi r^{*3}\,\Delta P}{3} = -\frac{16\pi\gamma^3}{3(\Delta P)^2} \tag{8}$$

where ΔP is the pressure difference between the inside of the nucleus and the surrounding phase, and the radius of the critical nucleus r^* is given by

$$r^* = \frac{2\gamma}{\Delta P} \tag{9}$$

For a crystalline nucleus formed from a liquid of the same composition, the pressure difference is equal to the free energy of fusion per unit volume at the temperature T of nucleation, and is approximately equal to

$$\Delta P \approx \frac{L}{vT_m}(T_m - T) \tag{10}$$

where T_m is the melting temperature, L is the heat of fusion, and v is the molar volume of the liquid.

Equation 8 is valid for the homogeneous nucleation of a phase. In actual practice nucleation is almost never homogeneous, but takes place on impurity particles, vessel walls, defects, or some other heterogeneity. These heterogeneities effectively lower the interfacial tension γ by a factor $f < 1$, so that

$$W^* = -\frac{16\pi(\gamma f)^3 v^2 T_m^2}{3L^2(T_m - T)^2} = -\frac{K_1}{(T_m - T)^2} \tag{11}$$

where K_1 is a coefficient approximately independent of temperature.

In condensed systems the rate of formation of fluctuations can be limited by diffusion or viscous flow in the liquid. This transport limitation is particularly important in viscous systems such as glasses. It can be taken into account by adding a transport factor to Eq. 7 for the rate of nucleus formation. Then

$$I = K_2 \exp\left(\frac{W^* - Q}{RT}\right) \tag{12}$$

where Q is the activation energy for the transport process and K_2 is assumed independent of temperature. Without the transport limitation there is a very sharp increase in nucleation rate at some critical temperature below the melting temperature because of the $(T_m - T)^2$ factor in the exponential. However, with the transport factor there is a maximum in nucleation rate with decreasing temperature that can be quite broad, depending on the relative size of the two exponential terms.

Very similar considerations apply to nucleation of a liquid phase from a homogeneous liquid solution of two components. It is assumed that the nucleus forms with the equilibrium composition of one liquid phase and grows by accretion of material from the remaining matrix liquid. The composition of the nucleating phase, that is, the one forming discrete particles, is the equilibrium phase of composition furthest from the overall composition of the liquid. As an example, consider nucleation in lithium silicate glasses whose equilibrium phase-separation diagram is shown in Fig. 6. In a homogeneous glass containing 26 mole % LiO_2 and 74 mole % SiO_2

the silica-rich phase nucleates in the lithium-rich matrix below about 800°C because the equilibrium amount of the silica-rich phase is less than that of the lithium-rich phase. An electron micrograph of phase separation in this glass after heat treatment at 500°C, Fig. 10, shows silica-rich particles as raised circles because they etch more slowly in hydrofluoric acid than the lithium-rich matrix.

In nucleation in a two-component system the driving force ΔP for nucleation is a function of composition and is not given simply by Eq. 9. The expression for the driving force is (Ref. 28, p. 250)

$$\Delta P = G_1 - G_m - (c_1 - c_m)\left(\frac{\partial G}{\partial c}\right)_{c_m} \tag{13}$$

where G_1 is the free energy of the nucleating phase of composition c_1, and G_m is the free energy of the matrix of composition c_m (normally the bulk composition). The interfacial energy γ is also a function of composition. In a simple nearest-neighbor treatment Becker[30] found that the interfacial tension was the sum of two terms, one a structural contribution independent of composition and the other proportional to $(c_1 - c_m)^2$.

Hammel studied nucleation rates of silica-rich particles in a 13 mole % Na_2O, 11 mole % CaO, 76 mole % SiO_2 glass in detail.[31] The number of particles he measured as a function of time are shown in Fig. 11 for several different nucleation temperatures. The particles were measured from electron micrographs down to a radius of about 30 Å; since the radii of critical nuclei were in this size range he apparently measured virtually all of the nucleated particles. Figure 11 shows that a constant rate of nucleation was achieved only after a period of time; this transient is discussed in more detail below. These constant rates are plotted as a function of nucleation temperature in Fig. 12.

In order to compare his results with the theory for homogeneous nucleation, Hammel measured the radius r^* of the critical nucleus at temperatures just below the miscibility gap. He made this measurement by holding a sample containing particles of a certain size in a temperature gradient and observing the temperature at which they neither grew nor dissolved, this then being the temperature at which the size was the critical size. If ΔP is proportional to the undercooling below the miscibility gap ΔT, a plot of $1/r^*$ against temperature should be a straight line; Hammel found this relationship. Then r^* at lower temperatures can be found by extrapolating the data at higher temperatures. To calculate the work W^* from Eq. 8 and r^* it is necessary to know ΔP. Hammel calculated ΔP from the miscibility curve and various equations; the equation of Lumsden[32] seemed to give the best fit. As discussed in the last section such calculations are rather uncertain. Hammel used the resulting value of W^*, and $Q = 4750°\text{K} =$

Fig. 10 Electron micrograph of phase separation in a 26 mole % Li_2O, 76% SiO_2 glass after 16 hr at 500°C. Replica of fracture surface etched 5 sec in 2% HF. Magnification 22,400 ×.

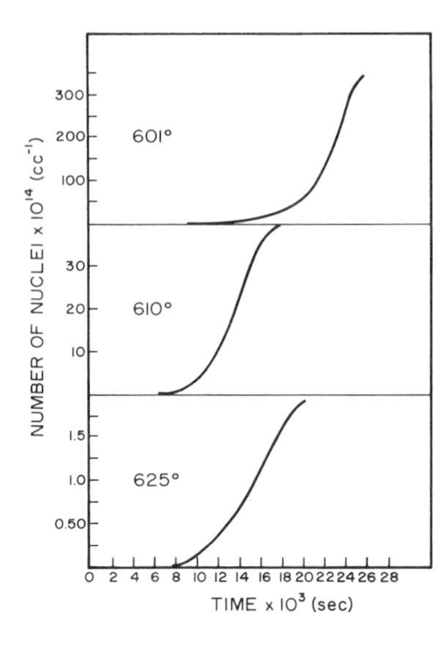

Fig. 11 Numbers of crystals in a 13 mole % Na_2O, 11% CaO, 76% SiO_2 glass as a function of time, measured by Hammel.[31]

99.5 kcal/mole as estimated from growth experiments described in the next section, together with a value of K_2 calculated from an equation of Turnbull[28], to compute the nucleation rates shown in Fig. 12. The absolute values of calculated and measured rates are surprisingly close, perhaps fortuitously so in view of uncertainties in ΔP and K_2. The temperature dependence does not agree as well, probably because of inaccurate values of ΔP and consequently W^*. It is also not certain that the value of Q from the growth measurements is the correct one for nucleation. Other workers found lower values of Q, as described in the next section; a lower value of Q would give poorer agreement on the absolute values, but better agreement on the temperature dependence than shown in Fig. 12. In any event the agreement between experimental and calculated nucleation rates, which is exceptional for nucleation experiments, provides evidence that the nucleation is homogeneous and supports the theory of nucleation outlined above.

The transient shown in Fig. 11 before reaching constant nucleation rate was also found by Burnett and Douglas[10] in a study of nucleation in a glass containing 12.5 mole % Na_2O, 12.5 mole % CaO, and 75 mole % SiO_2. These authors found that nucleation did not occur in their glass until it was at least 27°C below the miscibility temperature of 687°C. They studied nucleation by holding their glass at some lower temperature and then growing the nucleated particles at a temperature of 680°C where no more nucleation should occur. In the early stages of nucleation before a constant

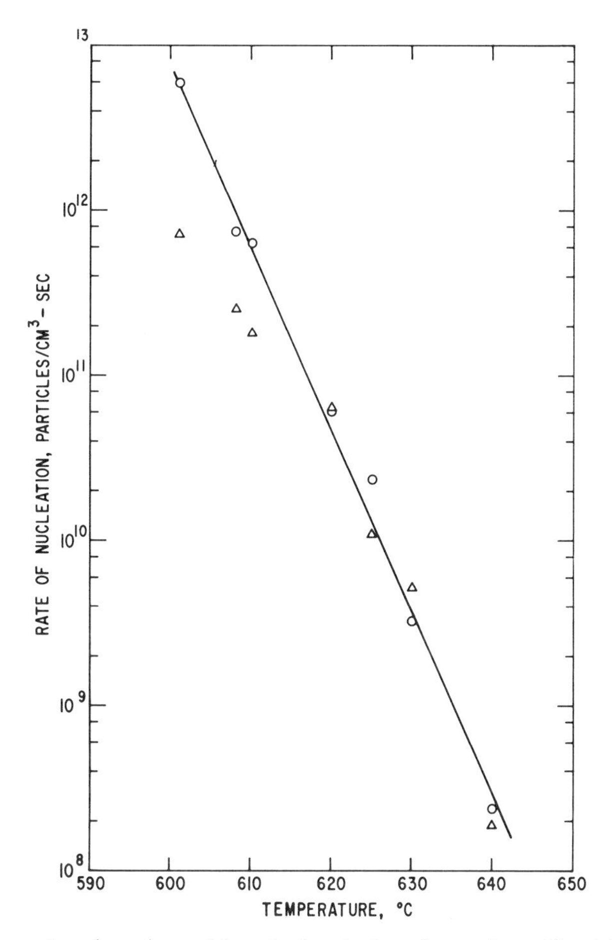

Fig. 12 Temperature dependence of the rate of nucleation of amorphous silica-rich particles in a 13 mole % Na_2O, 11% CaO, 76% SiO_2 glass. ○, measured; △, calculated by Hammel.[31]

rate of nucleation was achieved the number of particles N followed the equation

$$N = K_3 \exp\left(\frac{-\tau}{t}\right) \tag{14}$$

so that the nucleation rate is given by

$$I = K_4 \exp\left(\frac{-\tau}{t}\right) \tag{15}$$

where K_3, K_4, and τ are parameters independent of time. Equations of this form were first derived by Zeldovich and subsequently by many other authors. Such a dependence of particle size occurs because the steady-state distribution of embryo sizes that is assumed in deriving Eq. 7 has not been set up yet, because of the slow diffusion in the condensed state. The parameter τ is related to the concentration of diffusing species and its diffusion coefficient.

GROWTH

A particle of a second phase that has nucleated from a homogeneous multicomponent phase grows by accretion of material from the parent phase; the particle can have a different composition from the initial material. In this "classical" nucleation and growth treatment it is assumed that the particle has the equilibrium composition. The question of particle composition is discussed further in the next section. The isothermal rate of growth of the particle can be controlled either by diffusion of material in the parent phase or by a reaction at the particle-matrix interface. In diffusion controlled growth the concentration of diffusing material in the matrix at the particle-matrix interface is equal to the equilibrium solubility C_e of this material in the matrix. The profile of diffusing material in the matrix is shown in Fig. 13*a*. The initial concentration of material in the matrix at a large distance from the particle is C_∞, and the concentration of the material in the particle is C_p. In interface-controlled growth diffusion is rapid and there are only small concentration gradients in the matrix, so the interface concentration is close to C_∞, as shown in Fig. 13*b*. In diffusion control the rate of growth of a spherical particle of radius R is[33,34]

$$\frac{dR}{dt} = \alpha \frac{D}{R} \tag{16}$$

where α depends only on concentrations. In dilute solution $\alpha = (C_\infty - C_e)/(C_p - C_e)$. The radius grows proportional to the square root of time. In interface-controlled growth the radius grows proportional to time:

$$R = \frac{(C_\infty - C_e)Gt}{C_p - C_\infty} \tag{17}$$

where G is an interface-growth coefficient. When both processes are important the rate of growth in dilute solution is[35]

$$\frac{dR}{dt} = \frac{(C_\infty - C_e)DG}{(C_p - C_r)(D + GR)} \tag{18}$$

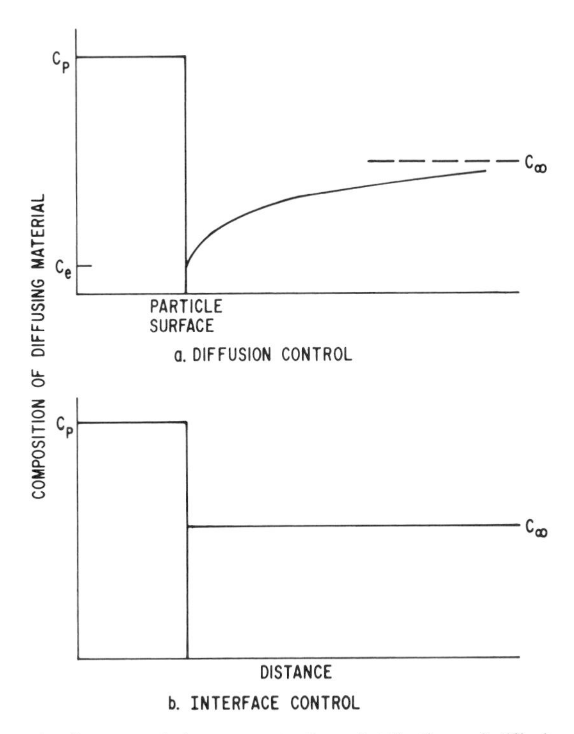

Fig. 13 Schematic diagram of the concentration distribution of diffusing material when growth is controlled by diffusion in the matrix (*a*) or by an interface process (*b*). C_e is the equilibrium concentration of solute in the matrix, C_p is the solute concentration in the particle, and C_∞ is the initial solute concentration in the matrix and the concentration far from the particle.

in which C_r is the concentration at the particle-matrix interface and is intermediate between C_∞ and C_e.

The equations above are valid only for an isolated particle. If a number of particles compete for the diffusing material, the growth rate eventually decreases to zero. An approximate equation for this competitive effect is

$$W = 1 - \exp(-bt^n) \tag{19}$$

where W is the fraction of total growth and b is a coefficient depending on growth mechanism, temperature, and concentration. For diffusion-controlled growth of a constant number of spherical particles $n = 3/2$; for interface control $n = 3$.

A more accurate equation for the diffusion-controlled growth of a constant number of competing spheres in dilute solution is[35]

$$\frac{(C_\infty - C_e)Dt}{3(C_p - C_e)R_f^2} = \frac{1}{2}\ln\left[\frac{W^{2/3} + W^{1/3} + 1}{(1 - W^{1/3})^2}\right] - \sqrt{3}\tan^{-1}\left(\frac{\sqrt{3}W^{1/3}}{2 + W^{1/3}}\right) \quad (20)$$

in which R_f is the final radius of the particles. This equation holds very well for the diffusion-controlled growth of gold particles in glass.[36]

At longer growth times larger particles grow and smaller particles dissolve to decrease the total particle surface. This "Ostwald Ripening" or coarsening is often observed in phase-separated glasses, and occurs after the matrix has almost reached the equilibrium composition. The mathematical treatment for this process has been given by Lifshitz and Slyozov[37] and Wagner[38] based on earlier work by Greenwood.[39] The driving force for coarsening is provided by the higher solubility of the smaller particles, resulting in a transfer of material from them to the larger particles. The mean radius R of the particles is given by

$$R^3 - R_0^3 = \frac{8\gamma C_e D V_m t}{9RT} \quad (21)$$

in which R_0 is the mean particle radius when time $t = 0$, γ is the particle-matrix interfacial tension, D is the diffusion coefficient of material in the matrix, V_m is the molar volume of the particles, R is the gas constant, and T is the temperature. Thus the mean particle radius increases as $t^{1/3}$.

Hammel[31] studied the growth of phase-separated silica particles in a soda-lime silicate glass with 13 mole % Na_2O and 11 mole % CaO, the same composition used for his nucleation studies described above. He found that the particle radius increased proportionally to the square root of time, as required for diffusion-controlled growth. He found an activation energy of growth of 94.5 kcal/mole, close to the activation energy of about 110 kcal/mole for viscous flow in glasses close to this composition, from diffusion coefficients calculated from Eq. 16. The solution was not dilute, so α was a function of composition as tabulated by Frank.[34] Hammel assumed that silica was the diffusing species. Burnett and Douglas[10] also found that the particle size was proportional to the square root of time in the early stages of growth of silica particles in soda-lime glass.

In the later stages of growth, after the glass matrix is close to the equilibrium composition, the particles in phase-separated glasses grow by the coarsening mechanism of Eq. 21, as shown by proportionality of the particle radius to the cube root of time. From such measurements Burnett and Douglas caculated an activation energy for diffusion of 59 kcal/mole for their glass containing 10 mole % Na_2O, 10% CaO, and 80% SiO_2, which is considerably smaller than the activation energy found by Hammel[31] for diffusion in a similar glass. The absolute values of the diffusion coefficients in the two studies were similar at 640°C: Hammel found $2.7(10)^{-15}$ cm^2/sec for growth, and Burnett and Douglas $0.6(10)^{-15}$ cm^2/sec for coarsening and $1.5(10)^{-15}$ cm^2/sec for growth. If Burnett and Douglas had used Hammel's measured value of 4.6 for the interfacial tension instead of 10, their value for coarsening would be $1.3(10)^{-15}$ cm^2/sec.

It is interesting to compare this diffusion coefficient with that calculated from the viscosity η and the Stokes-Einstein equation:

$$D = \frac{kT}{6\pi\eta R} \tag{22}$$

where R is the particle radius and k is the Boltzman's constant. For a D of $2.7(10)^{-15}$ at 640°C where the viscosity is about $2(10)^{11}$ P, the radius of the diffusing entity is 10^{-11} cm. Thus diffusion is apparently not directly related to viscous flow, in spite of the similarity between Hammel's diffusive activation energy and that for viscous flow. The diffusion coefficient for lattice oxygen in a soda-lime glass, as measured by Kingery and Lecron,[40] is about 10^{-13} cm^2/sec at 640°C, with an activation energy of 66 kcal/mole, so perhaps the rate of particle growth is related to diffusion of lattice oxygen, even though it must involve transport of silica.

Activation energies for diffusion have been calculated for other glasses from coarsening studies (composition in mole %): 10 Na_2O, 10 Li_2O, 80 SiO_2, 47 kcal/mole;[12] 1 K_2O, 26 Li_2O, 73 SiO_2, 71 kcal/mole;[41] 3 PbO, 97 B_2O_3, 82 kcal/mole;[42] 19 PbO, 76 B_2O_3, 5 Al_2O_3, 73 kcal/mole.[43] In all of these studies, as well as that of James and McMillan,[7] the particle radius was proportional to the cube root of aging time, as required by Eq. 21.

The presence of impurities such as water can have a considerable effect on the rate of growth of phase-separated particles,[44] apparently because of their disruption of the silica network and consequent more rapid diffusion.

MORPHOLOGY AND SPINODAL DECOMPOSITION

In the above studies the volume fraction of particles was small, and they grew as individual spheres. When the volume fractions of the two separating phases are not much different, the phases often separate into an interconnected structure, as shown in Fig. 1 (see also Ref. 44a). Two explanations of this morphology have been advanced. The first explanation is that phase separation occurs by the mechanism of spinodal decomposition[45,46] described below. The second is that particles nucleate and grow by the classical process described in the last two sections, but then agglomerate to the connected structure because of their high volume fraction.[48] Each of these views has attracted enthusiastic partisans, although neither has been solidly established by experiments. In this section both viewpoints are discussed so that the reader may choose between them.

The theory of spinodal decomposition is based on the free energy diagram for a two-phase system shown in Fig. 8*b*. Vital to the theory is a region in which $\partial^2 G/\partial C^2$ is negative, where G is free energy of the system and C is its composition. In such a region there is no barrier to phase growth, since any small fluctuation leads to a reduction in free energy. From this condition Cahn derived the mechanism of spinodal decomposition.[45] In this mechanism the transformation proceeds by a continuous change in the compositions of growing phases, whereas their extent remains constant. The composition changes take place in a regular three-dimensional array with a characteristic wavelength and grow until the compositions of the two phases reach the equilibrium values. This mechanism contrasts with the usual nucleation and growth process described above in which regions of constant composition (the equilibrium composition) nucleate and grow in size until the matrix also reaches the equilibrium value.

Cahn showed that the spinodal mechanism leads to an interconnected structure in an isotropic system such as one containing two liquid phases.[47] Cahn and Charles[46] suggested that this mechanism was therefore the reason for the interconnected structure in glasses, since a computer solution of the spinodal equations gives a cross-section very similar to that found in glasses, such as shown in Fig. 1. The region of spinodal decomposition as estimated from various models agrees qualitatively with the compositions at which an interconnected structure is observed in glasses, except that this structure is usually not observed at temperatures just below the consolute temperature even for the consolute composition. This result may be caused by rapid agglomeration of the interconnected structure.

Several studies of the earliest stages of phase separation in the interconnected region have been made with small-angle x-ray scattering and compared to the spinodal equations.[49–51] The height of the peak of scattered intensity as a function of angle is related to the composition difference between phases, whereas the angle at which the intensity is a maximum is related to the scale of the separation. Zarzycki and Naudin found that the angle of maximum intensity remained about constant with increase in aging time and height of the peak, as required by the spinodal mechanism, but this behavior was found only at the earliest aging times, as shown in Fig. 14. It would be very desirable to have electron microscopic evidence to confirm these deductions from the small-angle x-ray work; until they become available the evidence for the spinodal mechanism from this experimental technique remains tentative.

In support of the agglomeration mechanism of Haller, some electron

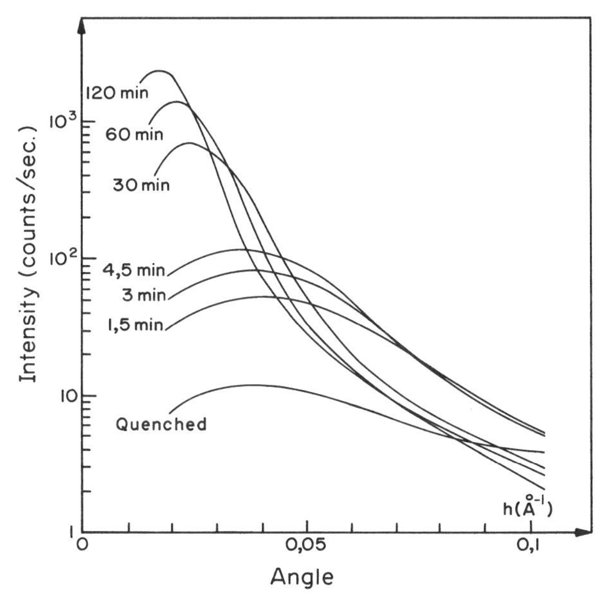

Fig. 14 Small-angle x-ray diffraction spectra of a glass containing 76 wt. % B_2O_3, 19% PbO, and 5% Al_2O_3, quenched from 1150°C and then heated at 450°C. From Ref. 42.

micrographs have shown discrete particles agglomerating to an interconnected structure as they were aged.[52,53] In Fig. 15 a transmission electron micrograph of a sodium-barium silicate glass is shown, taken by A. Turkalo of the General Electric Research and Development Center. This composition is close to that at the consolute temperature for the pseudobinary system SiO_2-$BaO \cdot Na_2O$. The glass, cooled rapidly from the melting temperature of about 1400°C, showed phase separation into discrete droplets of silica-rich phase surrounded by about an equal volume of sodium barium silicate phase, as shown in Fig. 15. After heating this glass for 202 min at 601°C an interconnected structure developed, as shown in Fig. 16. Thus it is possible for an interconnected structure to develop from one with discrete particles; however, it is not proven that this is the way all interconnected structures in phase-separated glasses are formed.

The presence of a region in which $\partial^2 G/\partial C^2$ is negative for phase-separating glasses has not been established experimentally. Various theories of binary solutions, such as the one described above, lead to a free energy function with a continuous derivative, but this may be merely a mathematical convenience.

Fig. 15 Transmission electron micrograph of a glass containing 88 mole % SiO_2, 6% Na_2O, and 6% BaO, quenched from 1400°C. Magnification 64,000 ×.

Fig. 16 Transmission electron micrograph of the same glass shown in Fig. 15, aged 202 min at 601°C after quenching. Magnification 60,000 ×.

Goldstein[53a] showed that the solute flux to the "joining points" of two adjacent spheres growing by diffusion becomes quite low as the interspherical distance decreases, and thus he concluded that the spheres would not coalesce during growth. Hopper and Uhlmann give other possible mechanisms leading to coalescence in spite of Goldstein's results.[54]

More experiments, particularly during the early stages of phase separation in the spinodal regions, are needed to decide by what mechanism the interconnected structure is developed.

REFERENCES

1. W. E. S. Turner and F. Winks, *J. Soc. Glass Tech.*, **10**, 102 (1926).
1a. M. E. Nordberg and H. P. Hood, U.S. Patent 2, 106, 744 (1934).
2. S. M. Ohlberg, H. R. Golub, J. J. Hammel, and R. R. Lewchuk, *J. Am. Ceram. Soc.*, **48**, 178, 331 (1965).
3. R. H. Doremus and A. M. Turkalo, *Science*, **164**, 418 (1969).
4. E. M. Levin, C. R. Robbins, and H. F. McMurdie, *Phase Diagrams for Ceramists*, American Ceramics Society, Columbus, Ohio, 1964; Supplement, 1969.
5. F. P. Glasser, J. Warshaw, and R. Roy, *Phys. Chem. Glasses*, **1**, 39 (1960).
6. Y. P. Gupta and U. D. Mishra, *J. Phys. Chem. Solids*, **30**, 1327 (1969).
7. P. F. James and P. W. McMillan, *Phys. Chem. Glasses*, **11**, 59, 64 (1970).
8. R. R. Shaw and D. R. Uhlmann, *J. Am. Ceram. Soc.*, **51**, 377 (1968).
9. R. J. Charles and F. E. Wagstaff, *J. Am. Ceram. Soc.*, **51**, 16 (1968).
10. D. G. Burnett and R. W. Douglas, *Phys. Chem. Glasses*, **11**, 125 (1970).
11. M. Yamane and T. Sakaino, *J. Am. Ceram. Soc.*, **51**, 178 (1968).
12. Y. Moriya, D. Warrington, and R. W. Douglas, *Phys. Chem. Glasses*, **1**, 19 (1960).
13. D. G. Burnett and R. W. Douglas, *Disc. Far. Soc.*, **50**, 200 (1970); *Phys. Chem. Glasses*, **12**, 117 (1971).
14. W. Haller, D. H. Blackburn, F. E. Wagstaff, and R. J.. Charles, *J. Am. Ceram. Soc.*, **53**, 34 (1970).
14a. G. Srinivasan, I. Tweer, P. B. Macedo, A. Sarkar, and W. Haller, *J. Noncryst. Solids*, **6**, 221 (1971).
14b. O. V. Mazarin, M. V. Streltsina, and A. S. Totesh, *Phys. Chem. Glasses* **10**, 63 (1969).
15. M. L. Hair and D. Chapman, *J. Am. Ceram. Soc.*, **49**, 651 (1966).
16. M. Tomozawa, in *Advances in Crystallization and Nucleation in Glasses*, American Ceramic Society, 1971, p. 41.
16a. J. H. Simmons, P. B. Macedo, A. Napolitano, and W. K. Haller, *Disc. Far. Soc.*, **50**, 155 (1971).
17. L. Shartsis, H. F. Shermer, and A. G. Bestul, *J. Am. Ceram. Soc.*, **41**, 507 (1958).
18. W. Vogel and K. Gerth, *Glastech. Ber.*, **31**, 15 (1958); *Silikattech.*, **9**, 353 (1958).
19. E. Plumat, *J. Am. Ceram. Soc.*, **51**, 499 (1968).
20. A. H. Cottrell, *Theoretical Structural Metallurgy*, Edward Arnold, London, 1948, pp. 153 ff.
21. J. W. Christian, *The Theory of Transformations in Metals and Alloys*, Pergamon, London, 1965, pp. 172 ff.
22. B. E. Warren and A. G. Pincus, *J. Am. Ceram. Soc.*, **23**, 301 (1940).
23. E. M. Levin and S. Block, *J. Am. Ceram. Soc.*, **40**, 95, 113 (1957); **41**, 49 (1958).
24. H. Rawson, *Inorganic Glass-Forming Systems*, Academic, New York, 1967, p. 123.

25. R. J. Charles, *Phys. Chem. Glasses*, **10**, 169 (1969).
26. R. J. Charles, *J. Am. Ceram. Soc.*, **50**, 631 (1967).
27. M. Volmer, *Kinetik der Phasenbildung*, Steinkopf, Dresden, 1939.
28. D. Turnbull, in *Solid State Physics*, Vol. 3, Academic, New York, 1956, pp. 225 ff.
29. Ref. 21, pp. 377 ff.
29a. J. W. Gibbs, *Scientific Papers*, Dover, N.Y., 1961, p. 254.
30. R. Becker, *Ann Phys.*, **32**, 128 (1938); see also Ref. 28, p. 270.
31. J. J. Hammel, *J. Chem. Phys.*, **46**, 2234 (1967).
32. J. Lumsden, *Thermodynamics of Alloys*, Institute of Metals, London, 1952, p. 335.
33. C. Zener, *J. Appl. Phys.*, **20**, 950 (1949).
34. F. C. Frank, *Proc. Roy. Soc.*, **201A**, 586 (1950).
35. H. L. Frisch and F. C. Collins, *J. Chem. Phys.*, **21**, 2158 (1953).
36. R. H. Doremus, in *Symp. on Nucleation and Crystallization in Glasses and Melts*, American Ceramic Society, Columbus, Ohio, 1962, p. 119.
37. E. M. Lifshitz and V. V. Slyozov, *J. Phys. Chem. Solids*, **19**, 35 (1961).
38. C. Wagner, *Z. Electrochem.*, **65**, 581 (1961).
39. G. W. Greenwood, *Acta Met.*, **4**, 243 (1956).
40. W. D. Kingery and J. A. Lecron, *Phys. Chem. Glasses*, **1**, 87 (1960).
41. R. A. McCurrie and R. W. Douglas, *Phys. Chem. Glasses*, **8**, 132 (1967).
42. J. Zarzycki and F. Naudin, *Phys. Chem. Glasses*, **8**, 11 (1967).
43. F. Naudin and J. Zarzycki, *C.R.*, **266**, 729 (1968).
44. N. J. Kreidl and M. S. Maklad, *J. Am. Ceram. Soc.*, **52**, 508 (1969).
44a. R. J. Charles, *J. Am. Ceram. Soc.*, **47**, 559 (1964).
45. J. W. Cahn, *Acta Met.*, **9**, 745 (1961).
46. J. W. Cahn and R. J. Charles, *Phys. Chem. Glasses*, **6**, 181 (1965).
47. J. W. Cahn, *J. Chem. Phys.*, **42**, 93 (1965).
48. W. Haller, *J. Chem. Phys.*, **42**, 686 (1965).
49. J. Zarzycki and F. Naudin, *C.R.*, **265**, 1456 (1967); *J. Noncryst. Solids*, **1**, 215 (1969); **5**, 415 (1971).
50. G. F. Neilson, *Phys. Chem. Glasses*, **10**, 54 (1969).
51. M. Tomozawa, R. K. MacCrone, and H. Herman, *Phys. Chem. Glasses*, **11**, 136 (1970).
51a. N. S. Andreev, G. G. Boiko, and N. A. Bokov, *J. Noncryst. Solids*, **5**, 41 (1970).
52. T. P. Seward, D. R. Uhlmann and D. Turnbull, *J. Am. Ceram. Soc.*, **51**, 634 (1968).
53. J. F. MacDowell and G. H. Beall, *J. Am. Ceram. Soc.*, **52**, 17 (1969).
53a. M. Goldstein, *J. Cryst. Growth*, **4**, 594 (1968).
54. R. W. Hopper and D. R. Uhlmann, *Disc. Far. Soc.*, **50**, 166 (1971).
55. V. I. Aver'yanov and E. A. Porai-Koshits, *The Structure of Glass*, Vol. 5, Consultants Bureau, New York, 1965, p. 63.
56. J. J. Hammel, paper given at the 7th International Congress on Glass, Brussels, 1965.
57. V. I. Aver'yanov and E. A. Porai-Koshits, *The Structure of Glass*, Vol. 6, Part 1, Consultants Bureau, New York, 1966, p. 98.

5

Crystallization

Glass is metastable with respect to crystalline phases at temperatures below its equilibrium liquidus temperature, which for alkali silicate glasses is usually higher than 700°C. Glasses can form because the rate of crystallization is low, as discussed in Chapter 2.

Surface crystallization of glass can lead to serious problems in glass manufacture because of the resultant changes in glass properties, such as viscosity and coefficient of thermal expansion. High stresses resulting from nonuniform contraction on cooling can cause fracture of the piece. Thus glass technologists have always been careful to avoid surface crystallization by not holding their glasses at temperatures where crystals grow rapidly in them. The proper conditions to avoid surface crystallization were found empirically, with the help of limited knowledge from phase diagrams. Recent experimental and theoretical studies on the crystallization of glass should be helpful in predicting crystallization rates in new compositions, and also in defining the limiting times and temperatures to which a glass piece can be subjected.

Opal glasses were among the first to be used for decoration, and have continued in this use down to the present. Their translucency derives from light scattered from internal crystals, resulting either from incomplete melting or from crystals that grew in the glass as it cooled. These glasses are often used as glazes on decorative or practical objects; however, their low strength has limited wider application.

Some of the few really new materials developed recently are uniformly crystallized glasses or "glass-ceramics." These materials were first developed at the Corning Glass Works,[1,2] and are made by special heat treatment of particular glass compositions to give a fine, uniform dispersion of crystals in the glass. These partially crystallized glasses can have many superior properties, such as high strength and impact resistance, low bulk and surface electrical conductivity, low dielectric loss, low chemical reactivity, low coefficient of thermal expansion, and a range of optical properties from clear to completely opaque or white resulting from light scattering from the crystals. Such glasses have been studied intensively since the original Corning discovery, and many companies now market them. Table 1 lists a few commercially available products with their enhanced properties and uses. Among other important uses not listed in the table are bearings, heat exchangers, reactor control rods, sealing, vacuum tube envelopes, substrates for electronic circuits, and capacitors. McMillan discusses these and other uses in his book entitled *Glass-Ceramics*,[3] as well as many other aspects of these materials.

Table 1 Properties and Uses of some commercial Glass-Ceramics

Glass type and company	Crystals	Properties	Uses
Corning 9606	$2MgO \cdot 2Al_2O_3 \cdot 5SiO_2$ (Cordierite)	Low expansion, transparent to radar	Radomes
Corning 9608	β-Spodumene	Low expansion, low chemical reactivity	Cookware
Owens-Illinois Cer-vit	β-Quartz	Very low expansion	Telescope mirrors
General Electric Re-X	Li_2O-$2SiO_2$	Low electrical conduction high strength	Insulators

In order to grow, a crystal must first nucleate; thus nucleation is discussed in the first succeeding section. A short section on morphology of crystals in glasses follows, and then rates of growth of crystals are treated.

NUCLEATION

The basic equations for the rate of nucleation were given in the last chapter. There it was shown that the nucleation rate I is

$$I = K \exp\left(\frac{W^*}{kT}\right) \tag{1}$$

where k is Boltzmann's constant, the effect of composition and temperature T on K is small, and the work of critical nucleus formation W^* for a spherical isotropic nucleus is

$$W^* = -\frac{16\pi\gamma^3}{3(\Delta P)^2} \tag{2}$$

γ is the interfacial tension and ΔP is the pressure difference between the interior of the nucleus and the surrounding phase, and is often called the driving force for nucleation or the free energy change per unit volume going from matrix to critical nucleus. An approximate equation for ΔP for a crystalline nucleus forming in a liquid or glass is

$$\Delta P = \frac{L}{vT_m}(T_m - T) \tag{3}$$

where T_m is the melting or liquidus temperature, L is the heat of fusion, and v is the molar volume of the liquid.

Therefore two important parameters in determining the nucleation rate are the interfacial tension γ and the driving force ΔP. Anything effectively lowering γ, such as impurity particles or defects, increases the rate of nucleation, and a lower temperature increases ΔP and consequently I.

Consider the formation of a nucleus of phase β from phase α. If there is an impurity particle present the effective interfacial tension γ_e for nucleation of β on the impurity particle is

$$\gamma_e = \gamma_{\beta i} - \gamma_{\alpha i} \tag{4}$$

where $\gamma_{\beta i}$ and $\gamma_{\alpha i}$ are the interfacial tensions between the phase and the impurity particle. Equation 4 results because the formation of unit area of β phase on the impurity particle displaces unit area of α phase on it, so the effective change in interfacial energy is the difference as shown. In most cases γ_e is less than $\gamma_{\alpha\beta}$, the interfacial tension between the two phases, so that an impurity particle enhances the nucleation rate. A free surface also can enhance the nucleation rate by Eq. 4 where the surface tensions are between α and β and the surrounding vapor or air. However, it is less likely in this case that γ_e would be less than $\gamma_{\alpha\beta}$.

In surface crystallization impurities are the most important source of nucleation. On cooling from the melting temperature, dust particles adhere to the glass surface, giving preferred nucleation. Other impurities such as alkalis can corrode the surface, leading to lower glass-air surface tension and enhanced nucleation. The centers of cristobalite crystals growing on a vitreous silica surface when analyzed with the electron microprobe show impurities such as calcium or sodium. Thus the glass surface should be protected from dust and other impurities to prevent surface nucleation of crystals.

A variety of methods has been used to nucleate uniform crystallization in glass. Fine metal particles grown in the glass can act as centers for crystallization. The platinum metals in concentrations of 0.001 to 0.1% catalyze crystallization in lithium silicate,[4] lithium aluminum silicate, and alkali barium silicate glasses.[5] The metal is added as a compound; it decomposes during glass melting and forms a very fine dispersion of metal particles, typically about 50 Å in diameter.[4] The noble metals, copper, silver, and gold, also act as nucleating agents in various lithium and barium silicate glasses, especially lithium aluminum silicates[6]. The noble metal can be added to the glass batch either as finely divided metal or as a compound. It dissolves in the glass in the ionic form during glass melting, and remains in this form if the glass is cooled quickly.[7] If the glass contains some antimony or arsenic oxide, the noble metal precipitates on reheating. The particles nucleate continuously at a constant rate;[8] the efficacy of antimony and arsenic as nucleating agents apparently is related to their having two oxidation states in the glass, but the exact mechanism of nucleation is uncertain.

The noble metal particles can also be nucleated by ultraviolet light or x-rays.[9] Cerium oxide in the glass improves the photosensitive nucleation of the noble metals.[10] Again the efficacy of the cerium as a nucleating agent is apparently related to its two oxidation states in the glass, and to its absorption band at 0.314 μm.[12] One mechanism proposed was that the radiation caused electrons to be ejected from Ce^{3+} ions and the electrons then reacted with the noble metal ions to form atoms, which agglomerated to a nucleus of a growing metal particle.[7,8] However, it is possible to reduce gold ions to atoms with hydrogen in a photosensitive glass, and yet these atoms do not form particles without irradiation.[11] Thus the nucleation mechanism is more subtle than a simple reduction process, and apparently involves an activated cerium ion[12] as a nucleating agent. The noble metal ions must be reduced at some stage of the growth process, but this reduction can occur either before or during growth.[11]

Lithium metasilicate crystals formed in different configurations on metal particles in photosensitive glasses can be etched out to make useful shapes. This "chemical machining" is possible because the lithium metasilicate is much more soluble in hydrochloric acid than is the base glass. Tolerances of ± 50 μm are possible in Cornings' "Fotoform" glass by this method. After machining, the remaining glass can also be crystallized to give better strength and stability at high temperatures. These glasses have been used in printed circuit boards, dielectrics, and fluid amplifiers.

In a lithium aluminum silicate glass lithium metasilicate particles did not nucleate until gold particles reached a size of 80 Å.[13] The reason for this size limitation is not clear. The lithium metasilicate is apparently metastable because the glass is in the region where tridymite and lithium disilicate are the stable crystalline phases.

Chance impurity particles can nucleate crystals in glass. Figure 1 illustrates a cluster of lithium disilicate crystals in a 26 mole % Li_2O, 74%

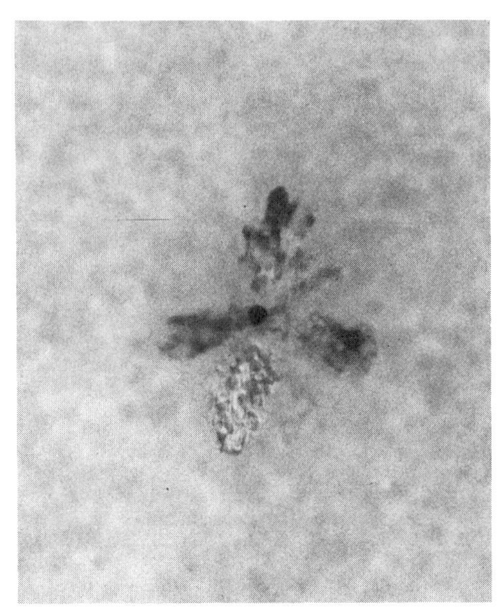

Fig. 1 Transmission electron micrograph of a cluster of lithium disilicate crystals growing in a 26 mole % Li_2O, 74% SiO_2 glass, heated 243 min at 500°C. Unetched thinned section. Magnification 34,000 ×.

SiO_2 glass containing no added nucleating agent.[14] The crystals are growing from an impurity particle. The composition of the particle was not identified, but it has flattened faces, implying that it is crystalline. Such a particle was found at the center of every cluster observed in this glass.

Phase separation has been suggested as a precursor to uniform crystallization in glass. Often phase separation gives one phase that is more easily crystallized than the bulk glass. For example, in a sodium borosilicate glass phase separation leads to one silica-rich phase and another sodium borosilicate phase containing less silica than the glass as a whole. This sodium borosilicate phase crystallizes during heat-treatment at relatively low temperatures, as shown in Fig. 2.[15]

From a study of a glass containing 28.6 mole % Li_2O and 71.4% SiO_2 Nakagawa and Izumitani[16] deduced that phase separation and crystalline nucleation were independent phenomena. However, in a further examination of binary lithium silicate glasses of composition from 27.4 to 32.5 mole % Li_2O, Tomozawa (to be published) concluded that in the early stages of phase separation crystalline nucleation was accelerated. He suggested that the diffusion zone around the growing silica particles might be a favorable region for crystalline nucleation.

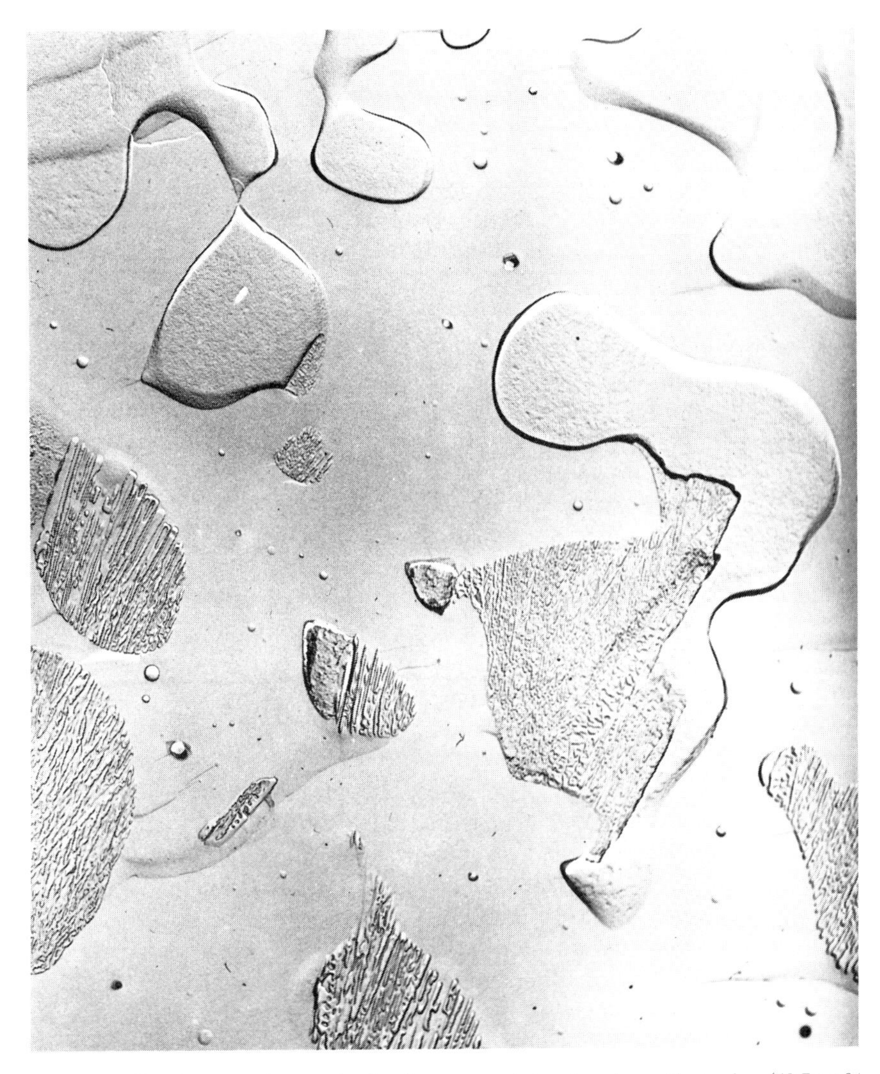

Fig. 2 Replica electron micrograph of a phase-separated sodium borosilicate glass (49.7 wt.% SiO_2, 45.5% B_2O_3, and 4.9% Na_2O) showing crystals growing in the borate-rich phase.

Rogers found that the activation energy for nucleation in some alkaline earth aluminosilicate glasses is above 100 kcal/mole, which is close to the activation energy for viscous flow in these glasses.[17] Consequently, he concluded that nucleation requires major rearrangement of the glass structure in addition to diffusion; however, the activation energy for diffusion of the critical constituents (probably including silica) could be close to that for viscous flow.

NUCLEATING AGENTS

The addition of "nucleating agents," such as titanium dioxide, phosphorous pentoxide, zirconium dioxide, and fluorides, to certain glasses leads to enhanced rates of uniform crystalline nucleation. The mechanism of nucleation by these agents is uncertain in most cases, although much work has been done on them and many mechanisms proposed.

The most widely used nucleating agent is titanium dioxide, TiO_2. This agent was used in much of the early work on glass-ceramics in such glasses as lithium aluminum silicates and magnesium aluminum silicates.[2] TiO_2 is quite soluble in silicate glasses and lowers their viscosity considerably. Zirconium dioxide, ZrO_2, has been suggested as a nucleating agent in various multicomponent lithium silicate glasses[18] because it is structurally similar to TiO_2; however it is less soluble than TiO_2 and does not lower the viscosity as much.

Phosphorous pentoxide, P_2O_5, has been used successfully as a nucleating agent in various lithium silicate and magnesium aluminum silicate glasses by McMillan and his co-workers.[3] Small amounts of P_2O_5 can influence crystalline nucleation rates strongly.

A number of other oxides have been used as nucleating agents.[3,18] Among these are chromium, vanadium, tungsten, molybdenum, iron,[19] zinc, and nickel. All the oxide nucleating agents except the last two have metallic ions with a high valence and field strength. This correlation may have some connection with the importance of field strength in bringing about phase separation, as described in the preceding chapter.

Fluorides have been used as nucleating agents in alkaline earth and alkali aluminosilicate glasses[3,20] and soda lime glasses.[21] This use follows from the traditional application of fluorides in opal glasses and glazes.

Mixed cadmium sulfide and selenide acts as a nucleation catalyst in various glasses[3]; the glass must be melted under reducing conditions to prevent oxidation of the sulfur or selenium, and the resulting glass ceramic is yellow or orange.

The heat-treatment of glasses containing nucleating agents is often carried out in two stages. First the glass is given a "nucleation treatment" at a temperature just above the annealing temperature; it is then crystallized at a higher temperature. The separation between these two processes is not sharp since some crystallization probably occurs during the nucleating treatment. A schematic graph of the rates of nucleation and growth is given in Fig. 3. This graph must be applied with caution since many crystallized glasses do not follow such a scheme.

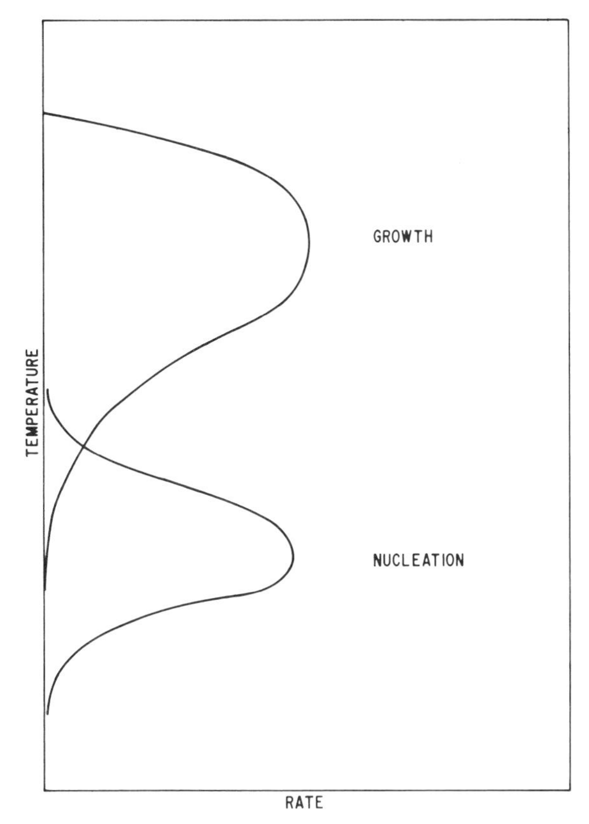

Fig. 3 Schematic representation of nucleation and growth rates of crystals in glasses.

At least three different mechanisms have been proposed for the action of nucleating agents. They can be classified as agent crystallization, catalysis of phase separation, and reduction of interfacial tension. Each is discussed in turn.

As discussed in the last section, metal particles can act as nuclei for uniform crystallization, so the first idea was that nucleating agents act by precipitating from the glass as fine crystals during the nucleation heat-treatment; they then serve as sites for crystallization of the main crystalline phase at the higher "growth" temperature. However, for most nucleating agents there is little or no evidence supporting this mechanism. Cadmium sulfide or selenide probably crystallizes as a uniform colloidal dispersion in glass that could catalyze nucleation. A fluorosilicate phase crystallizes first in alkaline earth aluminosilicate glasses,[20] but it is not certain that this phase nucleated the major phase. For the common nucleating agents such as TiO_2 and P_2O_5 there is not much evidence that a crystalline phase containing them is the first phase to precipitate during heat-treatment to form a glass ceramic. Crystals containing these agents usually grow at temperatures much higher than the nucleating temperature in glasses containing them as nucleation catalysts.

Extensive work on phase separation and uniform crystallization in glass started at about the same time. Perhaps this is the reason that a mechanism involving phase separation was invoked to explain the action of nucleation catalysts in uniform crystallization! As shown in the last section, one of two amorphous phases separating from a glass can be more favorable for nucleation and growth of crystals than the parent glass, so the proposal that nucleating agents lead to phase separation is a natural one. The tendency of metal ions of high field strength to favor phase separation as well as uniform crystallization also supports this explanation, although this similarity may be a coincidence. One good nucleating agent, P_2O_5, apparently does not separate into stable phases in binary P_2O_5-SiO_2 glasses,[22] as one would expect from the high field strength of the phosphorous ion in it.

The sequence of phase separation and subsequent crystallization in a magnesium aluminosilicate glass containing titania has been found in one of the few definitive studies of the mechanism of a nucleation agent.[23] The sequence of events was followed with light scattered from inhomogeneities in the glass, which indicated that the glass separated into two phases upon cooling from its melting temperature. Treatment at about 750°C increased the anisotropy of the separated particles, which was interpreted as resulting from crystallization in them. X-ray diffraction showed the presence of magnesium titanate, $MgO \cdot TiO_2$. Subsequent treatment at higher temperatures leads to the major crystalline phases of cordierite ($2MgO \cdot 2Al_2O_3 \cdot 5SiO_2$) and cristobalite. Thus in this glass the sequence is phase separation, then crystallization in the titania-rich phase, whose crystals then act as nuclei for the major crystalline phases. Apparently the titania does not act as a catalyst for phase separation, but rather concentrates in one of the separated phases in which it crystallizes uniformly to very fine crystals, which then catalyze further growth.

In some uniformly crystallized glasses there is good evidence that phase separation does not precede crystallization of the major phases, even if nucleating agents are present. Barry et al. found that β-eucryptite and $Li_2O \cdot Al_2O_3$ crystals formed in various lithium aluminum silicate glasses containing TiO_2 before any crystalline titanate phases were present.[24] In Fig. 4 are shown lithium disilicate crystals that grew in a multicomponent

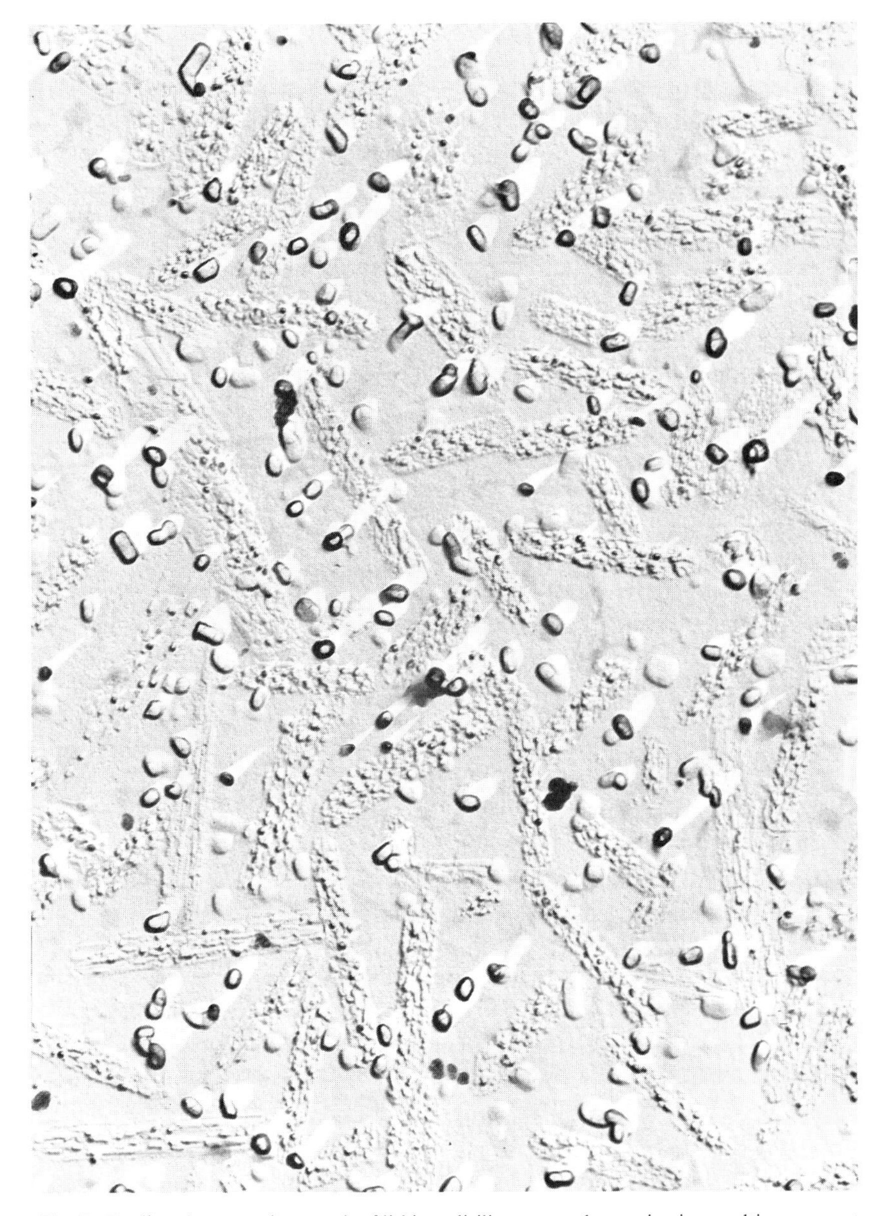

Fig. 4 Replica electron micrograph of lithium disilicate crystals growing in a multicomponent lithium silicate glass, heated 1 hr at 654°C and 4 hr at 850°C. Magnification 26,000 ×.

lithium disilicate glass which showed no phase separation either just after quenching or after various nucleating and crystallizing treatments[15]. Thus in many, if not most, glass-ceramics the nucleating agent does not enhance phase separation.

If not crystallization of the agent or enhanced phase separation, then what is the mechanism of the nucleating agents in increasing the rate of nucleation? After rejecting other possibilities, Hillig[25] suggested that perhaps these agents lower the interfacial tension γ between the crystal and glass, thus increasing the rate of nucleation by Eqs. 1 and 2. It is very difficult to obtain measurements of the interfacial tension, so there is no direct evidence for this mechanism. It may be that ions of high field strength can act as "surface active agents" to lower the interfacial energy. Barry et al. have extended this idea in developing a detailed model for the enhancement of crystalline nucleation by titania in lithium aluminum silicate glasses.[24] They argue that the Ti^{4+} ions associate with nonbridging oxygen ions, causing these ions and alkalis to concentrate at the edge of domains enclosing bridging oxygens. On the basis of this model Barry et al. explain some effects of composition on nucleation rate in their glasses. Their model is quite speculative, but at least it shows that the postulate of reduction of interfacial tension by nucleation agents should be considered seriously.

In a binary lithium silicate glass containing no nucleating agent, crystals grow on fine crystalline impurity particles, as shown in Fig. 1. When phosphorous is added to this glass, many more crystals are formed, and they are smaller, as shown in Fig. 5.[14] There is other evidence that both TiO_2 and P_2O_5 lower the rates of growth of lithium silicate crystals in glasses.[24,26] It appears that the number of crystals increases as a function of aging time in the glass containing phosphorous; compare Figs. 5 and 6. In the same glass composition without phosphorous all crystals nucleated quickly, and the number of growing crystals was constant with time at 500°C. Thus the phosphorous addition apparently increases the rate of homogeneous nucleation of the crystals. The phosphorous has no apparent effect on the phase-separated structure of the glass, and there was no microscopic or x-ray evidence for phosphate crystals. Therefore again the role of the nucleating agent is to increase the driving force ΔP for nucleation or reduce the interfacial tension γ, most probably the latter.

Two more exotic mechanisms for enhancement of nucleation by catalysts have been proposed. Rogers and co-workers[17] suggested that iron oxide, Fe_2O_3, is effective as a nucleation catalyst in alkaline earth aluminosilicate glasses because it increases the number of aluminum ions in the glass that are coordinated to six rather than four oxygen ions. The aluminum ions in the crystalline spinel that grows in these glasses are in sixfold coordination, so the iron makes formation of the spinel easier. This mechanism is, of

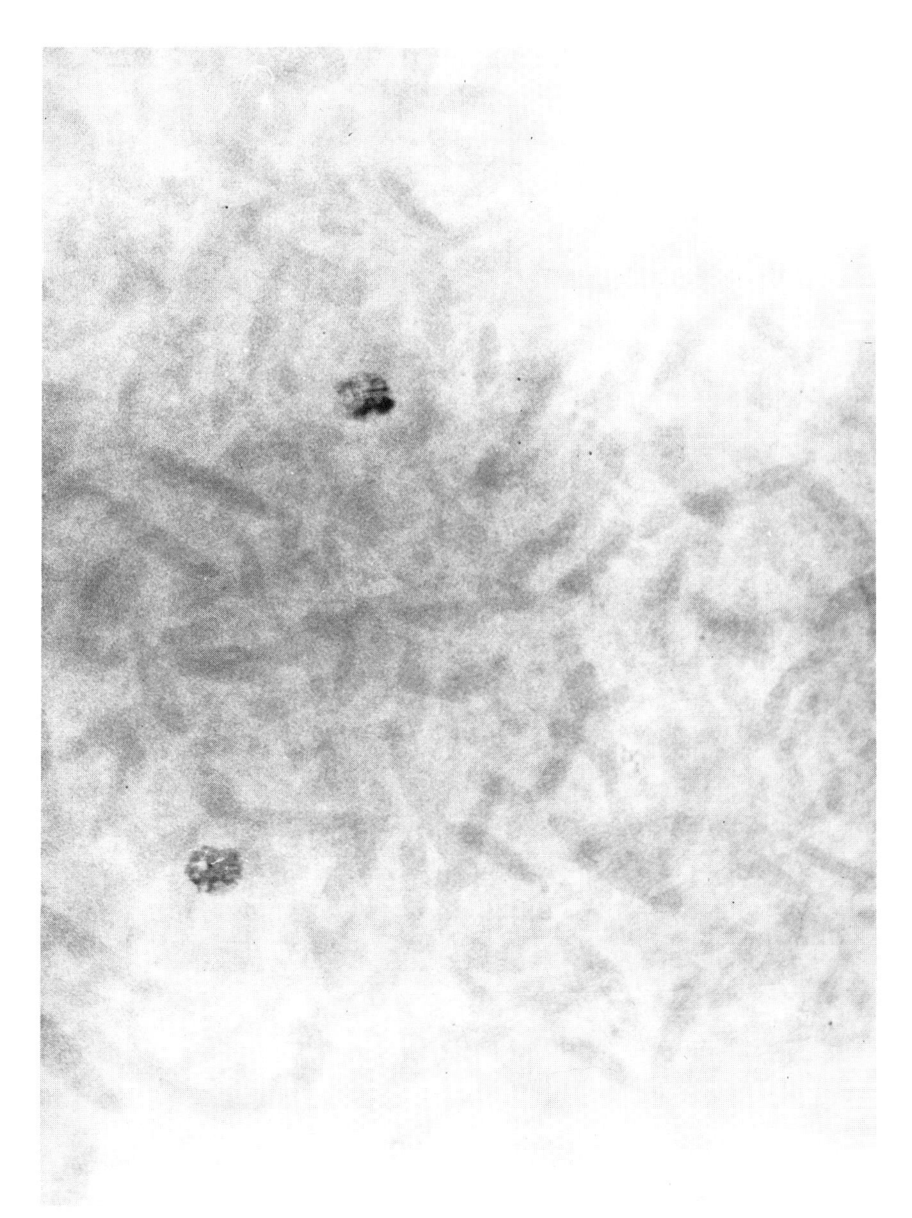

Fig. 5 Transmission electron micrograph of lithium disilicate crystals in a glass containing 71.7 mole % SiO_2, 27% Li_2O, and 1.3% P_2O_5, heated 243 min at 500°C. Unetched thinned section. Magnification 53,000 ×.

Fig. 6 Transmission electron micrograph of lithium disilicate crystals in a glass containing 71.7 mole % SiO_2, 27% Li_2O, and 1.3% P_2O_5, heated 1034 min at 500°C. Thinned section etched 4 sec in 2% HF.

course, specific to this particular system, and may be the reason why iron is effective as a nucleation catalyst here and not in other systems. Barry, Lay, and Miller proposed a "triggered" nucleation mechanism for TiO_2 in a lithium aluminosilicate glass.[27] They suggested that growing crystals with the β-quartz structure consume the major constituents of the glass but reject the titania. The increased titania in the matrix then leads to a higher nucleation rate in it, perhaps because the titania is surface-active, as mentioned above.

MORPHOLOGY

When crystals grow from a glass surface they often form a flat front progressing into the interior of the glass. However, this growing front can have an irregular shape and contain crystals of varied morphologies.[28]

In a binary lithium silicate glass the lithium disilicate crystals grew as spherical clusters, as shown in Figs. 1 and 7 in transmission electron micrographs.[14] The replica electron micrograph of Fig. 8 shows that these crystals grow in the lithium-rich matrix of the phase-separated glass, embedding the silica-rich droplets.

If a spherulite is defined as a spherical collection of crystals growing out of a common center, then these clusters of lithium disilicate crystals could be called spherulites. However, many so-called crystalline spherulites in polymers, liquid crystals, and minerals have a complex structure.[29,30] One characteristic feature of these complex spherulites is a crosslike figure in the polarizing microscope. These complex spherulites are often composed of helical, twisted strands of crystallites emanating from their center. In the early stages of growth the morphology is sometimes similar to a sheaf. No such morphologies were observed for the lithium disilicate crystals. The transmission pictures of Figs. 1 and 8 suggest that the clusters are composed of several single crystals of different orientations growing out of the center. Thus evidence for the similarity of these clusters to the more complex spherulites is lacking.

Barry et al.[24] found some polarization figures somewhat like those in polymers in composite crystals of lithium metasilicate, $Li_2O \cdot SiO_2$, and β-eucryptite in lithium aluminosilicate glasses. However, most of these figures are not crosslike, but show some radial bands and other irregularities. Their complexity is undoubtedly related to the filamentary interwoven structure of the composite crystals, which may be similar to that of other complex spherulites, but probably also has unique features. Dendritic composite crystals also grow in these glasses.

When about 5 mole % alumina is added to a lithium silicate glass, the morphology of the lithium silicate crystals growing in it is sharply changed

Fig. 7 Transmission electron micrograph of a cluster of lithium disilicate crystals in 26 mole % LiO_2, 74% SiO_2 glass, heated 1034 min at 500°C. Unetched thinned section. Magnification 30,000×.

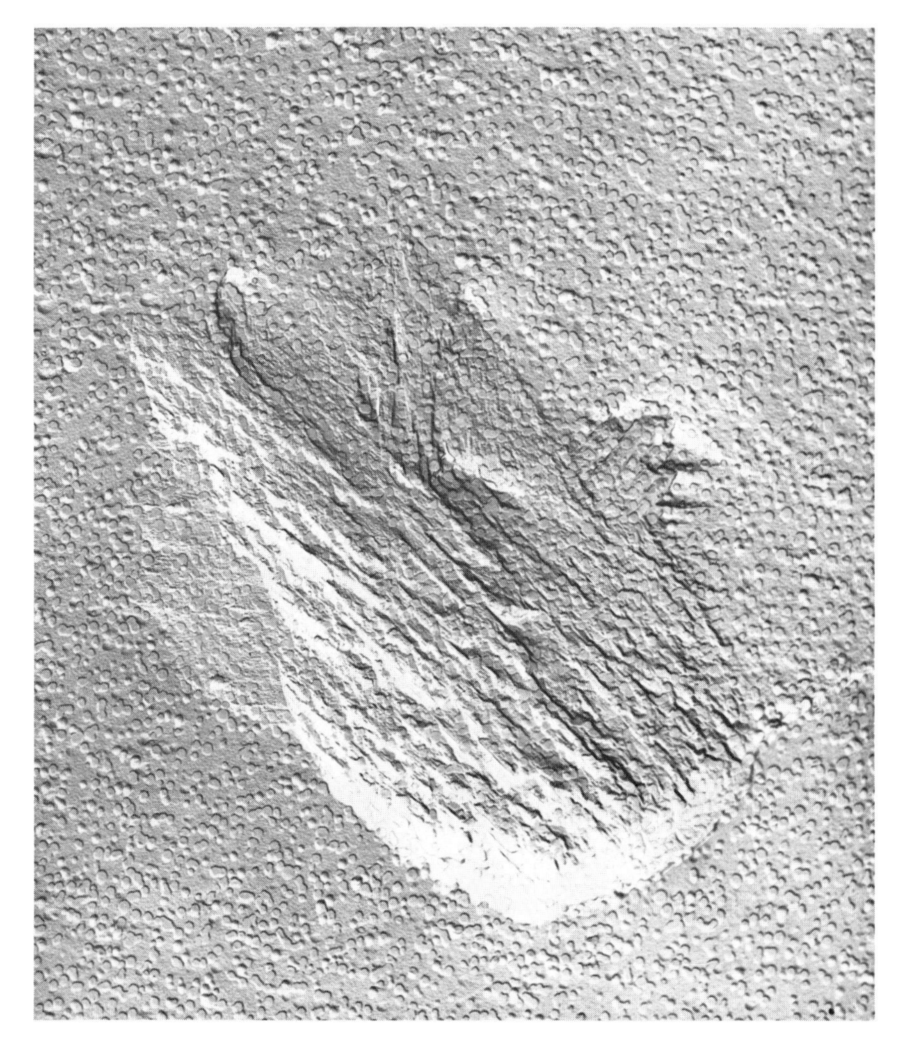

Fig. 8 Cluster of lithium disilicate crystals in a 26 mole % Li_2O, 74% SiO_2 glass, heated 16 hr at 500°C and then $1\frac{1}{2}$ hr at 550°C. Replica electron micrograph of a fracture surface etched 5 sec in 2% HF. Magnification 12,700 ×.

from spherical clusters to a laminar or micaceous structure, as shown by comparing Figs. 4 and 7. The reason for this change is not clear because the crystals apparently have the same composition in the two glasses.

In the sodium barium silicate system, barium disilicate crystals grow as spherulites at supercoolings of greater than 100°C, according to Burnett and Douglas.[31] At smaller supercoolings a platelet form of crystal grew either from the crucible wall or the glass-air surface. At a supercooling of about 100°C the platelets developed perturbations that grew into fibrils at a size of about 1 mm.

Burnett and Douglas claimed that their results agreed well with the model of Keith and Padden[30] for spherulitic growth. The two main conditions for this model are a medium of high viscosity and a matrix composition different from that of the crystal, so the growing phase expels some components as impurities. Fibrils growing from the surface of the spherulite can split into branches, at small angles to each other, that fill the space between fibrils. The precursor for the spherulite may be a single crystal, which becomes unstable at a certain size, after which it grows into a spherulite by fibrils and branching. Spherulitic growth in the soda-baria silica system does seem to fit this pattern, but in other inorganic glasses it is not followed by growing "spherulites."

When the barium disilicate spherulites were aged for long periods, the spherulites broke up into a lathlike structure.[31] The reason for this change in morphology is not clear; it could simply result from a reduction in surface energy from the spherulitic to lathlike form.

These examples show some of the possible morphologies for crystals in glasses; other types are known[6,17,32,33,33a] and new morphologies are being found regularly in this very active field of research.

GROWTH

The rate of growth of a crystal from a melt of the same composition, also called freezing or solidification, is controlled both by heat flow and a rearrangement process at the melt-crystal interface. It is difficult to be certain that heat flow is not a factor in experimentally observed crystallization rates. Here is a rough estimate of the crystallization velocities above which heat flow can be important. The heat flux J generated during freezing is

$$J = u A \rho L \tag{6}$$

where A is the area of the growing crystal-liquid interface, ρ is the density of the crystal, u is the rate of crystal growth, and L is the heat of fusion per unit mass of crystal. The heat flux also is proportional to the temperature gradient dT/dX:

$$J = AK\frac{dT}{dX} \tag{7}$$

where again A is the area of heat generation and K is the thermal conductivity. Equating Eqs. 6 and 7 gives a relation for the temperature gradient in terms of known quantities. For a binary alkali silicate glass, using the following parameters: $K = 3(10)^{-2}$ W/cm °K, $L\rho = 10^3$ J/cm^3, $u = 10^{-4}$ cm/sec, then $dT/dX \approx 3°$/cm. Thus in a bulk sample with dimensions in centimeters, the crystal-glass interface can be 3° higher than the temperature of the external face when the crystallization rate is 10^{-4} cm/sec. Therefore one might expect to find heat-flow effects in crystallization in glasses with growth rates higher than about 10^{-4} cm/sec.

In Table 3 of Chapter 2 maximum crystallization rates for a number of glasses are listed. Those rates for silica, germania, phosphorous pentoxide, and the organic polymer polyethylene adipate are well below 10^{-4} cm/sec, so it is probable that heat-flow effects are not important in crystallization in these materials. For the alkali disilicate and the borate glasses in the table the rates are above 10^{-4} cm/sec, so heat flow can be important in these glasses, depending on sample geometry.

As the temperature of a liquid is lowered under its freezing point its rate of crystallization first rises to a maximum and then decreases, as shown in Fig. 1 of Chapter 2 for fused silica. This maximum can be understood qualitatively as follows. As the temperature difference from the freezing temperature becomes greater, the driving force for crystallization is increased, increasing the rate. However, the rate of motion of the molecules becomes slower as the temperature is lowered, decreasing their rate of incorporation into the crystals. The net effect of these two competing processes is a maximum in the growth rate.

The rate of incorporation of units into the crystal is related to the viscosity η of the liquid. Thus a useful parameter for examining the temperature dependence of the crystallization or melting velocity u is its product with the viscosity $u\eta$. For the simple oxides SiO_2,[34] GeO_2,[35] and P_2O_5[36] $u\eta$ is proportional to the temperature difference ΔT between the melting and measuring temperature, both for melting and crystallization. Also Swift[28] found that devitrite crystals ($Na_2O \cdot 3CaO \cdot 6SiO_2$) grow in a soda-lime glass with constant $u\eta/\Delta T$ for melting and freezing. The composition of the glass was different from that of the crystals, but their rate of growth was constant with time, indicating that an interface process and not diffusion was controlling this rate. However, for binary oxide glasses, such as sodium[37,38,38a] and potassium[37] disilicates, lithium silicates,[39] barium,[40] strontium,[41] and lead[42] diborates, and for organic glasses, such as salol,[43] glycerol,[44] 1,2-diphenylbenzene,[45] 1,3,5-tri-α-naphthlbenzene,[46] and *o*-benzylphenol,[47] the term $u\eta$ is proportional to some higher power of the undercooling ΔT at small undercoolings. Furthermore the $u\eta$ values for crystallization of sodium disilicate glass are lower than $u\eta$ values for melting at comparable superheatings.[38a] At larger undercoolings the factor $u\eta/\Delta T$ becomes constant for some of these glasses, such as sodium and potassium disilicates, barium diborate, and salol. The glasses that do not show constant $u\eta/\Delta T$ at all undercoolings have growth rates above about 10^{-4} cm/sec and high entropies of fusion, whereas the simple oxide glasses with constant $u\eta/\Delta T$ at all temperatures for both melting and freezing have lower growth or melting rates and lower entropies of fusion. Devitrite growing in a soda-lime glass is an exception since its entropy of fusion is probably high, although it apparently has not been measured.

Values of $u\eta/\Delta T$ when they are constant with undercooling are given in Table 2. There is a report that devitrite grows at a rate of about

Table 2 Comparison of Measured and Calculated Rates of Crystallization of Glasses

Glass	Crystal	$u\eta/\Delta T$ in cm P/sec °K Measured	Calculated	Reduced entropy of fusion L/RT_m
	Constant $u\eta/\Delta T$ at all undercoolings			
SiO_2	Cristobalite	0.14	0.008	0.46
GeO_2	Hexogonal GeO_2	0.21	0.02	1.3
P_2O_5	Tetragonal P_2O_5	0.035	0.036	3.1
$17Na_2O$, $12CaO$ $2Al_2O_3$, $69SiO_2$	Divitrite $Na_2O \cdot 3CaO \cdot 6SiO_2$	0.008		
	Constant $u\eta/\Delta T$ at larger undercoolings only			
$K_2O \cdot 2SiO_2$	$K_2O \cdot 2SiO_2$	0.06		
$Na_2O \cdot 2SiO_2$	$Na_2O \cdot 2SiO_2$	0.05		3.7
Salol	Salol	$1.2(10)^{-4}$		>3

$2(10)^{-4}$ cm/sec from a glass of the same composition as the crystal, in contrast to the maximum rate of $2.8(10)^{-5}$ cm/sec found by Swift in a glass of different composition. With this increase all the $u\eta/\Delta T$ values for oxide glasses are within a factor of six of one another. Thus it is possible to estimate the order of magnitude of the crystallization velocity in an oxide glass from the viscosity and undercooling with $u\eta/\Delta T \approx 0.1$. This value may be lower for growth in a glass of different composition from that of the growing crystal, by comparison with the devitrite results and from the possibility of some mass transport limitations in the glass matrix. Diffusion-controlled growth and mixed diffusion- and interface-controlled growth were discussed in the preceding chapter.

Theories of varying complexity have been proposed to explain growth of crystals from the melt. However, their validity is still a matter of considerable controversy.[48,49] In this discussion a simple equation is first compared to the experimental results of glasses, and then some of the more esoteric ideas that have been invoked for crystallization from the melt are discussed.

Consider a crystal growing into a melt of the same composition. The rate of crystallization should be proportional to two factors: a gradient of a driving force and a coefficient that is the unit force needed to bring about the rearrangement of liquid to solid. Fluxes in kinetic processes are usually linearly related to two such factors; H. A. Wilson suggested such a relation for crystallization velocity many years ago.[50] The driving force for crystallization from the melt at a temperature T is the difference in Gibbs free energy between the liquid and crystal:

$$\Delta G \approx \Delta S_f(T_m - T) \tag{8}$$

where ΔS_f is the entropy of fusion at the melting point T_m, and is also equal to $-L/T_m$, where L is the heat of fusion. Equation 8 is valid when the heat capacities of the liquid and crystal are about equal. The velocity of crystallization u is the gradient in ΔG divided by a force f:

$$u = \frac{L(T_m - T)}{f\lambda T_m} \tag{9}$$

f is the force required to rearrange the amount of material from liquid to crystal needed to give unit velocity, and λ is the thickness of the transition layer between liquid and crystal. The force to move a particle of diameter d through a medium of viscosity η at unit velocity is $3\pi\eta d$. If this force is the same as the force f in Eq. 9, and d is about the same size as λ, then

$$u = \frac{L(T_m - T)}{3\pi\lambda^2\eta T_m} \tag{10}$$

Equation 10 has frequently been used to compare with solidification data. Equation 8 for the driving force should be reliable, and one would expect the rearrangement of molecules from liquid to crystal to be related to the viscosity. Thus the temperature dependence of crystallization velocity and the dependence on L and η should be given by Eq. 9. This expectation is fulfilled for those glasses in which $u\eta/\Delta T$ is constant, as shown in Table 2. Also shown in the table are values of $u\eta/\Delta T$ calculated from Eq. 10 using

$\lambda = 3$ Å. In most cases the agreement is poor, and there is no intelligible trend in the calculated values. These discrepancies reflect uncertainties in the geometrical factors in Eq. 10, in the use of the force-velocity equation, in the equating of λ and d, and in the meaning of λ. More theoretical and experimental work is needed to clarify these problems and to explain the cases in which $u\eta/\Delta T$ is not constant and is different for melting and freezing.

Impurities and defects can markedly modify crystallization rates in glasses. These effects are usually related to changes in viscosity of the glass. For examples, fused silica crystallized much more rapidly in the presence of water or oxygen than without them.[51] These gases are known to lower the viscosity of fused silica. Crystal growth in slightly reduced germania is faster than in stoichiometric material,[35] again paralleling a reduction in viscosity. Sodium impurity in germania also increases the growth rate; the effect may be intensified by concentration of the sodium ahead of the growing interface.[35] Various transition metal ions, such as iron, zinc, and vanadium, at concentrations of a few percent were found to increase crystallization rates in alkaline earth aluminosilicate glass, whereas chromium ions decreased the rate of crystal growth.[52]

Another equation for solidification can be derived from transition state theory. It is assumed that to cross the interface between liquid and crystal a molecule must acquire an "activation free energy" Q. The frequency of jumping from the liquid to the crystal is then $v_0 \exp(-Q/kT)$, where v_0 is some vibration frequency of the molecule. The frequency of jumping from crystal to liquid is $v_0 \exp[-(Q - \Delta G)/kT]$, where ΔG is the free energy difference between liquid and crystal as given by Eq. 8. The velocity of crystallization is then the net jumping rate times λ, the distance between liquid and crystal, or

$$u = \lambda v_0 \exp\left(\frac{-Q}{kT}\right)\left[1 - \exp\frac{\Delta G}{kT}\right] \tag{11}$$

If the diffusion coefficient in the liquid D is equal to $\lambda^2 v_0 \exp(-Q/kT)$, and is related to the viscosity by the Stokes-Einstein relation

$$D = \frac{kT}{3\pi\eta\lambda} \tag{12}$$

then

$$u = \frac{[1 - \exp(\Delta G/kT)]}{3\pi\lambda^2\eta} kT \tag{13}$$

If $\Delta G/kT$ is small Eq. 13 becomes the same as Eq. 10. Some authors[49,35,38a] have insisted that Eq. 13 with the exponential term is the "correct" form, and Eq. 10 should not be used. There is no experimental evidence to support this contention; in fact, data for the last three glasses in Table 1 fit Eq. 10 at higher undercoolings, whereas they do not fit Eq. 13. Thus the use of Eq. 10 or Eq. 13 at this stage of knowledge is a matter of taste.

From a nearest-neighbor bond model, such as described in the last chapter, Jackson[53] derived an expression for the difference in free energy between a molecule in the liquid and on the crystalline surface. From this equation he found that below a value of two for the reduced entropy of fusion $\alpha = L/RT_m$, the interface should be "rough." He surmised that on such a rough surface growth should be isotropic, whereas for α values larger than about three growth should be faceted; that is, certain low-angle lattice planes should have slow growth rates, and other lattice planes should have faster rates so that they rapidly grow out of existence. These morphological predictions are borne out remarkably well by experiment on glasses and other liquids. Phosphorous pentoxide and sodium disilicate, as well as most organic liquids, have $\alpha > 3$ and grow with faceted crystals, whereas silica and germania have lower reduced entropies of fusion and grow isotropically. This correlation does not seem to distinguish between types of kinetic behavior, however, because SiO_2, GeO_2, and P_2O_5 all have $u\eta/\Delta T$ constant for both melting and freezing.

Other treatments of the "roughness" of an interface have been made.[54–56] The critical evaluation of the applicability of these ideas to crystallization in glasses awaits more detailed examination of the glass-crystal interface.

The crystallization of organic polymers[57,58] is a specialized subject not treated in depth here. The rate of growth of polymer crystals is often controlled by a "secondary" nucleation process at the growing crystal-liquid interface. The crystallization velocity for this kind of process fits an equation of the form

$$u = C \exp\left(\frac{-B}{RT}\right) \tag{14}$$

where B and C are coefficients constant with temperature. Such an equation was found by Hillig to be valid for growth in the c-direction of ice.[59] Hillig interpreted this functionality to result from the continued nucleation of new ice layers on the growing crystal.

In growth of crystals from the vapor phase, steps on the crystal surface sweep across it to add material. This "layer spreading" mechanism was confirmed by experiments on the growth of filamentary crystals or "whiskers."[60] This mechanism also is important in crystal growth from the melt,[48] as shown by the elegant experiments of Sears on the melting and freezing of p-toluidine.[61] Whether this mechanism is important in all crystal growth from the melt, and particularly for glasses, is not certain.

REFERENCES

1. Anon., *Bull. Am. Ceram. Soc.*, **36**, 579 (1957).
2. S. D. Stookey, U.S. Patents 2,920,971 and 2,933,857, 1960.
3. P. W. McMillan, *Glass-Ceramics*, Academic, London, 1964.
4. G. E. Rindone, *J. Am. Ceram. Soc.*, **45**, 7 (1962).
5. British Patent No. 863,569, 1961; see Ref. 3, p. 52.
6. S. D. Stookey and R. D. Maurer, in *Progress in Ceramic Science*, J. E. Burke, Ed., Pergamon, New York, 1962, p. 77.
7. S. D. Stookey, *J. Am. Ceram. Soc.*, **32**, 246 (1949).
8. R. D. Maurer, *J. Appl. Phys.*, **29**, 1 (1958).
9. R. N. Dalton, U.S. Patent 2,422,472, 1947.
10. S. D. Stookey, U.S. Patent 2,515,275, 1950.
11. R. H. Doremus, in *Symp. on Nucleation and Crystallization in Glasses*, American Ceramic Society, Columbus, Ohio, 1962, p. 119.
12. J. S. Stroud, *J. Chem. Phys.*, **35**, 844 (1961); **37**, 836 (1962).
13. R. D. Maurer, *J. Chem. Phys.*, **31**, 494 (1959).
14. R. H. Doremus and A. M. Turkalo, *Phys. Chem. Glasses*, **13**, 14 (1972).
15. Courtesy of A. M. Turkalo, General Electric Research Development Center.
16. K. Nakagawa and J. Izumitani, *Phys. Chem. Glasses*, **10**, 179 (1969).
17. P. S. Rogers, *Mineral. Mag.* **37**, 741 (1970).
18. I. Sawai, *Glass Tech.*, **2**, 243 (1961).
19. P. S. Rogers and J. Williamson, *Glass Tech.*, **10**, 128 (1969).
20. S. Lyng, J. Markali, J. Krogh-Moe, and N. H. Lundberg, *Phys. Chem. Glasses*, **11**, 6 (1970).
21. S. P. Mukherjee and P. S. Rogers, *Phys. Chem. Glasses*, **8**, 81 (1967).
22. T. Y. Tien and F. A. Hummel, *J. Am. Ceram. Soc.*, **45**, 424 (1962).
23. R. D. Maurer, *J. Appl. Phys.*, **33**, 2132 (1962).
24. T. J. Barry, D. Clinton, L. A. Lay, R. A. Mercer, and R. P. Miller, *J. Mat. Sci.*, **4**, 596 (1969); **5**, 117 (1970).
25. W. B. Hillig, in *Symp. on Nucleation and Crystallization in Glasses and Melts*, American Ceramic Society, Columbus, Ohio, 1962, p. 77.
26. A. V. Phillips and P. W. McMillan, *Glass Tech.*, **6**, 46 (1965).
27. T. J. Barry, L. A. Kay, and R. P. Miller, *Disc. Far. Soc.*, **50,** 224 (1971).
28. H. R. Swift, *J. Am. Ceram. Soc.*, **30**, 165 (1947).
29. A. Keller, in *Growth and Perfection of Crystals*, R. H. Doremus, B. W. Roberts, and D. Turnbull, Eds., Wiley, New York, 1958, p. 499.
30. H. D. Kieth and F. J. Padden, *J. Appl. Phys.*, **34**, 2409 (1963).
31. D. G. Burnett and R. W. Douglas, *Phys. Chem. Glasses*, **12**, 117 (1971).
32. E. A. Porai-Koshits, Ed., *Catalyzed Crystallization of Glass*, Consultants Bureau, New York, 1964.
33. W. Vogel and K. Gerth, in *Symp. on Nucleation and Crystallization in Glasses and Melts*, American Ceramic Society, Columbus, Ohio, 1962, p. 11.

33a. A. G. Gregory and T. J. Veasey, *J. Mat. Sci.*, **6**, 1312 (1971).

34. F. E. Wagstaff, *J. Am. Ceram. Soc.*, **52**, 650 (1969).
35. P. J. Vergano and D. R. Uhlmann, *Phys. Chem. Glasses*, **11**, 30, 39 (1970).
36. R. L. Cormia, J. D. Mackenzie, and D. Turnbull, *J. Appl. Phys.*, **34**, 2239 (1963).
37. A. Leontewa, *Acta Physicochem. USSR*, **16**, 97 (1942).
38. W. D. Scott and J. A. Pask, *J. Am. Ceram. Soc.*, **44**, 181 (1961).

38a. G. S. Meiling and D. R. Uhlmann, *Phys. Chem. Glasses*, **8**, 62 (1967).

39. J. G. Morley, *Glass Tech.*, **6**, 77 (1965).

40. J. A. Laird and C. G. Bergeron, *J. Am. Ceram. Soc.*, **53**, 482 (1970).
41. S. R. Nagel and C. G. Bergeron, *Am. Ceram. Soc. Bull.*, **50**, 414 (1971) (abstr.).
42. J. P. DeLuca, R. J. Eagan, and C. G. Bergeron, *J. Am. Ceram. Soc.*, **52**, 322 (1969).
43. K. Neumann and G. Micus, *Z. Phys. Chem. N.F.*, **2**, 25 (1954).
44. M. Volmer and A. Maider, *Z. Phys. Chem.*, **154A**, 97 (1931).
45. R. J. Greet, *J. Cryst. Growth*, **1**, 195 (1967).
46. J. D. Magill and D. J. Plazek, *J. Chem. Phys.*, **46**, 3757 (1967); **45**, 3038 (1966).
47. G. Scherer and D. R. Uhlmann, *J. Cryst. Growth*, **15**, 1 (1972).
48. J. W. Cahn, W. B. Hilig, and G. W. Sears, *Acta Met.*, **12**, 1421 (1964).
49. K. A. Jackson, D. R. Uhlmann, and J. D. Hunt, *J. Cryst. Growth*, **1**, 1 (1967).
50. H. A. Wilson, *Phil. Mag.*, **50**, 238 (1900).
51. N. G. Ainslie, C. R. Morelock, and D. Turnbull, in *Symp. on Nucleation and Crystallization of Glasses*, American Ceramic Society, Columbus, Ohio, 1962, p. 97.
52. J. Williamson, *Mineral Mag.*, **37**, 759 (1970).
53. K. A. Jackson, in *Growth and Perfection of Crystals*, R. H. Doremus, B. W. Roberts, and D. Turnbull, Eds., Wiley, New York, 1958, p. 319.
54. W. K. Burton and N. Cabrera, *Disc. Far. Soc.*, **5**, 132 (1949).
55. J. W. Cahn, *Acta Met.*, **8**, 554 (1960).
56. H. J. Leamy and K. A. Jackson, *J. Appl. Phys.*, **42**, 2121 (1971).
57. L. Mandelkern, *Crystallization of Polymers*, McGraw-Hill, New York, 1964.
58. A. Kellar, *Rept. Prog. Phys.*, **31**, Pt. 2, 623 (1968).
59. W. B. Hillig, in *Growth and Perfection of Crystals*, R. H. Doremus, B. W. Roberts, and D. Turnbull, Eds., Wiley, New York, 1958, p. 350.
60. G. W. Sears, *Acta Met.*, **1**, 457 (1957).
61. G. W. Sears, *J. Chem. Phys.*, **23**, 1630 (1955).

Part Two

TRANSPORT

Those properties involving movement of molecules or electrons in glass are discussed in this part.

Viscosity and the glass transition are closely linked. In this discussion the glass transition is considered to result from the retardation of molecular motion. Stress relaxation is discussed in the chapter on the glass transition because it gives insight into changes in the transition region.

Diffusion of molecules and ions in glass is discussed in Chapters 8 and 9, respectively. Permeation of molecules requires their solubility in the glass, which is also discussed in Chapter 8. Electrical conductivity is closely linked to ionic motion, and is considered in Chapter 9. Ionic diffusion is responsible for certain electrical and mechanical losses in glass, which are discussed in Chapter 11. Electronic conduction predominates over ionic conduction in certain glasses, giving rise to semiconduction and switching and memory effects as discussed in Chapter 10.

6

Viscosity

The viscosity of a glass is one of its most important technological properties. It determines the melting conditions, the temperatures of working and annealing, fining behavior (removal of bubbles from the melt), upper temperature of use, and devitrification rate. The viscosities of different glasses vary enormously with composition and are strong functions of temperature.

When a shearing force is applied to a liquid, it flows, and the viscosity is a measure of the ratio between the force and rate of flow. If two parallel planes of area A, a distance d apart, are subjected to a tangential force difference F, the viscosity η is defined as

$$\eta = \frac{Fd}{Av} \tag{1}$$

where v is the relative velocity of the two planes. The unit of viscosity is grams per centimeter per second, which is called the poise (P). The viscosity of most common fluids such as water and organic liquids is about one hundredth of a P at room temperature.

The working point of a glass is defined as the temperature at which it has a viscosity of 10^4 P. At this temperature the glass can be readily formed or sealed. The softening point of a glass is the temperature at which it has a viscosity of $10^{7.6}$ P. At this viscosity a rod about 24 cm long and 0.7 mm in diameter elongates 1 mm/min under its own weight. The annealing point is the temperature at which the viscosity is $10^{13.4}$ P, and the strain point where it is $10^{14.6}$ P. The thermal expansion curve of an annealed glass begins to deviate considerably from linearity at a viscosity of about 10^{14} or 10^{15} P. At the annealing temperature a rapidly cooled silicate glass becomes reasonably strain-free in about 15 min.

The viscosities of several commercially important silicate glasses are compared in Fig. 1, as a function of temperature. The working, softening,

Table 1 Properties, Compositions, and uses of Important Commercial Silicate Glasses

Glass Type	Corning number	Kimble number	Approximate composition (wt. %)								Relative thermal Expansion $10^{-7}/°C$	Density (gms/cm^3)	Refractive index	Uses
			SiO_2	B_2O_3	Al_2O_3	CaO	MgO	PbO	Na_2O	K_2O				
Soda lime	0080	R-6	72.6	0.8	1.7	4.6	3.6		15.2		92	2.47	1.51	Windows, lamps
"Pyrex" boro-silicate	7740	KG-33	81	13	2				4		33	2.23	1.47	Headlamps, cooking, and labware
Soda boro-silicate	7050	K-705	68	24	1				7		46	2.24	1.48	Sealing
Alkali lead	0010	KG-1	77			1		8	9	5	93	2.86	1.54	Lamp tubing, sealing
Alkaline earth Alumino-silicate	1720	EZ-1	64	4.5	10.4	8.9	10.2		1.3	0.7	42	2.52	1.53	High temperatures

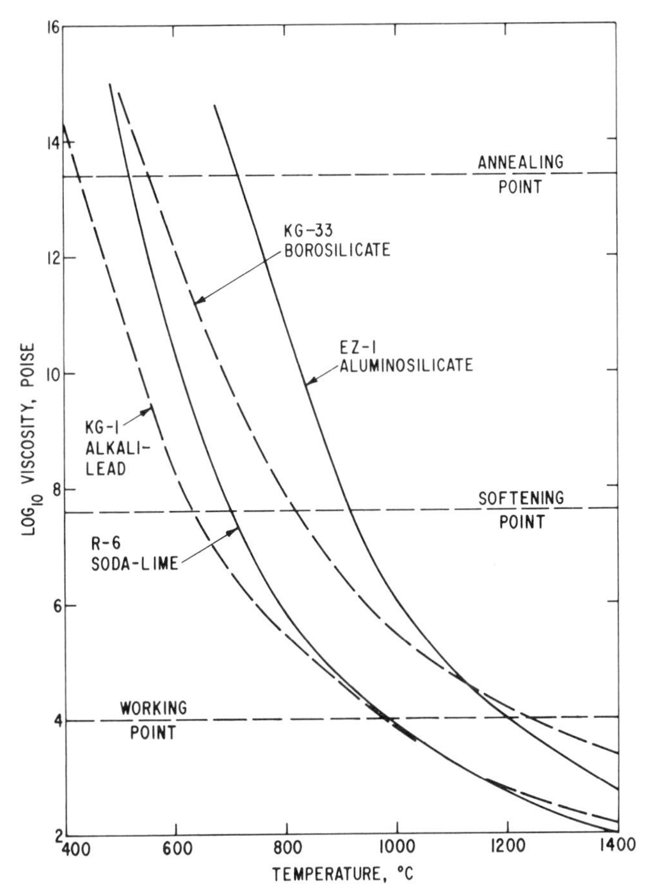

Fig. 1 Viscosity of some commercial silicate glasses listed in Table 1.

and annealing points are marked on the plot. Some other properties of these glasses and their approximate compositions are given in Table 1.

MEASUREMENT

Two methods are most frequently used to measure viscosity in glass: one at low viscosity and the other at high viscosity. At low viscosity (up to 10^8 P), the rotating cylinder or crucible method appears to be the most reliable. In this method the relative rate of rotation of two concentric cylinders is measured at a constant torque, and the viscosity is inversely proportional to the rate of rotation. Usually the viscometer is calibrated with liquids of known viscosity, but with care absolute measurements can be made. Higher viscosities are measured by the rate at which a glass rod elongates under a fixed force. The viscosity is then given by the formula:

$$\eta = \frac{Lmgf}{3\pi R^2 v} \tag{2}$$

where L is the length of the glass rod, R is its radius, m is the mass hanging on it, g is the gravitational constant, v is an instrumental reading proportional to the rate of elongation, and f is a calibration factor for the instrument.

Two problems arise in measuring the viscosity of viscous materials: one is the variation of viscosity with time, and the other is its variation with the force applied. Many investigators have found that the measured viscosity is not a function of applied force or velocity of flow,[1] implying that glass is a Newtonian liquid in which the flow rate is directly proportional to the shearing stress. However, Bartenev[2] measured the rate of flow of an alkali silicate glass rod at 655°C in tension at low shear stress and found that at stresses below 1 kg/cm^2 the flow deviated from Newtonian behavior. Above this stress the flow was linear with stress on a line passing through the origin, and as other measurements of the viscosity of glass have been made at stresses above this value the deviations were not previously noticed. Apparently there is some elastic deformation in the viscous glass at low stress before it flows.

Lillie studied the change of viscosity of glasses with time at the measuring temperature.[3] He found that above a viscosity of about 10^{12} P the viscosity of samples cooled quickly from a higher temperature initially appeared to be low and asymptotically approached some higher value with time of holding at the measuring temperature. The same asymptotic "equilibrium" viscosity was found when a sample was held at a lower temperature for a long time and then raised to the measuring temperature. The deviation of the initially measured viscosity from the equilibrium viscosity became progressively greater at higher viscosities until above about 10^{16} P it was not possible to measure the equilibrium viscosity even after very long times at the measurement temperature. Thus this viscosity of 10^{16} P should be considered as the upper limit of measurable values, and any viscosity measured higher than this cannot be considered an equilibrium value.

The variation of viscosity with time of a quenched sample involves a structural change from a state characteristic of a higher temperature to that of a lower temperature. The change is slow because the structural elements of the glass rearrange slowly, as reflected in the high viscosity. The exact structural nature of this rearrangement is uncertain. As a rough model one can think of the glass network as being progressively broken up as the temperature increases. When a glass sample is cooled quickly from a higher temperature, the network flows more easily because it has more broken bonds than at equilibrium at the lower temperature. As these bonds heal with holding at the lower temperature the viscosity increases to the equilibrium value. This model, as well as stress relaxation and viscoelasticity, will be discussed further in the next chapter on glass transition.

SIMPLE GLASS-FORMING OXIDES

As with many transport properties the viscosity η fits an Arrhenius-type equation over wide ranges of temperature:

$$\eta = \eta_0 \exp\left(\frac{Q}{RT}\right) \tag{3}$$

where Q and η_0 are temperature-independent coefficients called the activation energy and the preexponential factor, respectively.

Figure 2 shows the logarithm of viscosities of various simple glass-forming oxides as a function of reciprocal temperature. The activation energies for viscous flow of these glasses are given in Table 2. The data for silica and germania were carefully selected from those of many workers as being the most reliable. The viscosity is decreased by the presence of small amounts of impurities, as discussed below; this is one of the main reasons for errors.

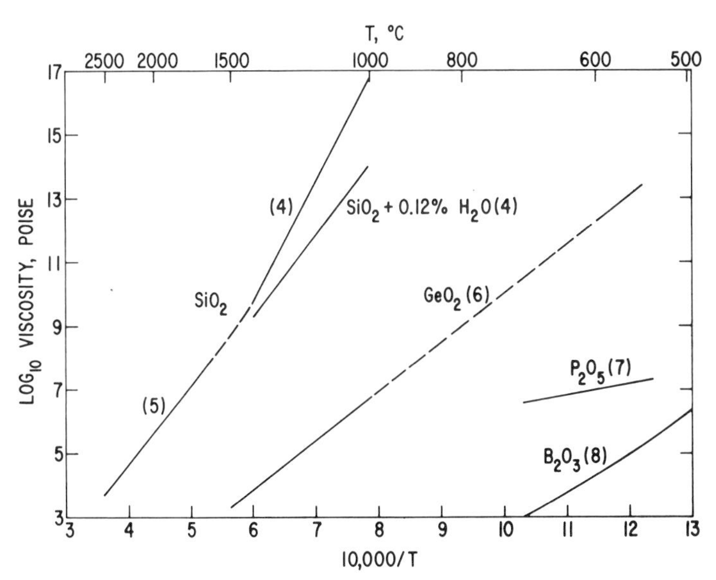

Fig. 2 Viscosities of various glass-forming oxides as a function of reciprocal temperature. Dashed lines show interpolations where data were not taken. Reference numbers are on the appropriate lines.

Hetherington et al.[4] showed that at lower temperatures silica must be held for long times at the measuring temperature before a reproducible value of viscosity can be measured. Hofmaier and Urbain[5] suggested that the higher activation energies found by many investigators at lower temperatures resulted from mixtures of liquid and crystals. However, Hetherington et al. took their samples from the center of a silica block, and since only surface crystallization occurs in a pure sample, they should have avoided crystallization. Hofmaier and Urbain claimed that the measurements of Fontana and Plummer[10] at low temperatures gave activation energies close to their high temperature values, but the absolute values of Fontana and Plummer agree closely with those of Hetherington et al., still implying a low activation energy at higher temperature.

Table 2 Activation Energies for Viscous Flow of Various Glasses

Glass	Activation energy (kcal/mole)	Temperature range (°C)	Refs.
Vitreous silica	170	1100–1400	4
	123	1600–2500	5
Vitreous germania	75	540–1500	6
Vitreous P_2O_5	41.5	545–655	7
Vitreous B_2O_3	83–12	26–1300	8,9,32

The data of Kurkjian and Douglas[6] on vitreous germania show a constant activation energy with a factor of 10^{10} change in viscosity. However, the data on vitreous silica definitely show a change in activation energy with temperature, and the results for B_2O_3 show a large change in activation energy with temperature. Also, multicomponent silicate glasses for which viscosity data are plotted in Fig. 1, and the organic glasses shown in Fig. 4, have changing activation energies for viscous flow as a function of temperature.

The strengths of oxygen-central atom bonds in the simple oxides such as SiO_2, GeO_2, P_2O_5, and B_2O_3 are about 100 kcal/mole. The very great difference in viscosity of these oxides shown in Fig. 2 is, therefore, quite surprising, since the mechanism of flow undoubtedly involves breaking of oxide bonds. The melting points of the crystalline oxides parallel the variations in viscosity: SiO_2 (cristobalite) 1710°C; GeO_2 (hexagonal),

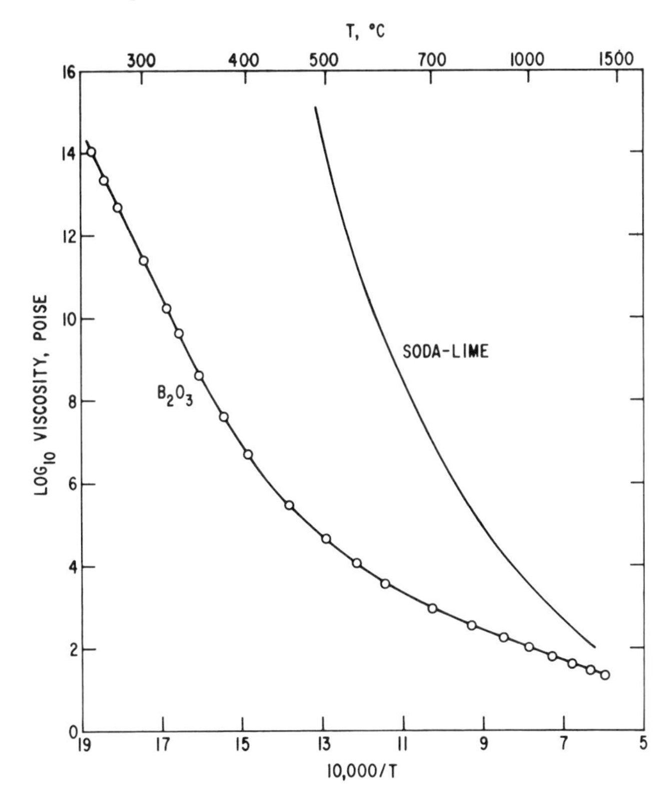

Fig. 3 Viscosity of boron trioxide[34] and a soda-lime glass[3] (70 wt. % SiO_2, 21% Na_2O, and 9% CaO) as a function of reciprocal temperature.

1116°C; P_2O_5, 580°C; B_2O_3, 450°C. The differences indicate that the detailed structure of the oxide network is also important in determining the ease of viscous flow. The linked-ring structure of B_2O_3, described in Chapter 3, is the probable reason for the low viscosity of this oxide, since the bonds between rings should be weaker than those in the rings. This structure also provides an explanation for the strong temperature dependence of the activation energy. The bonds between the rings should have a spectrum of strengths leading to a change in the effective activation energy for flow as a function of temperature. The P_2O_5 network possibly contains sheets of PO_4 tetrahedra, as mentioned in Chapter 3; each tetrahedron is bonded to three others instead of four as for SiO_2 and GeO_2. The lower viscosity and melting point of P_2O_5 may result from this structure.

Judged by their structure GeO_2 and SiO_2 should have a similar viscosity and melting point, since each apparently forms a random three-dimensional network in the rigid state (see Chapter 3). The great difference between both the absolute viscosity and its temperature dependence leads to doubt about the structural similarity. The random-network structure for vitreous silica is well established, as described in Chapter 3; thus there may be some subtle weakness in the germania network. Germania may also be more susceptible to impurity effects than silica, although there is no direct evidence for this contention.

The viscosities of the simple oxides decrease sharply with the addition of most impurity ions. Such impurities as water, in the form of —OH groups, and alkali oxides are particularly effective in lowering the viscosities. For example, the addition of 0.1 wt. % hydroxyl ion lowers the activation energy of flow for vitreous silica at temperatures from 1000 to 1400°C from 170 to 122 kcal/mole. The actual viscosity is lowered about three orders of magnitude at 1000°C by this addition, but is not too different from the value for water-free silica at 1400°C, as shown in Fig. 2. It is strange that the activation energy for silica containing water at lower temperatures is about the same as for water-free silica at higher temperatures, and extrapolates to about the same absolute values for the water-free material. Is this correspondence a coincidence? The answer is not clear. The addition of 0.165 mole % Na_2O to germania lowers the viscosity about a factor of 46 at 1000°C, and also causes a decrease in the activation energy of flow and a change in the activation energy with temperature.[6]

MULTICOMPONENT OXIDES

The addition of other oxides to silica invariably lowers its viscosity. The lowering is greatest with addition of alkali oxides and least with alkaline earth oxides and alumina. At higher temperatures the effect of alumina is less than for other oxides. Zinc and boric oxides among the multivalent oxides are particularly effective in lowering the viscosity of silicates. Morey[1] has collected data on the viscosity of many different silicate glasses and has given factors for the calculation of viscosity in soda-lime and soda-lead silicate glasses when other oxides are added to them.

The activation energy for viscous flow in multicomponent silicate glasses decreases as the temperature is increased.[3,11] At low temperatures where the viscosity is above 10^{12} P the activation energy for flow of sodium silicate glasses is about 100 kcal/mole, and it decreases to 50 kcal/mole or less at high temperatures where the viscosity is about 100 P. The value of 100 kcal/mole at low temperatures is much less than the low-temperature activation energy for flow in fused silica. The silicate lattice is broken up by the alkali oxide, making flow easier. With increasing temperature the activation energy decreases relatively more rapidly in sodium silicates than in fused silica, apparently because the lattice is already broken up by the alkali ions. A similar more rapid decrease of activation energy with temperature occurs in B_2O_3, again because of some weaker bonds.

The high-temperature activation energy of multicomponent silicates depends on the amount of oxides other than silica in the glass. For binary alkali silicates, this activation energy decreases from about 50 kcal/mole with 10% alkali oxide to 20 kcal/mole with 50% alkali oxide.[12] For binary alkaline earth silicate, the activation energy for viscous flow at high temperatures is about 40 to 50 kcal/mole for 25 to 50% alkaline earth oxide.[12]

Bockris and co-workers[12,13] have proposed a model of liquid binary silicates to explain these results and those of thermal expansion. At the orthosilicate composition (two moles of metal oxide per mole of SiO_2) the liquid consists of metal cations and SiO_4^{4-} anions. As more silica is added, the anions polymerize to larger units, making flow more difficult. Among the larger units in these anions are rings and chains of silica tetrahedra. "Islands" of three-dimensional silica in the melt are proposed in a related model.[14] While silicate melts must undoubtedly contain polymerized units of silica tetrahedra, there is no direct evidence for any of the particular anions proposed, so these models remain speculative.

The viscosity of B_2O_3 also decreases when other oxides are added, but much less than for SiO_2. This result is expected because of the lower viscosity of B_2O_3, resulting from some weaker bonds already present in the pure oxide. A summary of viscosities and other properties of borate melts is given by Mackenzie.[15]

Viscosity data on nitrate, sulfate, halide, and chalcogenide glasses have been collected by Rawson.[16] Especially curious is the viscosity of liquid sulfur, which is low at the melting point, decreases with increasing temperature up to 160°C, where it abruptly increases by a factor of about 10,000 over a temperature range of 20°C, and then decreases again at higher temperatures. At the melting point the liquid contains S_8 rings; at about 160°C these rings break up and polymerize to long chains, causing the rise in viscosity. At higher temperatures the chains decrease in length, again lowering the viscosity.[17,18]

Miller[19] has discussed and given references to earlier work on viscosities of glassy polymers.

Viscosity data on a number of organic glasses have been collected by Ling and Willard.[20] The activation energy for flow is constant with temperature above a viscosity of about 10^4 P. These activation energies are in the range from 12 to 35 kcal/mole. Since these measurements are at low temperatures (75 to 200°K), the viscosity drops very sharply as a function of temperature at these viscosities. For example, the viscosity of 2-methylpentane increases eight orders of magnitude in about 15°C. At higher temperatures where the

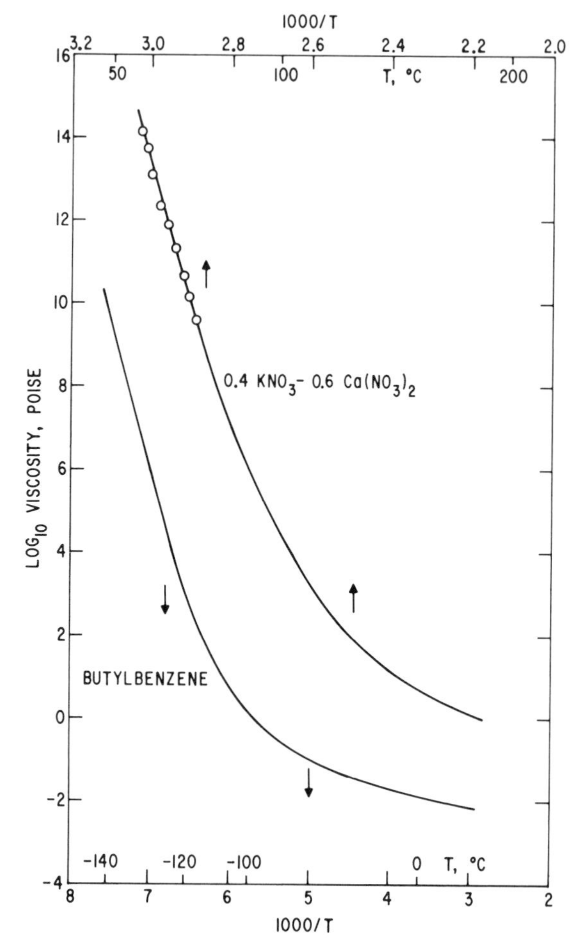

Fig. 4 Viscosities of a glass-forming fused salt, 0.4 KNO_3-0.6 $Ca(NO_3)_2$[34] and *N*-butylbenze.[20]

viscosity is above 10^4 P, the activation energy for flow decreases with increasing temperature.

The viscosities of four very different materials as measured over a wide viscosity range are shown in Figs. 3 and 4. In each case the Arrhenius relation for the exponential dependence of viscosity on the reciprocal of temperature (Eq. 3) is obeyed at the highest viscosities, and in some cases at the lowest, with an activation energy changing with temperature in between.

THEORIES

Various theories for the viscous flow of glass-forming liquids have been developed over the past decade. Unfortunately the validity of these theories is uncertain, and no satisfactory description of the viscosity of glasses has emerged, in spite of the agitation for different points of view. In this section some of the more popular ideas are summarized, but the complications of the subject cause doubt that a satisfactory and reasonably simple theory is on the way.

Theories for the viscosity of monomolecular liquids have the form of Eq. 3, where η_0 is a complicated function of various parameters, but usually varies much less with temperature than the exponential term, and to a first approximation Q is temperature-independent.[21] In a treatment based on the molecular theory of liquids, Q is equal to the mutual potential energy of a pair of molecules,[21] and in the transition state theory it is the height of the potential barrier to mutual rotation of neighboring molecules.[22] The mutual potential energy of molecules is at least calculable in principle, but the factors controlling the heights of potential barriers in all but the simplest cases are quite uncertain. Extension of these concepts to more complicated liquids, with strongly bonded molecules, is very difficult and has not been successfully accomplished. Thus other simpler approaches to the viscosity of glasses have been tried.

The empirical equation

$$\eta = \eta_0 \exp\left(\frac{B}{T - T_0}\right) \tag{4}$$

where η_0 and B are temperature-independent, has been found to fit viscosity data for glasses over certain temperature ranges.[23–25] This equation is the basis for two theories of viscosity: one based on "free volume" and the other on entropy.

In the simplest form of the free-volume treatment[26–29] the viscosity is given by the equation

$$\eta = \eta_0 \exp\left(\frac{B_1}{V - V_0}\right) \tag{5}$$

where V is the specific volume of the liquid and V_0 is its close-packed specific volume. If the expansion coefficient is independent of temperature, Eq. 5 is equivalent to Eq. 4, since then $V - V_0$ is proportional to $T - T_0$. This free-volume treatment can be thought of crudely as follows. At some temperature T_0, the molecules are packed together with the specific volume V_0, and in this condition no molecular motion is possible. As the temperature increases, some excess or free volume $V - V_0$ develops in the liquid, providing "space" for molecular motion and ultimately for flow. More elegant descriptions of this model are given by Cohen and Turnbull[28] and by Bueche.[29]

In the entropy approach[30] it is assumed that the liquid consists of regions that rearrange as units when they experience a sufficient fluctuation in energy. Thus there is a barrier in potential energy to the rearrangement of the units. The size of the units are functions of temperature and are determined by the configurational entropy of the liquid. As this entropy decreases, the sizes of the units increase until at zero configuration entropy the size is infinitely large and the liquid can no longer rearrange. Quantitatively

$$\eta = \eta_0 \exp\left(\frac{B_2}{TS_c}\right) \tag{6}$$

where S_c is the configurational entropy. This equation is equivalent to Eq. 4 when $S_c = \Delta C_p(T - T_o)/T$, and ΔC_p, the relaxational part of the specific heat, is constant with temperature.

In Eq. 4 a third parameter T_o is introduced to account for the changing activation energy with temperature. Thus, to establish more than empirical significance for the equation, one must find some significant correlation for values of T_o. Williams et al.[27] and Adams and Gibbs[30] found that T_o was about 50°C higher than the glass transition temperature Tg for many glasses. The glass transition temperature in these correlations was defined as the temperature at which there was a discontinuity in the coefficient of thermal expansion (see the following chapter for a discussion of transition temperature).

Goldstein compared the free-volume and entropy theories of viscosity and concluded that the entropy theory was superior,[31,32] chiefly because the free-volume theory cannot explain the pressure dependence of the glass transition temperature.

However, the most serious defect of any theory that results in Eq. 4 for the viscosity is that this equation is inconsistent with experimental data on glass-forming liquids at high viscosities. In a variety of different glasses the activation energy for viscous flow becomes temperature-independent at high viscosities (see Refs. 20, 33, and 34, and Figs. 3 and 4) just where Eq. 4 would predict it should vary most sharply. Equation 4 predicts an infinite viscosity at some low temperature; the experimental results indicate that the viscosity varies according to Eq. 3 at low temperatures. Equation 4 provides an empirical fit to data at intermediate temperature with some correlation between T_o and the temperature at which other properties change, but the theoretical significance of T_o for viscous flow is unclear. In at least one glass, GeO_2 (Fig. 2), T_o is zero, since Eq. 3 is valid over the whole viscosity range in which measurements were made. At high temperatures and low viscosities the activation energy is also constant for certain glasses (see Refs. 11 and 32, and Fig. 3). Thus a model for the viscosity of B_2O_3 has been proposed[32,34] in which the liquid has a "dissociated" state at high temperatures, and as the temperature is lowered agglomeration takes place until at low temperature and high viscosity the liquid is completely "associated." Such a simple structural picture at least is consistent with the experimental viscosities.

Goldstein has discussed the present theories for viscosities of liquids and finds them inadequate.[32] He outlined a picture of the flow process that could lead to the development of a more satisfactory theory, although much remains to be done to achieve this goal. In Goldstein's picture, flow at high viscosities (lower temperatures) is limited by potential barriers that are high compared to the thermal energy. As the temperature is raised to the point where the thermal energy is comparable to the heights of the potential barriers, the simple picture is no longer valid. Goldstein estimates that the picture may be valid down to viscosities as low as 10 P. He concludes that his model requires that no liquid can have a single relaxation time in shear, but must have a spectrum of relaxation times, as has been found experimentally from acoustic measurements at high viscosities.[35,36] However, at higher temperatures and lower viscosities the relaxation spectra approach those for a single relaxation time. A quantitative theory as derived from Goldstein's ideas is awaited with interest.

In conclusion, more work and thought are needed to develop a satisfactory description of viscous flow in liquids.

REFERENCES

1. G. W. Morey, *The Properties of Glass*, Reinhold, New York, 1954, pp. 140 ff.
2. G. B. Bartenev, in *Physics of Noncrystalline Solids*, J. A. Prins, Ed., North Holland Pub. Co., Amsterdam, 1965, p. 461; *The Structure and Mechanical Properties of Inorganic Glasses*, Wolters-Nordhoff Pub. Co., Groningen, 1970, pp. 141 ff.
3. H. R. Lillie, *J. Am. Ceram. Soc.*, **16**, 619 (1933).
4. G. Hetherington, K. H. Jack, and J. C. Kennedy, *Phys. Chem. Glasses*, **5**, 130 (1964).
5. G. Hofmaier and G. Urbain, in *Science of Ceramics*, Vol. 4, British Ceramics Society, 1968, p. 25.
6. C. R. Kurkjian and R. W. Douglas, *Phys. Chem. Glasses*, **1**, 19 (1960).
7. R. L. Cormia, J. D. Mackenzie, and D. Turnbull, *J. App. Phys.*, **34**, 2245 (1963).
8. G. A. Parks and M. E. Spaght, *Physics*, **6**, 69 (1935).
9. J. D. Mackenzie, *Trans. Far. Soc.*, **52**, 1564 (1956).
10. E. H. Fontana and W. A. Plummer, *Phys. Chem. Glasses*, **1**, 139 (1966).
11. G. S. Meiling and D. R. Uhlmann, *Phys. Chem. Glasses*, **8**, 62 (1967).
12. J. O. Bockris, J. D. Mackenzie, and J. A. Kitchener, *Trans. Far. Soc.*, **51**, 1734 (1955).
13. J. W. Tomlinson, M. S. R. Heynes, and J. O. Bockris, *Trans. Far. Soc.*, **54**, 1833 (1958).
14. J. O. Bockris, J. W. Tomlinson, and J. T. White, *Trans. Far. Soc.*, **52**, 299 (1956).
15. J. D. Mackenzie, in *Modern Aspects of the Vitreous State*, Vol. 1, J. D. Mackenzie, Ed., Butterworths, London, 1960, p. 188.
16. H. Rawson, *Inorganic Glass Forming Systems*, Academic, London, 1967.
17. R. E. Powell and H. Eyring, *J. Am. Chem. Soc.*, **65**, 648 (1943).
18. G. Gee, *Trans. Far. Soc.*, **48**, 515 (1952).
19. A. A. Miller, *J. Poly. Sci.*, **A2**, 1095 (1964); *J. Chem. Phys.*, **49**, 1393 (1968).
20. A. C. Ling and J. E. Willard, *J. Phys. Chem.*, **72**, 1918, 3349 (1968).
21. A. Bondi, *Rheology*, F. R. Eirich, Ed., Academic, New York, 1956, p. 321.
22. R. E. Ewell and H. Eyring, *J. Chem. Phys.*, **5**, 726 (1937).
23. H. Vogel, *Phys. Z.*, **22**, 645 (1921).

24. G. S. Fulcher, *J. Am. Ceram. Soc.*, **77**, 3701 (1925).
25. G. Tamman and W. Hesse, *Z. Anor. Allgem. Chem.*, **156**, 245 (1926).
26. A. K. Doolittle, *J. Apl. Phys.*, **22**, 1031 (1951); **23**, 236 (1952).
27. M. L. Williams, R. F. Landel, and J. D. Ferry, *J. Am. Chem. Soc.*, **77**, 3701 (1955).
28. M. H. Cohen and D. Turnbull, *J. Chem. Phys.*, **31**, 1164 (1959).
29. F. Bueche, *J. Chem. Phys.*, **30**, 748 (1959).
30. G. Adam and J. H. Gibbs, *J. Chem. Phys.*, **43**, 139 (1965).
31. M. Goldstein, *J. Chem. Phys.*, **39**, 3369 (1963); **43**, 1852 (1965).
32. M. Goldstein, *J. Chem. Phys.*, **51**, 3728 (1968).
33. P. B. Macedo and A. Napolitano, *J. Chem. Phys.*, **49**, 1887 (1968).
34. H. Tweer, N. Laberge, and P. B. Macedo, *J. Am. Ceram. Soc.*, **54**, 121 (1971).
35. P. Macedo and T. A. Litovitz, *J. Chem. Phys.*, **42**, 245 (1965); *Phys. Chem. Glasses*, **6**, 69 (1965).
36. R. Kano, G. E. McDuffie, and T. A. Litovitz, *J. Chem. Phys.*, **44**, 965 (1966).

7

Glass Transition

As a glass-forming liquid is cooled some of its properties change sharply in a narrow temperature range.[1] In Fig. 1 the specific volumes of a soda-lime silicate glass and glassy B_2O_3 are shown as a function of a temperature; an abrupt change in the slope of the curve, which is equal to the coefficient of volume expansion, occurs at a certain temperature. Other properties, such as the heat capacity (see Fig. 2 and Refs. 4 to 6) and electrical conductivity (Chapter 9), also change abruptly at about the same temperature. This temperature is called the glass transition or transformation temperature.

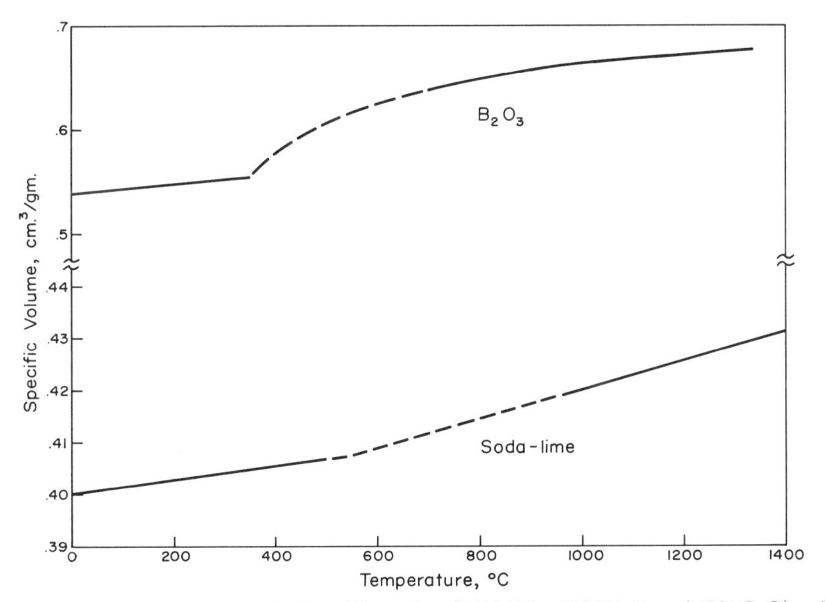

Fig. 1 Specific volumes of a soda-lime silicate glass (74% SiO_2, 16% NaO, and 10% CaO) and glassy B_2O_3 as a function of temperature.[2]

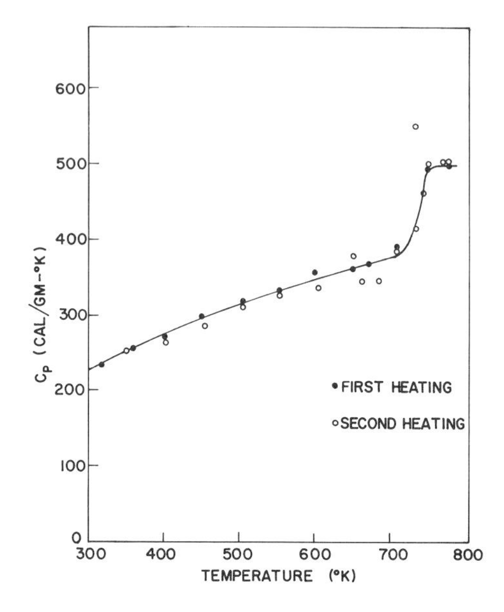

Fig. 2 Heat capacity of 0.15 Na_2O, 0.85 B_2O_3 glass as a function of temperature.[3]

It is significant that the viscosity does not show an abrupt change at the glass transition temperature. This result suggests that the viscosity determines the rate at which a property relaxes to some equilibrium value, but may not be related to the magnitude of property changes under different conditions. Therefore a theory for the viscosity probably will not result from consideration of changes in properties of the glass, but only from study of the details of the flow mechanism itself; hence the failure of the viscosity theories described in the last chapter that were based on the free volume or entropy of the glass.

Kauzmann[1] showed that glass transitions occur for many of the types of glasses listed in Chapter 2. It has been suggested that some amorphous solids do not show a glass transition temperature, but this result has not been established with any certainty.

It is reasonable to associate the glass transition temperature with the slowing down of rearrangements in the glass structure. When these rearrangements occur rapidly during the time of experimental measurements, the glass has the properties of a liquid; when the rearrangements are slow, the glass structure is "frozen" and it behaves like a solid. In fact, the specific volumes and their temperature dependencies of the glasses shown in Fig. 1 are quite close to those of the corresponding crystalline solids below the glass transition temperature. The rate of these molecular rearrangements should be closely associated with viscous flow, and indeed the glass transition temperature occurs at a viscosity of about 10^{13} P.

Thus the presence of a glass transition results from the slowing down of viscous relaxation in a liquid. Estimates of the relaxation time at a viscosity of 10^{13} P are in the range of minutes to a few hours, in reasonable accord with the usual times of experiments.

Since the transition temperatures determined from specific volume and heat capacity are nearly the same,[1] the molecular motions involved in changing these properties with temperature changes are probably similar. Alternatively, the same apparent transition temperatures for these two properties may result because the relaxation rate of glasses changes so rapidly with temperature. This rapid change is shown in the high activation energies for viscous flow at high viscosities, as described in the last chapter.

Other properties of glasses show changes at the glass transition temperature. There is some evidence that the elastic modulus of glasses changes sharply at the glass transition temperature,[7,8] although it is difficult to measure a meaningful value of the modulus much above the glass transition temperature. As this temperature is approached from below, the delayed elasticity of glasses increases rapidly, as shown in Chapter 11. This change is a manifestation of the increasing ease of molecular motion and consequent change in elasticity with time at a particular temperature. An increase in hydrostatic pressure apparently increases the transition temperature.[1] This result might be expected because of the reduction in relaxation rates with increasing pressure. Changes in the absorption spectra of nitrate glasses occur at the glass transition.

Kauzmann[1] showed that the dielectric relaxation of several organic glasses changes abruptly at the glass transition temperature. The dielectric relaxation times are about 30 min at the transition temperature, consistent with the estimates mentioned above. It is known that the molecular motions in dielectric relaxation of liquids are related to their viscosities, so a correlation of dielectric relaxation with transition temperature is expected. This sort of dielectric relaxation should not be confused with other dielectric losses in ionic glasses, including oxides. These losses are associated with ionic motion, as discussed in Chapter 11.

STRESS AND STRAIN RELAXATION

When a glass is cooled rapidly from a temperature above the transition region into this region, it retains some properties of the higher temperature, and these properties " relax " to those characteristic of the lower temperature as a function of time. An example of this process was given in the last chapter in describing the change of viscosity with time at high viscosity. The time to reach a stable or equilibrium state is, of course, longer the lower the holding temperature.

If a piece of glass is cooled rapidly from a temperature above the transition region, nonuniformities in its temperature during cooling lead to regions of differing specific volume and thus to stresses in the glass. These stresses can weaken the piece, and change its properties, so it is desirable to remove them by heating the piece at an appropriate temperature in the transition region. At this "annealing" temperature the volume differences and thus the stresses are removed as the volumes approach that volume characteristic of the holding temperature. The rate at which this annealing process goes on is very important in preparing glasses for use and so has been much studied. Good empirical relations for annealing have been devised, and in particular it has been found that the rate of annealing is closely related to the viscosity of the glass, hence the designation of the temperature where the viscosity is $10^{13.4}$ P as the annealing point, since at this viscosity most stresses are removed in about 15 min.

In contrast to this empirical knowledge, the relation of the rate of removal of strain and of the change of other properties in the transition region to the molecular structure of glass is only qualitative, and often not even that. A large number of measurements of property changes have been made, but correlation between changes of different properties is often difficult, and, in most cases, no universal mathematical functions were found to describe the results. Earlier studies were reviewed by Morey,[9] Goldstein,[10] and Kurkjian[8] (see also Refs. 11 to 13).

Tool[14] introduced the interesting idea of a fictive temperature as a parameter for characterizing the state of a glass. The fictive temperature of the glass is the temperature that corresponds to the state of the glass. For example, if a glass is rapidly cooled from a temperature above the transition region, it retains the properties characteristic of this temperature, which is therefore its fictive temperature. The fictive temperature changes as a function of time during annealing until it becomes the annealing temperature. The difficulty with this approach is that the same value of properties, such as the specific volume, can be attained in a glass by two quite different heat treatments, implying that a single parameter is not sufficient to characterize the state of a glass. Thus, whereas the idea of a fictive temperature has value in certain simple types of experiments and perhaps as a technological characterization, it is not sufficiently precise for a detailed description of the state of the glass sample (see also Ref. 15).

It is important to characterize the initial state of the glass carefully before studying the kinetics of property changes in the transition region. One way to start with a known state is to anneal the glass carefully at the temperature of the experiment. Then, if some force is imposed on the glass, such as a shearing stress, rates of changes under this force can be studied starting from a known condition. This procedure was followed by Kurkjian[8] in studying stress relaxation in a soda-lime silicate glass. Since relaxation of stress is the aim of annealing, this type of experiment gives insight into the annealing process.

To study the stress relaxation, Kurkjian rapidly twisted an annealed glass rod a certain amount with a known stress. The stress in the rod at this fixed twist and at a constant temperature was then measured as a function of time. The "instantaneous" initial stress So was used to normalize the stresses at different temperatures. When this was done, the stress relaxation at different temperatures in the transition range fitted on the same master curve, as shown in Fig. 3, where the normalized stress S/So is plotted as a function of

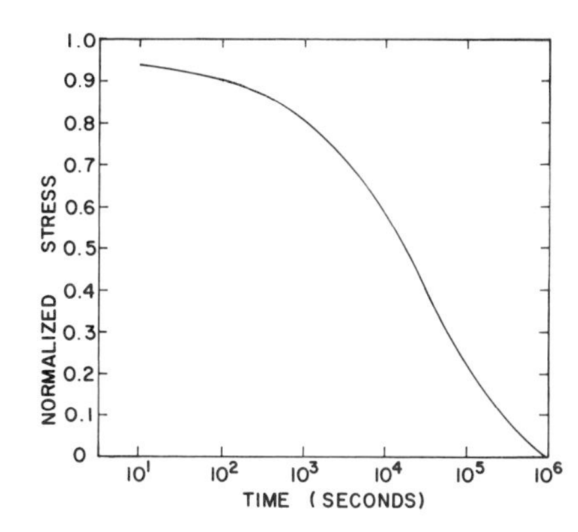

Fig. 3 Stress relaxation of a soda-lime silicate glass as a function of time.[5] Data at different temperatures are normalized to 473°C using an activation energy of 150 kcal/mole.

time. The existence of such a master curve shows that a single mathematical function describes the relaxation data at all temperatures. Kurkjian fitted his stress S as a function of time to a sum of six exponential terms of the form

$$S = \Sigma_i \varepsilon_i M_i \exp\left(\frac{-t}{\tau_i}\right) \tag{1}$$

where ε_i is the initial strain, and M_i is the modulus for a process with relaxation time τ_i. In these terms one can say that the spectrum of relaxation times was the same for the different temperatures of the experiment.

The relaxation spectrum of Eq. 1 gives a phenomenological description of the experimental results. The spectrum found by Kurkjian had a long tail at short times and cut off rather sharply at long times. The relationship of this spectrum to molecular structure and processes is not clear at this time, and should provide interesting material for further study. The importance of Kurkjian's work was to show that by proper attention to the initial conditions, the same mathematical function describes the results at different temperatures.

A further discussion of stress relaxation in glass together with a review of other mathematical functions and models used to describe this process has been given by Bartenev.[16]

One often wishes to follow the rate of change of a property from a nonequilibrium, or stabilized, state to the equilibrium condition. The difficulty here is to characterize the initial state of the glass, and this problem has not been satisfactorily solved. Some experiments of this kind have been discussed by Goldstein[10] and Douglas,[17] but much work remains to give even a phenomenological description of these changes.

REFERENCES

1. W. Kauzmann, *Chem. Rev.*, **43**, 219 (1948).
2. L. Merker, *J. Soc. Glass Tech.*, **43**, 179 (1959).
3. D. R. Uhlmann, A. G. Kolbeck, and D. L. DeWitte, *J. Noncryst. Solids*, **5**, 426 (1971).
4. V. E. Shnaus, C. T. Moynihan, R. W. Gammon, and P. B. Macedo, *Phys. Chem. Glasses*, **11**, 213 (1970).
5. C. T. Moynihan, P. B. Macedo, J. D. Aggarwal, and V. E. Schnaus, *J. Noncryst. Solids*, **6**, 322 (1971).
6. S. S. Chang and A. B. Bestal, *J. Chem. Phys.*, **55**, 933 (1971); **56**, 503 (1972).
7. G. S. Parks and J. D. Reugh, *J. Chem. Phys.*, **5**, 364 (1937).
8. C. K. Kurkjian, *Phys. Chem. Glasses*, **4**, 128 (1963).
9. G. W. Morey, *The Properties of Glass*, Reinhold, New York, 1954, p. 166.
10. M. Goldstein, in *Modern Aspects of the Vitreous State*, Vol. 3, J. D. Mackenzie, Ed., Butterworths, London, 1964, p. 90.
11. S. S. Spinner and A. Napolitano, *J. Res. Nat. Bur. Stand.*, **A70**, 147 (1966).
12. M. Goldstein and N. Nakoneczny, *Phys. Chem. Glasses*, **6**, 126 (1965).
13. J. S. Haggerty, Ph.D. Thesis, Massachusetts Institute of Technology, Cambridge, Mass., 1965.
14. A. Q. Tool, *J. Am. Ceram. Soc.*, **29**, 240 (1946).
15. H. N. Ritland, *J. Am. Ceram. Soc.*, **39**, 403 (1956).
16. G. M. Bartenev, *The Structure and Mechanical Properties of Inorganic Glasses*, Wolters-Nordhoff Pub. Co., Groningen, 1970, pp. 111 ff.
17. R. W. Douglas, in *Physics of Noncrystalline Solids*, J. A. Prins, Ed., North Holland Pub. Co., Amsterdam, 1965, p. 397.

8

Molecular Solution and Diffusion in Glass

Gases can dissolve molecularly in glass because of its open structure. If the gas molecule is small, it can diffuse rapidly in a simple glass such as fused silica. The permeability of fused silica and other glasses to helium has been known for many years; Barrer[1] reviewed earlier work back to 1900. The different permeabilities of glass for different gases can be used to separate and purify gas mixtures. Molecular diffusion is also important in "fining" or removal of bubbles from glass melts.

Gases such as hydrogen, oxygen, and water, which dissolve molecularly in glass, can also react with the glass network. These reactions affect certain properties of the glass, for example, optical absorption, viscosity, and electrical conductivity, and are important in fining glass. Such reactions are discussed below and in Chapter 13 on chemical reactions in glass.

In this chapter experimental results and theories of molecular solubility in glass are first discussed, followed by sections on the methods of measuring molecular diffusion in glass, the effects of temperature and molecular size on molecular diffusion, chemical reactions, the influence of glass composition on molecular diffusion, and theories for this diffusion. This chapter summarizes and extends parts of my earlier review on diffusion in glass.[2]

MOLECULAR SOLUBILITY OF GASES

Several different methods have been used to measure gas solubilities in glass. The simplest technique involves saturating the glass, usually in the form of fibers or powder, with gas at a certain temperature and then pumping out the gas at a higher temperature and measuring the amount given off.[3–5] The pressure drop of a gas in a vessel containing glass samples also gives a measure of the solubility.[6] The most common method[7–12] employed is to combine measurements of the permeation coefficient K and the diffusion coefficient D, since the solubility is equal to the ratio K/D, as shown in the next section.

The solubility is defined here as the ratio C_i/C_g, where C_i is the concentration of gas dissolved in the glass, and C_g is the concentration of molecules in the gas phase. This solubility is unitless, and is sometimes called the "coefficient of solubility," after Ostwald. Other definitions of solubility are S, the volume of gas at standard temperature and pressure dissolved per unit volume of glass per unit of external gas pressure, and the number of molecules of gas dissolved per unit volume of glass per unit external pressure. These solubilities are proportional to C_i; $S = (273/T)(C_i/C_g)$. The ratio C_i/C_g is preferred to the other definitions because it does not depend on external gas pressure, and because its temperature dependence gives directly the energy of interaction between the gas and the solvent (glass), as shown below, whereas the temperature dependence of S gives the enthalpy of interaction.

The solubilities of helium, neon, hydrogen, deuterium, and argon in fused silica are shown in Figs. 1–3. There are some rather surprising discrepancies

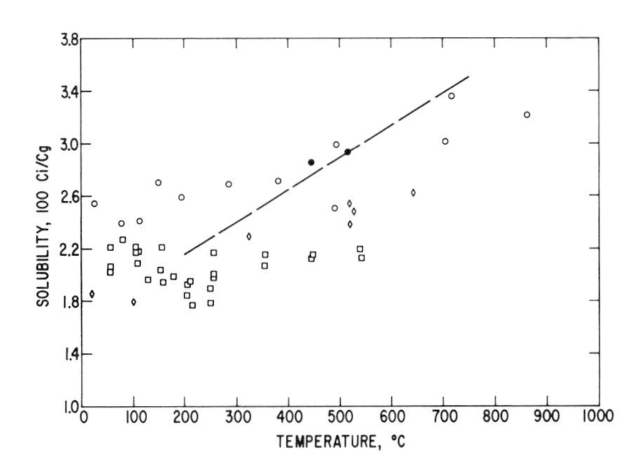

Fig. 1 Solubility ratio C_i/C_g for helium in fused silica. ●, Williams and Furguson[4]; ○, Swets, Lee, and Frank[7]; ◇, Woods and Doremus[5]; □, Shackelford[6]; —, Shelby.[11]

between results of different investigators, particularly for helium and hydrogen. The high values of hydrogen solubility found by Shackelford[6] may have resulted because he used a fused silica container, which can contribute an uncertain background. However, there are some striking generalities that can be made. The first is the small temperature dependence

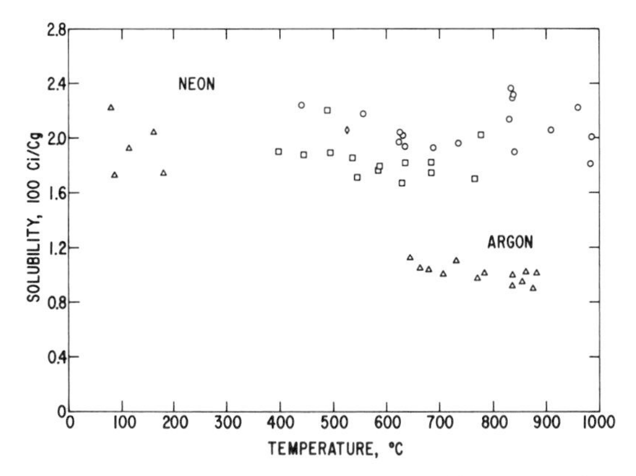

Fig. 2 Solubility ratio C_i/C_g for neon and argon in fused silica. ○, Frank, Swets, and Lee[8]; ◇, Woods and Doremus[5]; △, Perkins and Begeal[10]; □, Shackelford.[6]

of the solubility, which is constant within experimental error from 80 to 1000°C for neon and increases only slightly, if at all, with temperatures over this range for helium. There appears to be some increase in solubility of hydrogen at low temperatures, but the results of different investigators differ widely in this region, so that this increase may result from experimental problems. A second interesting finding is the similarity of the solubility of different gases. The rough average values for all investigators are 0.024 for helium, 0.019 for neon, and 0.03 for hydrogen. Argon is appreciably lower at

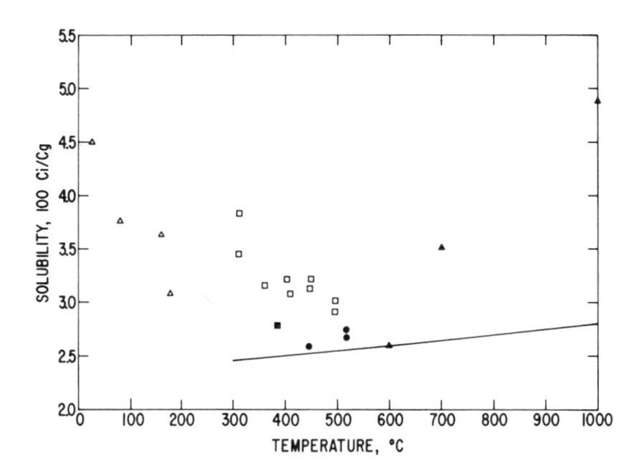

Fig. 3 Solubility ratio C_i/C_g for hydrogen and deuterium in fused silica. Hydrogen: ■, Wustner[3]; ●, Williams and Furguson[4]; □, Shackelford.[6] Deuterium: △, Perkins and Begeal[10]; ▲, Beauchamp and Walters[25]; —, Lee and Fry.[19]

0.01, which is the same as Norton's value of 0.01 for oxygen measured at about 1000°C.[12] These values may be compared with the molecular diameters of these gases in angstrom units: helium, 2.0; neon, 2.4; hydrogen, 2.15; oxygen, 3.2; argon, 3.2. The solubility C_i/C_g is constant with gas pressure up to a pressure of at least 1000 atm; Wustner measured the solubility of hydrogen at this pressure and found the same value as at 1 atm.[3]

Shelby[11a] found no appreciable difference in the solubilities of helium in several different types of fused silica with different impurity levels, for example, for a water concentration from a few parts per million up to 0.1%. He also found[13] little change of the solubility of helium in a series of SiO_2-TiO_2 glasses containing up to 10% TiO_2.

The solubility of helium and neon in the tridymite crystalline form of silica is about the same as in vitreous silica, but the solubility of these gases in the cristobalite crystalline form is about a factor of four lower than in tridymite or fused silica.[14,15]

Woods and Doremus[5] found that the solubility of helium and neon in germania (GeO_2) glass was about 0.013, or appreciably less than in fused silica.

Some results for solubilities of gases in Pyrex borosilicate glass (81% SiO_2, 13% B_2O_3, 4% Na_2O, 2% Al_2O_3) are summarized in Table 1. The results are similar to those for vitreous silica, although there appears to be a greater temperature dependence of helium solubility. The solubility of helium in Pyrex glass is lower than in vitreous silica, presumably because of the phase-separated structure of Pyrex as shown in Fig. 1, Chapter 4. The silica-rich matrix in Pyrex dissolves gas much as does vitreous silica, but occupies only part of the total glass volume, giving a lower effective

Table 1 Solubility of Gases in Pyrex Glass[a]

Temperature (°C)	Gas	Solubility C_i/C_g	Refs.
23	He	0.006	17
25	He	0.006	16
27	He	0.009	18
200	He	0.010	18
200	He	0.017	16
515	He	0.024	4
100–300	D_2	0.033	16
465	D_2	0.012	19

[a] Taken from Ref. 16.

solubility. In a borosilicate glass (60% SiO_2, 30.5% B_2O_3, and 9.5% Na_2O by weight) containing more borate and soda than Pyrex the solubility of helium at 300 to 500°C was 0.013 and of neon 0.010, showing that as the amount of the silica phase in the phase-separated structure of the borosilicate decreases, the molecular solubility decreases.[5]

Eschbach[20] measured solubilities of helium in four different borosilicate glasses and a lead glass; as the amount of network-formers ($SiO_2 + B_2O_3$) decreased, the solubility C_i/C_g decreased to as low as about 0.002 at 200°C for a lead glass. Eschbach also found a considerable decrease in helium solubility with temperature for two silica glasses whose complete compositions were not given. In addition he found[21] a low solubility for neon in Dusan borosilicate glass (76% SiO_2, 16% B_2O_3, 2% Al_2O_3, 6% Na_2O).

The solubility of helium in molten soda-lime glass was found by Mulfinger and Scholze[22] to be about 0.02 at temperatures from 1200 to 1500°C. These authors attributed the lower value, as compared to that for fused silica, to the presence of modifying sodium and calcium ions that "fill up" the space for gas solubility.

Two different models have been proposed for the molecular solubility of gases in glass. In one model, the gas is considered to dissolve in an inert material with a certain amount of free volume that is available for solution of gas.[23] This volume is constant with temperature and pressure as long as the glass is not appreciably affected by these variables.

The free volume is given directly by the ratio C_i/C_g. This relation can also be derived from statistical equations, such as those of Fowler and Guggenheim[24] for the chemical potential of an ideal gas μ_g and of this gas dissolved in a condensed phase μ_i:

$$\mu_g = RT \ln \phi + RT \ln C_g \tag{1}$$

$$\mu_i = -E + PV + RT \ln \phi + RT \ln C_i - RT \ln v_f \tag{2}$$

where ϕ is a function involving the translational and internal degrees of freedom of the molecules, $-E$ is a potential energy resulting from interactions between the dissolved gas and the solvent liquid, P is the total pressure, R is the gas constant, T is the absolute temperature, V is the partial molar volume of the solute gas, and v_f is the fractional free volume available to the dissolving gas. At equilibrium these two chemical potentials must be equal in order to give a relation for C_i/C_g in terms of the free volume. In deriving this relation it is assumed that the function ϕ is the same in the gaseous and dissolved states. This is equivalent to assuming that the dissolved molecule preserves three degrees of translational freedom. Then

$$\ln \frac{C_i}{C_g} = \frac{+E - PV}{RT} + \ln v_f \tag{3}$$

The total volume of the glass does not change appreciably as gas is dissolved in it; therefore $V = 0$, and the small temperature dependence of C_i/C_g shows that $E = 0$. Consequently, the free volume is equal to C_i/C_g, and is about 3% for fused silica.

This result can be compared with the free volume calculated by other methods. The free volume is often given by

$$V_f = 1 - \frac{V_0}{V} \tag{4}$$

where V is the bulk molar volume of the material, and V_0 is the volume of a mole of its molecules, these molecular volumes being calculated from the dimensions of the individual molecules. It is difficult to calculate the molecular volumes of the SiO_2 groups in fused silica. From the relative densities of fused silica (2.2) and coesite (3.0), a dense crystalline form of silica, one would expect a free volume of about 27%, far greater than the value calculated from the gas solubilities. Even using the density of quartz to calculate the molecular volumes gives a free volume of 13%. Therefore only a part of this kind of free volume in fused silica is available for solution of gas.

The free volume can be considered in terms of the detailed molecular structure of fused silica. It seems likely that the gas molecules dissolve in holes or interstices in the fused silica. Such interstices exist in cristobalite, which has a density close to that of fused silica. Molecular size, up to a molecular diameter of about 3 Å, appears to have a small effect on the gas solubility; then, if the oxygen and argon values are reliable, solubility decreases. Therefore it seems reasonable to assume that the average diameter of the interstices is about 3 Å. The total number of interstices for a free-volume fraction of 0.03 is then about $10^{21}/cm^3$. The largest gas concentration measured experimentally (by Wustner in hydrogen at 850 atm) is about $1.7(10)^{20}$ molecules/cm^3, which is still below the estimated number of interstices.

The discrepancy between the free volume calculated from the relative densities of fused silica and coesite or quartz and that for gas solubility is perhaps understandable in terms of the solution of gas molecules in interstices. There can be free volume in the fused quartz structure that is not in the form of interstices and thus is not available to the dissolving gas, either because it is not connected to the main network of interstices or because each part of it is too small to contain a gas molecule.

The number N_r of interstices of radius greater than the minimum radius r_0 can be estimated from the usual statistical equation:

$$\frac{N_r}{N} = \exp\left(\frac{-W}{RT}\right) \tag{5}$$

where N is the total number of interstices and W is the energy of formation of an interstice of radius r from one of radius r_0. An estimate for W comes from the equation of Frenkel[25] for the elastic energy to enlarge a spherical cavity of radius r_0 to a radius of r:

$$W = 8\pi G r_0 (r - r_0)^2 \tag{6}$$

The shear modulus is G, which is about $3(10)^{11}$ dynes/cm^2 for fused silica. Using $r_0 = 1.5$ Å, about 3% of the interstices have a diameter of 3.4 Å at 600°C. Thus a certain distribution of hole sizes above the minimum size is expected. The lower solubility of oxygen and argon may be caused by such a distribution, resulting in a smaller free volume being available to the larger molecules. It is possible that the distribution of hole sizes is fixed at higher temperatures where the viscosity of the silica is lower and is frozen in at lower temperatures.

From this description of gas solubility in fused silica one would expect the ratio C_i/C_g to decrease with increasing pressure above a certain pressure that is determined by the distribution of interstice sizes available for solution. This decrease should appear at a considerably lower pressure for gases with larger molecules, such as oxygen, nitrogen, and argon, than for helium, neon, and hydrogen, which have smaller molecules.

A second model for the molecular solubility of gas in glass has been developed from statistical mechanics.[6,14,26] For a dissolved molecule with three degrees of translational freedom, and localized in a site, the following equation was derived,[6,26] in the present notation:

$$\frac{C_i}{C_g} = N_s V_s \exp\left(\frac{E}{RT} - 1\right) \tag{7}$$

where N_s is the number and V_s is the volume of solubility sites. Since $N_s V_s$ is equivalent to v_f, Eq. 7 is the same as Eq. 3 except for a factor of e. However, all of these authors[6,14,26] preferred to consider the dissolved gas molecules as harmonic oscillators. Then for three degrees of vibrational freedom in the dissolved state, the following equation is found[6,14,26]:

$$\frac{C_i}{C_g} = \left(\frac{kT}{\nu}\right)^3 \frac{N_s}{(2\pi mkT)^{3/2}} \exp\left(\frac{E}{RT}\right) \tag{8}$$

Here m is the mass of a molecule of the dissolving gas, k is the Boltzmann constant, and ν is the vibrational frequency of the dissolved gas molecules. Equation 8 is an approximation valid when $(h\nu/kT) < 1$, where h is Plank's constant, and is equivalent to Eq. 3 if the vibrational frequency is considered to equal $(kT/2\pi m)^{1/2}(v_f)^{-1/3}$. Thus the two models are formally very similar, and present experimental results cannot readily distinguish between them. Shackelford[6] found that his vibration frequency ν was about inversely proportional to the molecular mass, giving a constant v_f for different atoms. Thus the use of one model or the other is a matter of taste; I prefer the free-volume model because of its simplicity and more direct physical interpretation. Beauchamp and Walters[15] and Perkins and Begeal[10] have also given reasons for considering molecules dissolved in glass as having translational rather than linear oscillator character. Theoreticians and statistical mechanics may prefer the vibration model.

In comparing results of the vibrational model with experimental data, ν and E were treated as adjustable parameters.[6,14,24] For fused silica, ν in the range from $3(10)^{12}$ to $7(10)^{12}$ vibrations/sec was found, which is reasonable. Values of E of about 1 to 3 kcal/mole were computed; Barrer and Vaughan[14] calculated similar values from a model of helium-oxygen interactions. In the free-volume model the gas-glass interaction is considered to be negligibly small, and changes in C_i/C_g resulting from different glass structures are attributed to changes in the volume available for solution, rather than to changes in the vibration frequency of dissolved molecules.

MEASUREMENT OF MOLECULAR PERMEATION AND DIFFUSION IN GLASS

The permeation and diffusion of a gas through glass have usually been measured in two different ways: with a glass membrane (in the form of a tube or a bulb) or in a powder. Thin membranes of many different glasses can be made by blowing or polishing, but sometimes it is difficult to seal them into a vacuum system, in which case the powder technique is useful.

Adsorption of a gas on the glass surface has been considered to be a necessary prelude to its solution and diffusion in the glass; however, the amount of gas dissolved in glass is proportional to the gas pressure at pressures far above that required for saturation of surface adsorption sites. Thus it seems likely that the gas molecules can dissolve directly in the glass without surface adsorption. Equilibrium between the gas and gas dissolved at the glass surface occurs rapidly for molecular solubility, and surface adsorption and reactions are not involved.

A glass membrane is mounted in a vacuum system to measure permeation and diffusion of a gas through it. First the spaces on both sides of the membrane are pumped out, and then the permeating gas is introduced on one side at the desired pressure. The amount of gas that appears on the other side of the membrane is then measured with time. The amount of gas that permeates the glass is kept so low that the drop in pressure on the inlet side is negligible and the pressure on the outlet side is small. Under these conditions the concentrations of gas dissolved in the membrane at its two surfaces are constant, being equal to some value C_i on the inlet side and zero on the outlet side. Since thin-walled tubes or bulbs of comparatively large diameter are used in these experiments the diffusion equation for a slab applies without appreciable error. The flux J of diffusing gas, which passes through unit surface area of the membrane of thickness L in unit time under the boundary conditions $C = C_i$ when $x = 0$, $C = 0$ when $x = L$, $C = 0$ when $t = 0$, is[27]

$$J = \frac{DC_i}{L}\left[1 + 2\sum_{n=0}^{\infty} \cos n\pi \exp\left(\frac{-Dn^2\pi^2 t}{L^2}\right)\right] \tag{9}$$

where t is the time after emission of the gas. Holstein[18] has shown that this equation can be transformed to

$$J = 2C_i \frac{D^{1/2}}{t} \sum_{n=0}^{\infty} \exp\left[\frac{-L^2(2n+1)^2}{4Dt}\right] \tag{10}$$

where the series converges rapidly for small t. For long times ($t \gg D/L^2$) these equations reduce to $J = DC_i/L$, which is the "steady-state" solution and implies a constant gradient throughout the membrane.

The measured pressure p of gas on the outlet side is related to J by the equation

$$J = \left(\frac{V}{ART_p}\right)\left(\frac{dp}{dt}\right)$$

where V is the volume on the outlet side where p is measured, A is the surface area of the glass membrane, R is the gas constant, and T_p is the temperature at which the pressure is measured. The results of permeation measurements in the steady state are usually expressed in terms of a permeation velocity K, which is defined as

$$K = \left(\frac{VL}{\Delta pA}\right)\left(\frac{dp}{dt}\right) = \frac{LRT_pJ}{\Delta p}$$

where Δp is the pressure difference across the membrane. The units of K generally used are cm^3 of gas at stp per cm^2 of glass area per sec per atm of gas pressure difference per cm of glass thickness. If the outlet pressure is negligible $p = C_g RT$, where C_g is the concentration of gas on the inlet side in moles/cm^3, and T is the temperature of the diffusion experiment. Then, since $J = DC_i/L$, $K = DT_pC_i/C_gT$, where T_p is 273°K. The solubility S is defined as the volume of gas (at stp) dissolved in unit volume of glass per atm of external gas pressure:

$$S = \frac{C_i T_p}{C_g T} \qquad \text{and} \qquad K = DS \tag{11}$$

The units of K are also cm^3/sec, since S is dimensionless. A discussion of the units of K and S is given by Rogers et al.[18] in an appendix to their article.

In the early stages of permeation D can be calculated from the measured change in pressure with time (dp/dt) in the outlet side. For this purpose Eq. 10 is useful, since for short times only the first term of the series is needed, so that a plot of log $[t^{1/2}\, dp/dt]$ against $1/t$ is linear, because J is proportional to dp/dt; the slope of the line is equal to $L^2/4D$, from which D can be calculated since L is known. This method was suggested by Rogers et al.[18] and gives the most accurate values of D from permeation data. The intercept of this line with the log $[t^{1/2}\, dp/dt]$ axis ($1/t \rightarrow 0$) can also be used to calculate the solubility S if D is known. When the outlet pressure p_0 is plotted against time, the curve becomes linear at long times (this is also the steady-state region), and the diffusion coefficient can as well be found from the intercept t_c of this line with the pressure axis from the relation $D = L^2/6t_e$, as shown by Barrer.[1]

If the gas on the inlet side is pumped out rapidly after the steady state is attained, the diffusion coefficient can be calculated from the rate at which the dissolved gas diffuses out of the membrane. The initial concentration for this desorption is given by $C = C_i(1 - x/L)$, and the boundary conditions are that $C = 0$ at both surfaces. The flux at the outlet is then[27]

$$J = \frac{2C_i D}{L} \sum_{n=1}^{\infty} \cos n\pi \exp\left(\frac{-Dn^2\pi^2 t}{L^2}\right) \tag{12}$$

Thus the diffusion coefficient and solubility can be calculated from permeability data in a variety of ways, and comparisons between the various calculations should give some idea of the reliability of the results.

In the powder method the diffusion coefficient is calculated from the amount of gas absorbed or desorbed from a powdered sample, and the solubility is found from the total amount of gas absorbed or desorbed per unit weight of glass. For desorption from spheres of radius R, the amount of gas desorbed Q is[5,14]

$$\frac{Q}{Q_\infty} = \frac{6(Dt)^{1/2}}{\pi R} - \frac{3Dt}{R^2} \tag{13}$$

where Q_∞ is the total amount of gas desorbed. A comparison of the experimental curve with Eq. 13 gives a measure of D when t, a, and Q_∞ are known.

EFFECT OF TEMPERATURE, PRESSURE, AND MOLECULAR SIZE ON MOLECULAR DIFFUSION IN GLASS

The most extensive measurements on diffusion of molecules in glass have been made on fused silica, since this glass is one of the most permeable to gases. Few measurements of permeation of gases other than helium or hydrogen have been made in other glasses.

The diffusion coefficient of helium through thin fused silica tubes was measured by the membrane technique from room temperature to 1000°C by Swets, Frank, and Lee.[7] In a plot of log D versus $1/T$ these authors found two straight line portions; however, a plot of log D/T versus $1/T$ gives a good straight line,[28] as shown in Fig. 4. These diffusion results are among

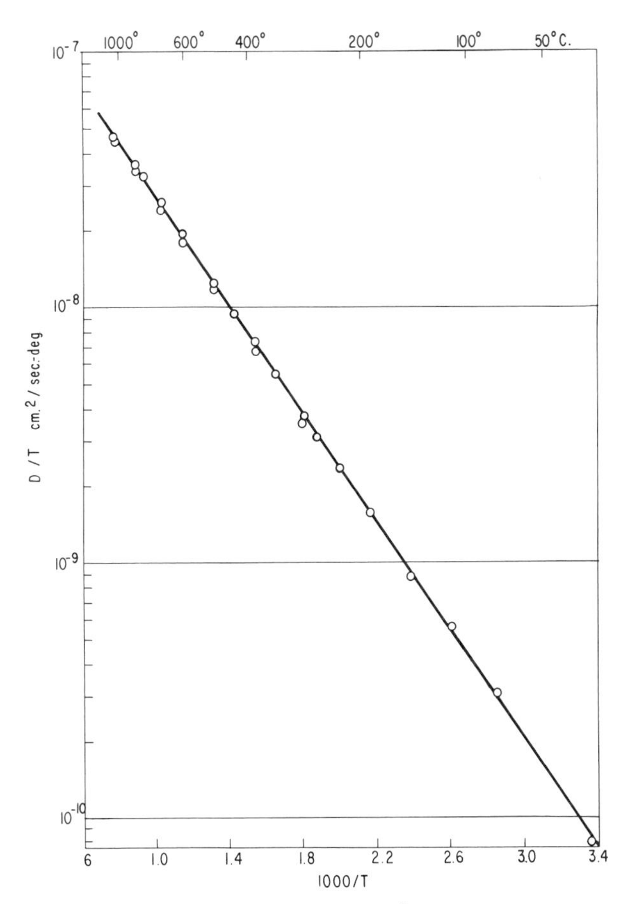

Fig. 4 Diffusion of helium in fused silica.[7]

the few taken over a wide enough temperature range and are of sufficient accuracy to determine whether a preexponential factor of temperature is needed in the Arrhenius equation:

$$D = D_0 \exp\left(\frac{-Q}{RT}\right) \tag{14}$$

Shelby[11b] confirmed in more extensive measurements that a preexponential factor of temperature T^n with $n = 1.06 \pm 0.11$ in this equation gives the best fit for helium and neon diffusing in fused silica. The data of Frank, Swets, and Lee[8] on neon diffusion in fused silica from 400 to 980°C are combined with those of Perkins and Begeal[10] in Fig. 5. Perkins and Begeal used a thin film of silica formed on porous Vycor for their membrane. The neon data also fit well on a log D/T versus $1/T$ plot and show distinct curvature on a log D versus $1/T$. Therefore it seems likely that the preexponential factor of temperature in Eq. 14 is valid for all molecular diffusion in fused silica and perhaps in other glasses as well.

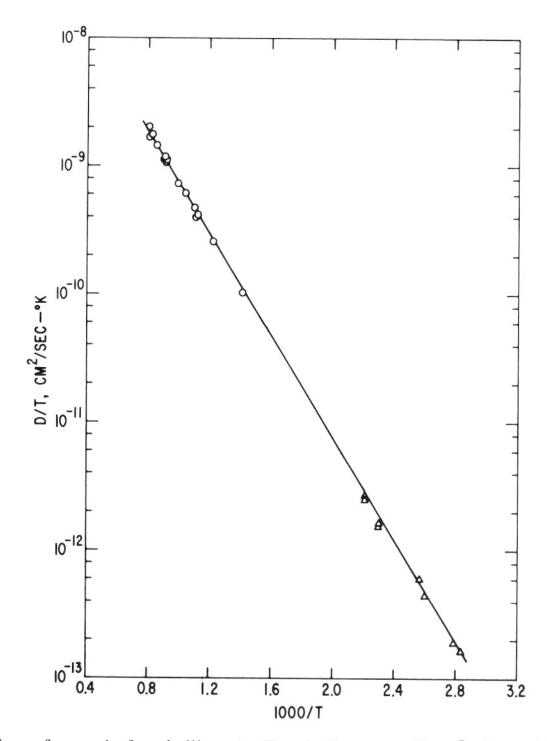

Fig. 5 Diffusion of neon in fused silica. ○, Frank, Swets, and Lee[8]; △, Perkins and Begeal.[10]

Table 2 Molecular Diffusion in Fused Silica

Molecule	Diameter (Å)	Diffusion coefficient (cm^2/sec) 25°C	1000°C	Activation energy Q' (kcal/mole)	Refs.
Helium	2.0	$2.4(10)^{-8}$	$5.5(10)^{-5}$	4.8	7
Neon	2.4	$5(10)^{-12}$	$2.5(10)^{-6}$	8.8	8,10
Hydrogen (deuterium)	2.5	$2.2(10)^{-11}$	$7.3(10)^{-6}$	8.5	9,10
Argon	3.2		$1.4(10)^{-9}$	26.6	10
Oxygen	3.2		$6.6(10)^{-9}$	25	12
Water	3.3		$\approx 3(10)^{-9}$	17	29,30
Nitrogen	3.4			26?	1,31
Krypton	4.2			≈ 46	10
Xenon	4.9			≈ 72	32

The size of the diffusing ion affects molecular diffusion in fused silica quite strongly, in contrast to the small dependence of solubility on size. The diffusion coefficients at 25 and 1000°C and the activation energies for diffusion for a number of molecules in fused silica are given in Table 2. The activation energies Q' have been calculated from an equation of the form

$$D = D'T \exp\left(\frac{-Q'}{RT}\right) \tag{15}$$

where Q' and D' are temperature-independent. This equation agrees well with experimental data and leads to a simple activation energy over the whole range of temperatures of the measurements, as mentioned previously. The molecular diameters in the table were estimated from the gaseous viscosities at 700°C; the water value was extrapolated. The diffusion coefficients and activation energies for helium, neon, and hydrogen have been measured by several different workers and are considered quite reliable. The results for argon were measured only by Perkins and Begeal,[10] but in view of the good agreement of their results on other gases with those of previous investigators, their results for argon should also be reliable. The diffusion coefficient for oxygen as measured by Norton should be dependable, but the activation energy was determined from only two different measurements and therefore is less certain. The activation energies for nitrogen and krypton were estimated from permeation measurements since no diffusion measurements have been made on these gases. The values for water are complicated by reaction with the silica lattice, as discussed later. The strong dependence of activation energy on molecular size is discussed further in the section on theories.

The diffusion coefficients of molecularly dissolved species show little dependence on gas pressure up to pressures ofat least 1000 atm. McAfee[17] found an anomalous increase in the permeation of helium at room temperature in bulbs of Pyrex borosilicate glass when they were subjected to a tensile stress close to the breaking stress. Laska and Doremus[32] confirmed this anomalous increase and by analyzing the permeation and uptake curves as a function of pressure concluded that it resulted from a decrease in the effective thickness of the bulbs, not from any change in the diffusion coefficient or the solubility of the helium. They speculated that this reduction in effective thickness of about 30% might result from fine cracks in the glass that were pinned before propagating all the way through the glass. This pinning is contrary to conventional ideas of glass fracture, in which a crack that starts to propagate continues through the sample to cause failure. The anomalous increase in permeation was not found for fused silica, nor was it present at 200°C in Pyrex. Perhaps internal voids in the borosilicate glass opened up at higher stress because of its phase-separated structure. More work is needed to test these ideas.

Annealing stressed borosilicate glasses reduces the rate of helium permeation in them, and increases the activation energy for diffusion.[33,34] Similar effects are found in ionic diffusion, as described in the next chapter; these changes probably result from the decrease in density of the glasses with annealing.

MOLECULAR DIFFUSION AND CHEMICAL REACTION

Experiments on the diffusion of water and oxygen in silica are complicated by reactions of these gases with the silica lattice. These reactions will be considered again in the chapter on chemical reactions in glass; here only their interference with diffusion is treated.

Water reacts with silica glass to form —OH groups in the following way:

$$H_2O + -\overset{|}{\underset{|}{Si}}-O-\overset{|}{\underset{|}{Si}}- = 2(-\overset{|}{\underset{|}{Si}}OH) \tag{16}$$

The rate at which these groups build up in fused silica in contact with water vapor has been studied extensively by Roberts and co-workers.[29,35–37] They favored a mechanism for diffusion of water into silica in which a proton jumps from an SiOH group to a neighboring Si—O—Si bridge, followed by a jump of the hydroxyl group. This mechanism is unlikely[30]; it requires a much higher activation energy than the 18 kcal/mole found by Moulson and Roberts[29] because the silicon-oxygen bonds that must be broken in this mechanism have energies of over 100 kcal/mole. If a lattice diffusion mechanism were valid one would expect a similar diffusion coefficient of water in crystalline quartz, but at 1000°C, where water diffuses readily into fused silica, there is no indication of its diffusion into quartz.[38]

Therefore a more likely model of water diffusion in fused silica is one in which the water molecules dissolve molecularly in the glass and subsequently react with the silicon-oxygen lattice by Eq. 16.[30] Crank (Ref. 27, pp. 121 ff.) has treated simultaneous diffusion and reaction in which the concentration of diffusing material is much lower than the concentration of reaction product. In this case diffusion takes place with an effective diffusion coefficient

$$D_e = 2D\left(\frac{C_r}{K^2}\right) \tag{17}$$

where C_r is the concentration of species that has reacted and remains fixed, D is the diffusion coefficient of dissolved species (molecular water), and K is an equilibrium constant. The relation between C_r, the concentration of SiOH groups in the present case, and C_i, the concentration of molecularly dissolved water, is

$$C_r = K(C_i^{1/2}) \tag{18}$$

since each water molecule reacts to form two SiOH groups. This square-root relationship between the amount of SiOH groups and C_i, and consequently the ambient water vapor pressure, was found by Moulson and Roberts.[29]

The effective diffusion coefficient of Eq. 17 is directly proportional to the concentration of C_r of SiOH groups; the solution of the diffusion equation for this dependence was found by Wagner[39] (see also Crank, Ref. 27, p. 162). This solution is compared with the profile of SiOH groups found by Roberts and Roberts[36] in Fig. 6. The fit is good, confirming the model of molecular diffusion and reaction of water. The activation energy and diffusion coefficient of water given in Table 2 were calculated from this comparison, assuming a solubility C_i/C_g for water of about 0.01 and using the value of K in Eqs. 17 and 18 calculated from the data of Ref. 29.

Oxygen molecules can also react and exchange with the oxygen in the silicon-oxygen lattice. In permeation of oxygen through a glass membrane this exchange is unnoticed, but in experiments involving exchange with oxygen gas of an isotropic ratio different from that of normal oxygen, this exchange becomes important. Several workers have studied this type of exchange reaction between oxygen isotopes and fused silica.[40–43] These results were not consistent with a simple diffusion process, and it was

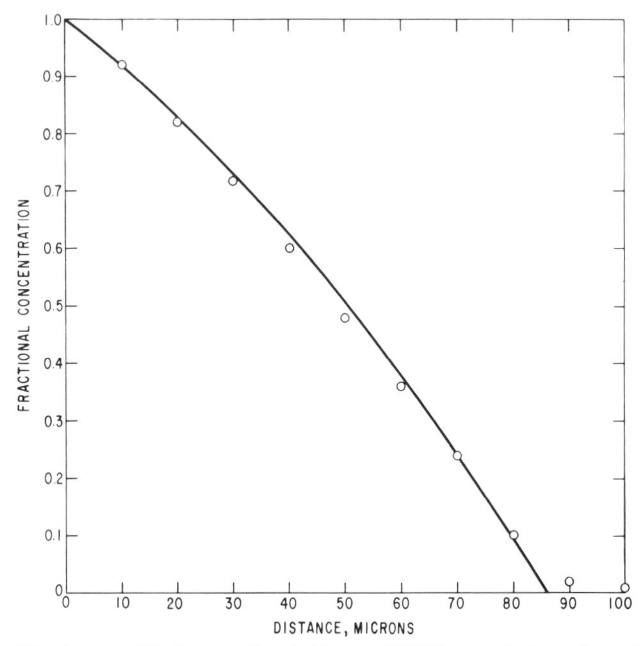

Fig. 6 Profile of water diffusing into fused silica at 1100°C. —, calculated from the diffusion coefficient of Eq. 17; ○, smoothed data from Ref. 36.

necessary to include a surface reaction to explain the time dependence of the isotropic ratio in the gas. Furthermore, the activation energies for diffusion were very different in each of the studies, although the coefficients measured were somewhat more similar. It seems likely that these difficulties can be explained by the assumption of molecular solution and diffusion of the oxygen molecules with subsequent reaction with the silica lattice. In this case no surface reaction is needed. Haul and Dumbgen[42] came close to this interpretation, since they found that the effective diffusion coefficient was proportional to the external oxygen pressure, and that a corrected diffusion coefficient, taking into account the total lattice concentration of oxygen, was nearly the same as the coefficient measured by Norton.[12] However, Haul and Dumbgen still used a surface, rather than a bulk, reaction to explain their results.

EFFECT OF GLASS COMPOSITION ON MOLECULAR DIFFUSION

Shelby[11] examined the diffusion of helium in a variety of types of fused silica with different amounts of impurities. For example, the hydroxyl concentration varied from 5 to 1300 parts per million, and one sample had

8 ppm of chlorine. The silicas were made in different ways: electrical or flame fusion of quartz powder, and oxidation of $SiCl_4$ in an oxyhydrogen or a water-free plasma. Within the relatively small error of his measurements, Shelby found no differences in the diffusion or solubility of helium in these glasses. Thus changes in impurities and methods of manufacture normally encountered for fused silica apparently do not effect molecular diffusion in it, in spite of some earlier reports of such effects.

The diffusion of helium and neon were measured in vitreous germania by Woods and Doremus,[5] and of helium in this glass by Shelby.[43a] Woods and Doremus found for helium and neon, respectively, activation energies of 4.9 and 8.9 kcal/mole and preexponential factors of $6.1(10)^{-6}$ and $5.6(10)^{-6}$ cm^2/sec. The activation energies were close to those for fused silica, without the preexponential temperature term, but the actual values of the diffusion coefficient at room temperature were $2(10)^{-9}$ for helium and $2(10)^{-12}$ for neon, thus about an order of magnitude lower than the helium value in Table 2, but only a factor of about 2.5 lower than the neon value. Shelby found an activation energy of 8.0 kcal/mole and a preexponential factor of about $1.5(10)^{-3}$ cm^2/sec for helium in germania with somewhat higher diffusion coefficients than Woods and Doremus. Apparently the larger germanium atom leads to a more "filled up" lattice than silica, even though the specific volume per mole of germania is somewhat larger than that of silica.

Shelby found that the diffusion coefficient of helium in vitreous B_2O_3 was about four times greater than for fused silica, although the activation energies for the two glasses were about the same.[43a]

Norton[44] reported a slightly higher permeability of Vycor glass (96% SiO_2, 4% B_2O_3) to helium than fused silica, and Lee and Fry[19] found a slightly higher diffusion coefficient for deuterium in Vycor than in fused silica. Apparently the Vycor process (see Chapter 4) leads to a glass with a slightly more open structure than that of fused silica.

Some diffusion measurements on gases in sodium borosilicate glasses and a titania silicate glass are summarized in Table 3. In two of the borosilicates the activation energy for diffusion is about the same as in fused silica, but as more sodium borate is added to the glass the absolute value of the diffusion coefficient decreases. These results are consistent with the phase-separated structure of the borosilicates; as the amount of the continuous silica phase decreases the effective area for diffusion decreases, but the activated step for diffusion remains the same. Eschbach's results for "Duran" borosilicate glass gave about the same activation energy for helium diffusion as in fused silica, but higher values for hydrogen and neon.[20,21] This difference results partly because of the higher temperatures of Eschbach's measurements, since no preexponential temperature correction was made, but even with this correction his activation energies for neon and hydrogen are high. In

Table 3 Molecular Diffusion in Sodium Borosilicate and Titania Silicate Glasses

Glass	Gas	Diffusion coefficient at 200°C (cm^2/sec)	Activation energy (kcal/mole)	Refs.
"Duran"	He	$4(10)^{-9}$	6.4	20
(76% SiO_2, 16% B_2O_3,	H_2	$1.2(10)^{-9}$	13.7	20
2% Al_2O_3, 6% Na_2O)	Ne	$2(10)^{-10}$ (extrapolated)	15.5	21
"Pyrex"	He	$4.7(10)^{-7}$	6.5	18
(81% SiO_2, 13% B_2O_3			5.8	16
4% Na_2O, 2% Al_2O_3)	D_2	$4.1(10)^{-9}$	11.2	16
Borosilicate	He	$5(10)^{-8}$	5.6	5
(60% SiO_2, 30.5% B_2O_3,	Ne	$7(10)^{-11}$	9.0	5
9.5% Ns_2O)				
Fused Silica	He	$8(10)^{-7}$	6.1	7
	Ne	$1.8(10)^{-9}$	9.0	10
	D_2	$6.5(10)^{-9}$	10.5	9
	He	$8(10)^{-7}$	4.8 *T* in	7
10% TiO_2, 90% SiO_2	He	$1.2(10)^{-6}$	4.5 D_0	13

the titania glass the diffusion coefficient is higher and the activation energy is lower than for helium in fused silica, indicating that the addition of titania leads to easier diffusion. This result is evidence that the titania groups substitute for silica groups in the network structure rather than acting as network modifiers, in which case one would expect a decrease in diffusion coefficient and an increase in activation energy as described later.

In many studies of the effect of composition on molecular transport in glass only the permeation rates, and not the diffusion coefficients, were measured. The results of Norton[44] for the permeation of helium in different glasses are shown in Fig. 7. The compositions of some of these glasses are given in Table 1, Chapter 6. As more network modifying ions are added, the permeation rate decreases and the activation energy increases. Two later more extensive studies of the permeation of helium through silicate glasses, particularly borates, gave similar results.[45,46] Shelby (unpublished) analyzed the results of these latter two authors by plotting the permeation rate at 300°C as a function of the sum of concentrations of network formers (SiO_2, B_2O_3, and P_2O_5) in these glasses, as shown in Fig. 8. There is a good correlation between the amount of network former, or alternatively the

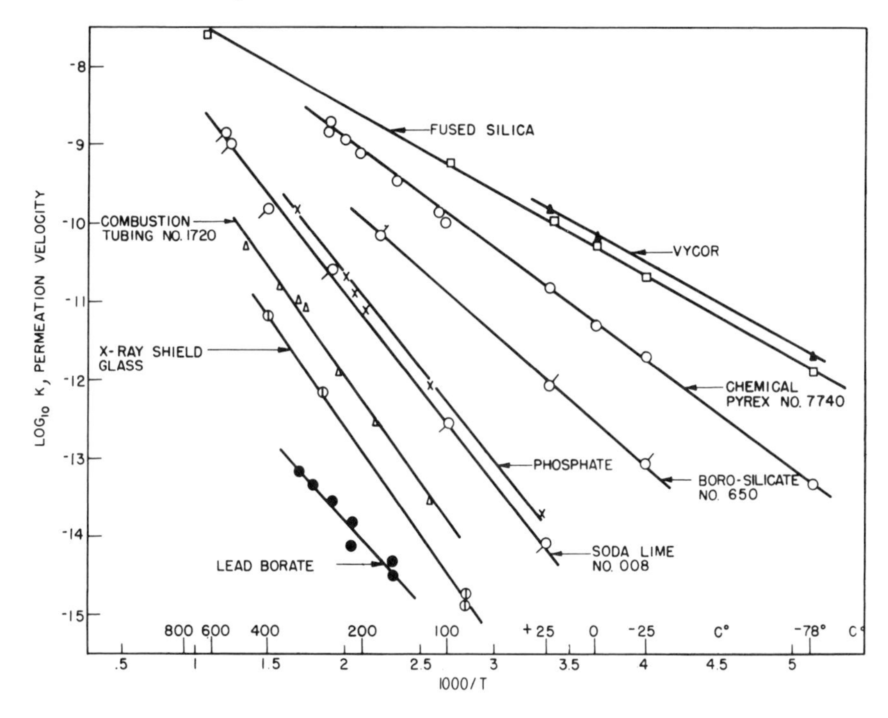

Fig. 7 Permeation rates for helium in various glasses as a function of temperature, measured by Norton.[44] Permeation units are cm^3 of gas at STP per sec per cm^2 area per mm thickness per cm Hg pressure.

amount of network modifier ions, and the permeation rate. In this comparison aluminum ions were considered to be network modifiers instead of network formers, although it is known that in some glasses aluminum can substitute for silicon as a network former. However, particularly in the alkaline earth aluminosilicates the permeation data show that alumina groups tend to reduce permeation of helium in the same way as ions that break up the silicon-oxygen network. Possibly in these glasses the alumina groups act as network breakers. Alternatively, the alumina groups may block molecular diffusion even if they are part of the glass network, since each alumina group must be accompanied by a charge-compensating cation. In any event the correlation shown in Fig. 8 should be valuable for predicting the permeation of helium in a glass in which it has not been measured, and indicates that each mole of network-breaking ions affects the permeation of helium in about the same way.

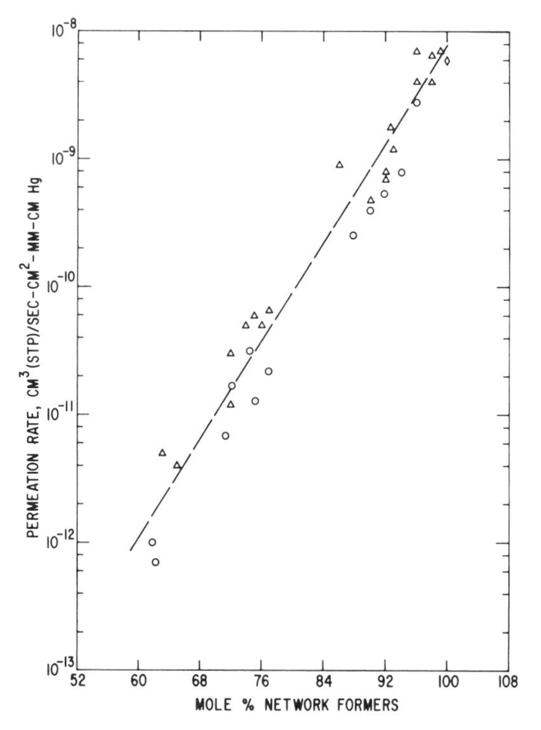

Fig. 8 Permeation rates for helium in various glasses at 300°C as a function of the % of network former ($SiO_2 + B_2O_3 + P_2O_5$) in the glass. △, Altemose[45]; ○, Vostrov et al.[46]

The data of Eschbach[20] on diffusion and solubility of helium in six different silicate glasses indicate that a good part of the changes in permeation rates shown in Fig. 8 are caused by changes in solubility. At 60% network former Eschbach found a diffusion coefficient about twenty times smaller than that of fused silica, whereas the permeation rate was about 6000 times smaller. Shelby (to be published) also found that both the diffusion coefficient and solubility of helium in sodium silicate glasses decreased substantially as the concentration of silica decreased.

Other measurements of molecular diffusion in silicate glass have been made in various ways. Reynolds measured the diffusion of argon in a potash-calcium aluminosilicate glass,[47] and found a diffusion coefficient of about 10^{-11} cm^2/sec at 400°C with an activation energy of 42 kcal/mole. Greene and co-workers measured the rate of contraction of oxygen bubbles in various molten glasses.[48,49] I calculated diffusion coefficients from their data, as shown in Table 4.[50] The oxygen is probably diffusing into the glass

Table 4 Diffusion Measurements from Contracting Bubbles in Molten Glass

Gas	Glass	Diffusion coefficient at 100°C (cm^2/sec)	Q	Refs.
O_2	Ba-Al-Alkali Silicate	$8(10)^{-7}$	24	48,49,50
	Soda-lime silicate	$1.4(10)^{-7}$	53	48,49,50
	Borosilicate	$2.8(10)^{-6}$	32	48,49,50
He	Soda-lime silicate	$5(10)^{-5}$	13.4	51
Ne	Soda-lime silicate	$3(10)^{-6}$	13.4	51
H_2O	Soda-lime silicate	?	36	52

in molecular form in these experiments, but the process may be complicated by reaction of the oxygen with constituents of the glass. Frichat and Oel measured the diffusion of helium and neon in molten soda-lime glass from the rate of contraction of gas bubbles in the glass.[51] These results are also shown in Table 4. They found the same activation energy for diffusion of the two gases, but the helium diffusion coefficient was about 18 times higher. Thus there appears to be a somewhat different mechanism of diffusion in molten glass than in rigid glass. Nemer[52] measured the diffusion of water from contracting bubbles in a molten soda-lime glass. His absolute values of the diffusion coefficients were unreliable, and again reaction of the water with the glass may have been important.

Barton and Morain[53] measured the permeation of hydrogen in various silicate glasses from the rate of reduction of silver in the glass by hydrogen penetrating into it. They found that in soda-lime glasses the calcium was more effective in reducing hydrogen permeation than sodium in glasses with the same silica content. Furthermore, the smaller the alkaline earth ion in glasses with the same sodium-alkaline earth-silica compositions, the higher was the permeation rate of hydrogen, except in barium and strontium glasses, which showed about the same rate of hydrogen permeation. These results indicate that hydrogen permeation would not show as good a correlation with only network concentration as does helium, as shown in Fig. 8, perhaps because the larger hydrogen molecule is more strongly influenced by changes in the size and charge of network modifying ions.

The rate of growth of small gold particles in glass can be controlled by diffusion of gold atoms in the glass to the growing particle.[54,55] The activation energy for diffusion of gold atoms in a glass containing 72% SiO_2, 23% Na_2O, 4% Al_2O_3, and 1% ZnO changes in the glass transition region, but the diffusion coefficient does not show a sudden jump as it does for ionic diffusion, as shown in the next chapter.

THEORIES FOR MOLECULAR DIFFUSION

In a random-walk treatment of diffusion, the diffusion coefficient has the form

$$D = \gamma \overline{\lambda^2} \Gamma \tag{19}$$

where γ is a geometrical constant, $\overline{\lambda^2}$ is the average of the squares of step lengths, and Γ is the average number of steps per unit time. An expression for Γ is given by the transition state theory of kinetic processes[2]:

$$\Gamma = \frac{kT}{h} \exp\left(\frac{\Delta S}{R} - \frac{\Delta H}{RT}\right) \tag{20}$$

where k, h, and R are Boltzmann's, Plank's, and the gas constant, respectively, and ΔS and ΔH are the entropy and enthalpy of activation. ΔH is thus equal to the activation energy Q' of Eq. 15.

A method of calculating the activation energy for molecular diffusion was proposed by Anderson and Stuart.[56] They assumed that this activation energy is the energy required to deform the silica network enough to allow the atom to pass from one interstice to another. The interstices were considered to be connected by "doorways" of average radius r_D; the activation energy is then the elastic energy required to enlarge the doorway to the radius r of the diffusing molecule, assuming it to be incompressible. This energy was estimated from the equation (Eq. 6)

$$Q' = 8\pi G r_D (r - r_D)^2 \tag{21}$$

derived by Frenkel[25] for the elastic energy to dilate a spherical cavity from the radius r_D to r. G is the elastic modulus of the glass. From Eq. 21 the square root of the activation energy should be a linear function of the diameter of the diffusing molecule. Figure 9 shows this functional dependence for the data on molecular diffusion in fused silica recorded in Table 2. The doorway diameter is equal to 1.1 Å from the diameter extrapolated to zero activation energy. The elastic modulus calculated from Eq. 21 and the slope of the line in Fig. 9 is about $1.2(10)^{11}$ dynes/cm^2, which is less than the value of $3(10)^{11}$ dynes/cm^2 for fused silica. This difference may result because the strain energy in the loose network structure of silica is less than in a close-packed liquid, for which Eq. 21 was derived. In any event the functional dependence of activation energy on molecular diameter, as given

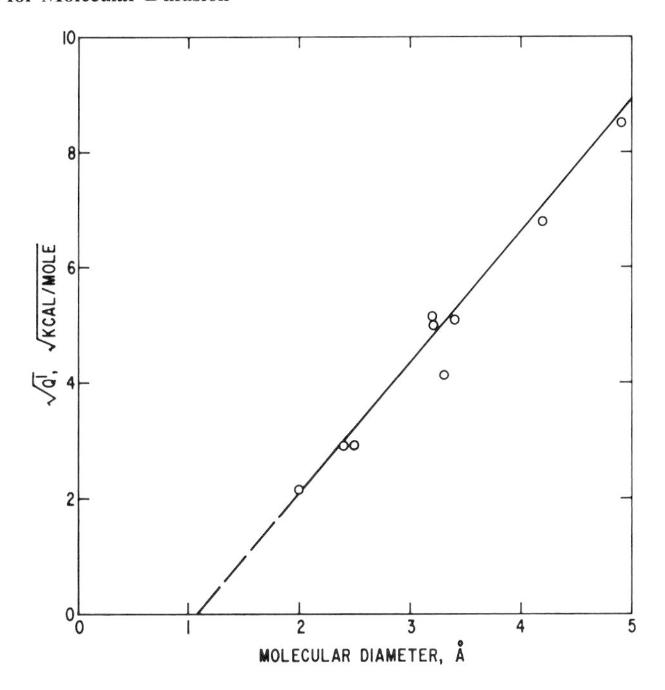

Fig. 9 Square root of the activation energies for molecular diffusion in vitreous silica as a function of molecular diameter. Data from Table 2.

in Eq. 21, describes well the activation energies for diffusion of molecules in fused silica, and the doorway diameter calculated from the equation is reasonable.

The entropy of activation can be estimated from experimental D' values (Eq. 15) and Eqs. 20 and 21. If $\gamma = \frac{1}{6}$ and $\gamma \approx 5.5$ Å for molecular diffusion in fused silica, the entropies of activation are negative and become more negative as size of the diffusing molecule increases. Perkins and Begeal[10] found the same result with a slightly different preexponential factor in Eq. 20.

Zener[57] has shown that the entropy change ΔS should be proportional to the negative of the temperature coefficient of the elastic modulus of a substance if diffusion in it involves only straining its lattice. The elastic moduli for fused silica increase with temperature, in contrast to the decrease found for ordinary silicate glasses and metals; this positive temperature coefficient is consistent with the negative values of ΔS.

If it is assumed that $\lambda \approx 5.5$ Å for other glasses, then $\Delta S/R$ is negative for the diffusion of helium in Pyrex in the temperature range 25 to 350°C; in this range the elastic moduli of Pyrex increase with temperature. The values of $\Delta S/R$ are positive for the diffusion of gold in a soda-alumina glass and argon in a potash-lime glass, and the elastic moduli of these glasses decrease with increasing temperature. These correlations probably cannot be extended to higher temperatures because in the annealing range the bulk elastic moduli will not be directly related to molecular transport, since the glass structure is changing with temperature. Thus there is the correct qualitative correlation between ΔS and the temperature dependence of elastic moduli, which is in agreement with the suggestion that the main contribution to the free energy of activation of molecular diffusion in these glasses involves the elastic strain of the glass network.

REFERENCES

1. R. M. Barrer, *Diffusion in and Through Solids*, Cambridge University Press, Cambridge, England, 1941, pp. 117 ff.
2. R. H. Doremus, in *Modern Aspects of the Vitreous State*, Vol. 2, J. D. Mackenzie, Ed., Butterworths, London, 1967, pp. 1 ff.
3. H. W. Wüstner, *Ann. Phys.*, **46**, 1095 (1915).
4. G. A. Williams and J. B. Ferguson, *J. Am. Chem. Soc.*, **44**, 2160 (1922).
5. K. N. Woods and R. H. Doremus, *Phys. Chem. Glasses*, **12**, 69 (1971).
6. J. F. Shackelford, Ph.D. Thesis, University of California, 1971; see also J. F. Shackelford, P. L. Studt, and R. M. Fulrath, *J. Appl. Phys.*, **43**, 1619 (1972).
7. D. E. Swets, R. W. Lee, and R. C. Frank, *J. Chem. Phys.*, **34**, 17 (1961).
8. R. C. Frank, D. E. Swets, and R. W. Lee, *J. Chem. Phys.*, **35**, 1451 (1961).
9. R. W. Lee, *J. Chem. Phys.*, **38**, 448 (1963).
10. W. G. Perkins and D. R. Begeal, *J. Chem. Phys.*, **54**, 1683 (1971).

11a. J. E. Shelby, *J. Am. Ceram. Soc.*, **55**, 61 (1972).

11b. J. E. Shelby, *J. Am. Ceram. Soc.*, **54**, 125 (1971).

12. F. J. Norton, *Nature*, **191**, 701 (1961).
13. J. E. Shelby, *J. Am. Ceram. Soc.*, **55**, 195 (1972).
14. R. W. Barrer and D. E. W. Vaughan, *Trans. Far. Soc.*, **63**, 2275 (1967).
15. E. K. Beauchamp and L. C. Walters, *Glass Tech.*, **11**, 139 (1970).
16. H. M. Laska, R. H. Doremus, and P. J. Jorgensen, *J. Chem. Phys.*, **50**, 135 (1969).
17. K. B. McAfee, *J. Chem. Phys.*, **28**, 218 (1958).
18. W. A. Rogers, R. S. Burlitz, and D. Alpert, *J. Appl. Phys.*, **25**, 868 (1954).
19. R. W. Lee and D. L. Fry, *Phys. Chem. Glasses*, **7**, 19 (1966).
20. H. L. Eschbach, in *Advances in Vacuum Science and Technology*, E. Thomas, Ed., Pergamon, London, 1960, p. 373.
21. H. L. Eschbach, R. Jaeckel, and D. Miller, *Z. Naturforsch.*, **18a**, 434 (1963).
22. H. O. Mulfinger and H. Scholze, *Glastech. Ber.*, **35**, 466, 495 (1962).
23. R. H. Doremus, *J. Am. Ceram. Soc.*, **49**, 461 (1966).
24. R. H. Fowler and E. A. Guggenheim, *Statistical Thermodynamics*, Cambridge University Press, Cambridge, England, 1939, pp. 81, 373.
25. J. Frenkel, *Kinetic Theory of Liquids*, Oxford University Press, London, 1946, pp. 10–11.
26. P. L. Studt, J. F. Schackelford, and R. M. Fulrath, *J. Appl. Phys.*, **41**, 2777 (1970).
27. J. Crank, *Mathematics of Diffusion*, Oxford University Press, London, 1956, pp. 47 ff.

28. R. H. Doremus, *J. Chem. Phys.*, **34**, 2186 (1961).
29. J. Moulson and J. P. Roberts, *Trans. Far. Soc.*, **57**, 1208 (1961).
30. R. H. Doremus, in *Reactivity of Solids*, Mitchell, DeVries, Roberts, and Cannon, Eds., Wiley, 1969, p. 667.
31. J. Johnson and R. Burt, *J. Opt. Soc. Am.*, **6**, 734 (1922).
32. H. Matzke, *Phys. Stat. Solidi*, **18**, 285 (1966).
33. P. L. Smith and N. W. Taylor, *J. Am. Ceram. Soc.*, **23**, 139 (1940).
34. L. C. Walters, *J. Am. Ceram. Soc.*, **53**, 288 (1970).
35. T. Drury and J. P. Roberts, *Phys. Chem. Glasses*, **4**, 79 (1963).
36. G. J. Roberts and J. P. Roberts, *Phys. Chem. Glasses*, **5**, 26 (1964); **7**, 82 (1966).
37. I. Burn and J. P. Roberts, *Phys. Chem. Glasses*, **11**, 106 (1970).
38. A. Kats, "Hydrogen in Alpha-Quartz," Thesis, Delft, 1961.
39. C. Wagner, *J. Chem. Phys.*, **18**, 1229 (1950).
40. R. Haul and G. Dumbgen, *Z. Electrochem.*, **66**, 636 (1962).
41. E. W. Sucov, *J. Am. Ceram. Soc.*, **46**, 14 (1963).
42. W. C. Hagel and J. D. Mackenzie, *Phys. Chem. Glasses*, **5**, 113 (1964).
43. E. L. Williams, *J. Am. Ceram. Soc.*, **48**, 191 (1965).
43a. J. E. Shelby, *J. Appl. Phys.*, **43**, 3068 (1972).
44. F. J. Norton, *J. Am. Ceram. Soc.*, **36**, 90 (1953).
45. V. O. Altemose, *J. Appl. Phys.*, **32**, 1309 (1961).
46. G. A. Vostrov and O. J. Bolchakov, *Pribory i Technika Eksperimenta*, No. 2, 112 (1966).
47. M. B. Reynolds, *J. Am. Ceram. Soc.*, **40**, 395 (1957).
48. C. H. Greene and R. F. Gaffney, *J. Am. Ceram. Soc.*, **42**, 271 (1959).
49. C. H. Greene and J. Kitano, *Glastech. Ber.*, **32K**, V, 44 (1959).
50. R. H. Doremus, *J. Am. Ceram. Soc.*, **43**, 655 (1960).
51. G. H. Frischat and H. J. Oel, *Glastech. Ber.*, **38**, 156 (1965); *Phys. Chem. Glasses*, **8**, 92 (1967).
52. L. N. Nemer, *Glass Tech.*, **10**, 176 (1969).
53. J. L. Barton and M. Morain, *J. Noncryst. Solids*, **3**, 115 (1970).
54. R. D. Maurer, *J. Appl. Phys.*, **29**, 1 (1958).
55. R. H. Doremus, in *Symp. on Nucleation and Crystallization in Glasses and Melts*, M. K. Reser, Ed., American Ceramic Society, Columbus, Ohio, 1962, p. 119.
56. O. L. Anderson and D. A. Stuart, *J. Am. Ceram. Soc.*, **37**, 573 (1964).
57. C. Zener, in *Imperfections in Nearly Perfect Crystals*, Wiley, New York, 1952, p. 289.

9

Electrical Conductivity and Ionic Diffusion

In many applications glass must serve as an electrical insulator or conductor, and therefore understanding of its electrical conductivity is important. Ionic transport plays a role in other applications and in manufacture of glass. Dissolution in water and weathering of glass result from interionic diffusion, and rates of chemical reactions in glass melting are often controlled by diffusion. Ionic motion in the amorphous silica layers on silicon devices can interfere with their performance and must often be suppressed.

In most oxide glasses the electrical conductivity results from ionic motion. In certain special compositions containing multivalent oxides, such as vanadium pentoxide or iron oxide, the conduction is electronic, as discussed in the next chapter. Most chalcogenide glasses, those containing pure or combined sulfur, selenium, or tellurium, are also electronic conductors; it is their semiconducting and switching properties that have excited interest in these glasses. The "salt"-type glasses of halides, nitrates, sulfates, and aqueous solutions are ionic conductors. Organic glasses can be either electronic conductors or show ionic conduction resulting from impurities.

The ionic conductivity of virtually all oxide glasses results from the transport of monovalent cations. In most commercial glasses the conducting ion is sodium. Faraday's law is found to hold for these glasses, and a number of electrolysis experiments,[1,2] reviewed in detail in earlier publications,[3,4] have established the ionic nature of the conduction process. Lithium ions also are quite mobile in oxide glasses. Potassium and hydrogen ions sometimes carry current, although their mobility is usually lower than that of sodium and lithium. Even in glasses with no nominal addition of monovalent ions, the conductivity results from transport of monovalent cations. In fused silica, electrolysis experiments show that sodium and lithium ions are the conducting species, even though they are present only in quantities of a few parts per million.[5]

The conduction mechanism in various "alkali-free" silicate and borate glasses containing lead, aluminum, and alkaline earth ions is uncertain. Conduction does not result from sodium ion motion in sodium-lead silicate glasses containing less than 5% mole Na_2O, 50% SiO_2, and the balance PbO, as shown from electrolysis measurements.[6] In a 50 mole % BaO, 50% SiO_2 glass the diffusion coefficient and concentration of sodium were too low to give the measured conductivity,[7] as calculated from the Einstein equation, given in the section below on the relation between conductivity and diffusion. The electrical conductivity of 50 mole % CaO, 50% SiO_2 glasses was insensitive to sodium concentration in the range from 45 parts per million to 1.26 mole % Na_2O.[8] Thus it appears that alkali ions do not carry the current in any of these glasses. The diffusion coefficient of oxygen in calcium aluminum silicate glasses is apparently too low to account for their conductivity[9]; in calcium aluminum borate glasses the contribution of oxygen is uncertain.[9,10] In lead silicate and borate glasses the diffusion coefficient of oxygen gives too high a conductivity from the Einstein equation, and the activation energy for oxygen diffusion, 11 kcal/mole, is much less than the 24 kcal/mole for conduction.[11] It therefore seems likely that these experiments measure the diffusion of molecular oxygen and its reaction with the lattice, as described in the last chapter, and that the diffusion of lattice oxygen is much slower.

The electrical conductivity of lead silicate,[6] calcium silicate,[12] and barium aluminum borate[13] glasses was affected by the presence of —OH groups. Thus it seems possible that hydrogen ions are carrying the electrical current in these glasses.[6,12,13] In most silicate glasses the mobility of hydrogen ions is about three to four orders of magnitude lower than the mobility of sodium ions,[14,15] but it is possible that the mobility of protons is higher in the alkaline earth, lead, and aluminosilicate glasses. It is known that alumina groups lead to relatively less tightly bound hydrogen ions in silicate glasses,[15] so a relatively high hydrogen ion mobility is particularly likely in aluminosilicate glasses.

In barium silicate glasses with 30 and 50 mole % barium oxide, Evstrop'ev and Kharyuzov[7] found that the diffusion coefficient of barium ions and the electrical conductivity obeyed Einstein's equation; however, in calcium silicate glasses the diffusion coefficient of calcium is much too low to account for the electrical conductivity, and apparently hydrogen ions carry the current.[12] The diffusion coefficient of lead in lead silicate glasses gives about the right value of conductivity from the Einstein equation; however, electrolysis experiments show that the lead ion is not conducting.[6] It is also possible that electrons carry part or all of the current in these glasses; thus work is needed to clarify the conduction mechanism in these alkali-free glasses.

This chapter contains sections on the following subjects: methods and pitfalls in measuring ionic conductivity and diffusion; the relation between conductivity and diffusion; the effect of temperature, stress, and glass composition on ionic transport in glass; interionic diffusion in glass; diffusion of nonconducting ions; and, finally, theories of ionic transport.

Earlier reviews on these subjects have been written by Morey,[3] Stanworth,[16] Stevels,[17] Owen,[4] and myself.[15,18,19] Winchell collected results of many diffusion experiments in crystalline, glassy, and molten silicates.[20]

MEASUREMENT OF ELECTRICAL CONDUCTIVITY AND IONIC DIFFUSION

One might expect that the measurement of electrical conductivity in glass would be simple and straightforward; however, there are many pitfalls to be avoided, and it is possible that many measurements reported in the literature are not especially reliable.

Two methods have been used to measure electrical conductivity of solid ionic conductors: one technique uses direct current (dc) and the other uses alternating current (ac). In a dc measurement, problems with electrodes are frequent. A space charge is often set up in the glass because of partial blocking of the ionic current by the electrodes; then the current decreases rapidly with time, and its value must be extrapolated to zero time if an accurate value of conductivity is desired. To avoid this electrode polarization some authors have used electrodes of sodium amalgam or sodium metal. However, since the conducting species are sodium ions, not atoms, oxidation and reduction reactions must occur at both electrodes if space charges are to be avoided, and if these reactions are slow, uncertainties appear in the measurements. To avoid these electrode problems an alternating current is usually used, of a frequency from 10^3 to 10^6 cycles/sec. Silicate glasses show dielectric losses at these frequencies, however, so care must be taken to make the measurements over a wide frequency range. If a constant sample resistance is found over several decades of frequency, one can be reasonably certain that an accurate value of conductivity is being measured. Such constancy is usually not found. The result of these difficulties is that many experimental measurements of electrical conductivity of glasses that have been reported in the literature are not accurate to more than about $\pm 30\%$ in absolute value. Relative measurements, for example, of temperature dependence, are more reliable because the same proportional error tends to occur under different conditions.

A method that avoids these problems is to return to a dc measurement and to use electrodes of fused salts.[21,5] In this way electrode reactions and frequency effects are eliminated. A simple experimental arrangement consists of a glass tube partially filled with molten salt in a molten salt bath. A variety of metallic electrodes can be inserted into the bath; if small currents are used, contamination of the bath by electrode reactions is minimized. The molten salt has a much higher conductivity than the glass, so nearly all the potential drop is across the glass. Care must be taken that the bath contains the ion conducting in the glass, usually sodium, in sufficient quantity to prevent depletion, and that other possible conducting ions, such as lithium, must be present in low enough quantity so that they do not cause any change in the conductivity of the glass by exchanging with the conducting ions in it. One difficulty with this method is to find stable salt melts for wide ranges of temperatures. Alkali nitrates have low melting points but decompose above about 450°C. Many other salts are hygroscopic; that is, they absorb moisture from the atmosphere, so that many of them can be used only in a dry atmosphere. Any fused salt should be purged with dry gas to remove residual water. In some salts such water is tenaciously held, and they must be heated for long times at high temperatures to remove it.

Other problems occur in the preparation of the glass samples. In addition to avoiding obvious conducting impurities, if one wishes to steady a particular ion, it is especially important to prevent contamination with water. For example, McGee (to be published) has shown a considerable difference in the conduction behavior of sodium silicate glasses that were melted in air and under vacuum. Glasses melted in air show a sharp change in the dependence of resistance and in the activation energy on concentration at about 30% Na_2O, but in vacuum-melted glasses with lower water contents these breaks or sharp changes no longer appear, and the conductivity is lower. These results are surprising in view of the low mobility of hydrogen ions exchanged for sodium ions in sodium silicate glasses[14]; it is possible that either the hydroxyl groups change the rate of transport of sodium ions, or that a small concentration of hydrogen ions has a high mobility and contributes to the conduction process. The latter situation would be in conflict with the many electrolysis experiments that show all the current being carried by sodium ions. In any event conduction and diffusion of vacuum-melted glasses should be studied further.

Stresses in glass can increase the conductivity up to a factor of 10. These stresses are removed by holding the glass at its annealing temperature for about $\frac{1}{2}$ hr and then cooling it slowly (1°/min). All results reported here are on annealed glasses unless otherwise stated.

The specific resistivity ρ in ohm-cm is defined as the resistance of a sample of unit area A and unit thickness d, so $\rho = AR/d$, where R is the resistance of a platelike sample. The conductivity σ is the reciprocal of the resistivity: $\sigma = 1/\rho$.

Ionic diffusion can be studied with radioactive tracers; such diffusion, called "tracer diffusion," takes place without chemical changes or ionic gradients in the sample. The tracer can either be deposited on the sample surface from a solution to provide a concentrated source of tracer, or the tracer concentration at the sample surface can be maintained constant by a large dilute source such as a fused salt.

The diffusion coefficient D is defined by the equation

$$J = D\frac{\partial c}{\partial x} \tag{1}$$

where J is the flux of diffusing species and $\partial c/\partial x$ is its gradient of concentration c in the x direction. From the equation of continuity

$$\frac{\partial c}{\partial t} = \frac{\partial(-J)}{\partial x} = D\frac{\partial^2 c}{\partial z^2} \tag{2}$$

where t is the time and D is independent of concentration.

If at time zero an amount Q of substance per unit area is added to one face of a semiinfinite solid containing none of this substance, then the solution of Eq. 2 is[22]

$$c = \frac{Q}{\sqrt{\pi D t}} \exp\left(-\frac{x^2}{4Dt}\right) \tag{3}$$

where c is the concentration of diffusion substance at a distance x from the face after diffusion for a time t. To calculate the diffusion coefficient D from Eq. 3 only the slope of the log concentration against x^2 plot for a certain time of diffusion is needed, since Q and t are constants for these conditions. The boundary condition of Eq. 3 applies when material containing a radioactive or isotropic tracer of high specific activity is deposited on a solid. The concentration profile in the solid after diffusion for a certain time can be determined by removing successive layers of the solid and analyzing for the tracer with a counter or mass spectrometer.

Under certain conditions Frischat found profiles of sodium tracer in glasses, measured as described in the preceding paragraph, that deviated from Eq. 3 close to the sample surface.[23] However, McVay and Farnum were able to eliminate similar effects by making the diffusion anneal in argon instead of atmospheric air.[23a] It seems likely that the deviations resulted from introduction of hydrogen ions by reaction and ion exchange with water in the air. These results are further evidence of the possible interference of water in diffusion and conductivity measurements in glass, as mentioned above.

If the solute concentration at the face of a semiinfinite solid that contains no solute initially is maintained at c_i, then with D constant

$$c = c_i \operatorname{erfc}\left(\frac{x}{2\sqrt{Dt}}\right) \tag{4}$$

where

$$\operatorname{erfc} z = \frac{2}{\sqrt{\pi}} \int_z^{\infty} e^{-y^2}\, dy$$

The total amount of material M that has diffused into the solid after a time t is

$$M = 2c_i\left(\frac{Dt}{\pi}\right)^{1/2} \tag{5}$$

The functional form of Eq. 5 is preserved even if D is a function of the concentration of the diffusing material. This is so because the solute concentration c depends only on the variable $y = x/\sqrt{D_i t}$, where D_i is the diffusion coefficient when $c = c_i(x = 0)$. Then the flux J_0 crossing the surface $x = 0$ is

$$J_0 = -D\left(\frac{\partial c}{\partial x}\right)_{x=0} = -\frac{\sqrt{D_i}}{t}\left(\frac{\partial c}{\partial y}\right)_{y=0}$$

The derivative is not a function of x, t, or D_i, since y does not appear in it. The resulting expression for M is

$$M = \int_0^t J_0\, dt = -2\sqrt{D_i t}\left(\frac{\partial c}{\partial y}\right)_{y=0}$$

Therefore M is proportional to $\sqrt{t}$ regardless of the functional relation between D and solute concentration. This relation can usually be deduced from the profile of diffusing material. To determine this profile the concentration of diffusing material as a function of penetration distance into the glass after a certain time of diffusion must be found. To do this layers of glass can be removed, either by etching with acid, by grinding, or by sectioning, and the amount of diffusing material in each layer determined by chemical or tracer analysis.

If the electrical resistance of a glass is changed by the diffusing species this change can sometimes be used to follow the progress of diffusion. Consider a solid sample of fixed cross-sectional area with flat faces and thickness L. The resistance R per unit cross-sectional area of the sample measured between its faces can be expressed as

$$R = \int_0^L \frac{dx}{\sigma}$$

where σ is the specific conductivity of the glass at the distance x from one face. If the penetration of diffusing material is limited to a thin layer of the sample throughout the diffusion process, there is a constant concentration c such that

$$R = \int_0^x \frac{dx}{\sigma} + R_0$$

throughout the diffusion, when x is the distance corresponding to c and R_0 is the initial resistance of the sample when it contains no solute. Since the penetration of diffusing material is limited to a thin layer, the equations for a semiinfinite solid apply. Thus, if the concentration of diffusing material is held constant at one face at c_i, the reduced solute concentration c/c_i at a distance x from the face is a function of $y = x/\sqrt{D_i t}$ only, in which D_i is the diffusion coefficient of solute at the concentration c_i. Then the above equation can be transformed to

$$R - R_0 = 2\sqrt{D_i t} \int_0^y \frac{dy}{\sigma} \tag{6}$$

Since y is constant with time, because c is constant, and σ varies only with y for a particular diffusion run at constant temperature, the integral is a constant throughout a diffusion experiment, and the resistance increment $R - R_0$ is proportional to $\sqrt{t}$. This equation was derived without assuming a relation between resistivity and concentration, or between the diffusion coefficient and concentration. To relate $R - R_0$ to D_i under conditions of changing temperature the Einstein equation (Eq. 11) can be used for σ.

RELATION BETWEEN ELECTRICAL CONDUCTIVITY AND IONIC DIFFUSION

If the electrical current in a glass is carried by a single ionic species, the electrical conductivity of the glass is related to the diffusion coefficient of the ion by the Einstein equation. This equation can be derived as follows:[24] the driving force for ionic transport in the x direction is the negative of the gradient in total chemical potential ("electrochmical potential"):

$$\frac{\partial \mu}{\partial x} = \frac{RT}{c}\frac{\partial c}{\partial x}\frac{\partial \ln a}{\partial \ln c} + ZFE \tag{7}$$

where a is the thermodynamic activity, c is the concentration, Z is the ionic charge, F is the faraday, E is the electrical potential gradient, R is the gas constant, and T is the temperature. A generalized mobility can be defined as the average ionic velocity v per unit driving force. Then since the flux J of diffusing species is equal to cv, the mobility u^* for tracer diffusion is given by

$$u^* = \frac{-v}{\dfrac{RT \partial c}{c \; \partial x}} = \frac{-J}{RT\dfrac{\partial c}{\partial x}} = \frac{D}{RT} \tag{8}$$

since there is no electrical potential gradient in tracer diffusion, and the term

$$\frac{\partial \ln a}{\partial \ln c} = 1 + \frac{c}{\gamma}\frac{\partial \gamma}{\partial c} \tag{9}$$

is unity because there is no gradient in activity coefficients. The tracer-diffusion coefficient is D. The electrical mobility u_e in an experiment measuring the electrical conductivity σ is

$$u_e = \frac{-v}{ZEF} = \frac{\sigma}{Z^2F^2c} \tag{10}$$

The Einstein equation is found by equating the electrical and diffusive mobilities:

$$\sigma = \frac{Z^2F^2Dc}{RT} \tag{11}$$

In other derivations of this equation thermodynamic expressions for a system in equilibrium are used, but these expressions are unsatisfactory, since in a transport experiment the system is not in equilibrium. Deviations from the Einstein equation in aqueous solutions because of the distortion of the electric field around a moving ion demonstrate this deficiency. The present derivation is based on a simple physical assumption, the equivalence of electrical and diffusive mobilities, and a deviation from the Einstein equation shows when this assumption is not valid.

In glasses, Eq. 11 is often not obeyed exactly; to give equality σ must be multiplied by a factor f of less than one. In sodium silicate glasses values of f calculated from the data of Johnson[25] for diffusion and Seddon et al.[26] for electrical conductivity vary from 0.74 for 12 mole % Na_2O to 0.25 for 36% Na_2O. Haven and Verkerk[27] found values from 0.51 for 13% Na_2O to 0.40 for 30% Na_2O. Terai[28,30] found f values between 0.17 and 0.3 for various sodium aluminosilicate and aluminophosphate glasses, and concluded that there was a slight increase in f with temperature. Terai reviewed other calculations of f in sodium aluminosilicates and found considerable disagreement between different workers. Fitzgerald[29] found f values between 0.3 and 0.5 by measuring the electrical conductivity of the same glasses for which Johnson[25] measured the diffusion coefficient.

However, some measurements on other glasses have found good agreement with Eq. 11 and no need for an f-factor. My result of $f = 1$ for a commercial soda-lime silicate glass was confirmed by interdiffusion experiments, as described below.[24] A comparison of diffusion coefficents[31] and electrical conductivity[5] of vitreous silica also showed good agreement with Eq. 11. Furthermore, Garfinkel[32] found agreement with this equation for Pyrex borosilicate glass. Thus it seems possible that there is some peculiarity in the measurements leading to f-factors lower than one. A speculative possibility for a cause for high conductivities might be water in the glass, as described in the preceding section on experimental methods. Further work is needed to establish the reasons for f values lower than one.

In spite of the problems mentioned in the paragraph above the Einstein equation is valuable in comparing results of conductivity and diffusion experiments. Thus the temperature dependences of these two processes are accurately related in Eq. 11, and so are other relative changes caused by stress, ion size, and glass composition. In the following discussions of these effects no distinction needs to be made between conductivity and diffusion, since the trends apply to both processes.

DEPENDENCE OF IONIC TRANSPORT ON TEMPERATURE AND PRESSURE (STRESS)

The electrical resistivity of a glass containing about 26.5 wt. % Na_2O and 73.5% SiO_2 is shown in Fig. 1 as a function of reciprocal temperature. The data were taken by Babcock[33] at higher temperatures and Sedden et al.[26] at lower temperatures on glasses of almost the same composition. Sedden et al. annealed their glass carefully and cooled it slowly to get reproducible measurements. The temperature dependence shown in the figure parallels that found by Johnson[25,18] for diffusion of sodium ions in binary sodium silicate glasses (see Fig. 6). A similar temperature dependence for diffusion or conductivity is found for other alkali silicate glasses, both binary and multicomponent, that have been measured. For example, the temperature dependence of diffusion coefficients for Pyrex borosilicate glass shows a break near the transition temperature.[34]

At lower temperatures the resistivity fits an Arrhenius-type equation:

$$\rho = \rho_0 \exp\left(\frac{Q}{RT}\right) \tag{12}$$

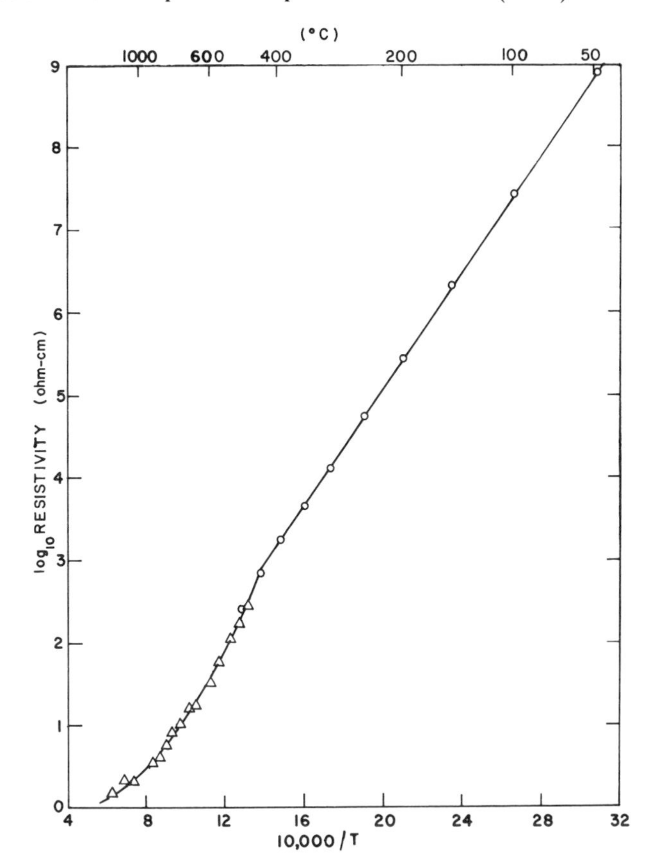

Fig. 1 Electrical resistivity of a 26.5 wt. % Na_2O, 73.5% SiO_2 glass as a function of reciprocal temperature. ○, Seddon et al.[26]; △, Babcock.[33]

with a preexponential factor ρ_0 and activation energy Q independent of temperature. This dependence shows the relative stability of the glass structure below the transition temperature. As the temperature approaches that where structural rearrangements become possible, the resistivity drops more rapidly with increasing temperature, leading to a second temperature region where Eq. 12 is valid, with a higher activation energy than at lower temperatures. As the glass structure is further broken up with increasing temperature, the effective activation energy continuously decreases. At temperatures above the glass transition temperature, Eq. 12 may be obeyed over short temperature ranges, but log ρ versus $1/T$ is actually curved. For the glass in Fig. 1 the activation energy at low temperatures is 16.0 kcal/

mole, about 23 kcal/mole just above the glass transition temperature, and about 8 kcal/mole from 1100 to 1500°C.[35]

The intersection of the lines on the log ρ versus $1/T$ plot at low and intermediate temperatures has been used to determine a transition temperature. However, this temperature of about 430°C for the glass in Fig. 1 is considerably lower than the temperature of about 490°C, where the viscosity is 10^{13} P, which is the more conventional transition temperature. Thus it appears that ionic transport changes character at a lower temperature than thermal expansion, the more usual property used as a criterion for the transition temperature.

The use of Eq. 12 for the resistivity implies, from Eq. 11, a factor of temperature in the preexponential term for the diffusion coefficient in an Arrhenius-type equation. However, the experimental data for diffusion and conductivity in glasses are not accurate enough or taken over a wide enough range of temperature to tell whether a preexponential factor of temperature is needed for either property. Thus activation energies for both diffusion and conductivity are reported for Arrhenius equations with no preexponential temperature factor. The result is that activation energies for diffusion are somewhat (about 10%) lower than those for electrical conductivity.

The electrical conductivity and viscosity of glasses have been related by various empirical formulae.[3,33,36] However, it seems unlikely that there is any quantitative relation between these properties because their temperature dependence is very different. The viscosity shows no unusual changes in the transition region, as discussed in Chapter 6, whereas ion transport is markedly affected at temperatures in the transition region. Furthermore, the Stokes-Einstein equation relating viscosity η and the diffusion coefficient D of a particle of radius r:

$$D = \frac{RT}{6\pi\eta r} \tag{13}$$

gives ionic radii orders of magnitude too small for alkali ions diffusing in silicate glasses. Therefore, although both viscosity and ionic transport reflect changes in the glass structure with temperature, there is no general quantitative relation between them.

The temperature dependence of ionic transport in vitreous silica appears to be different from that in other silicates containing more transporting ion, perhaps because the transport properties of silica in the rigid state can be measured over such a wide temperature range. The diffusion coefficient of sodium in vitreous silica made from fusion of natural quartz crystals is shown in Fig. 2, from the measurement of Frischat.[31] A further discussion of diffusion of sodium in vitreous silica is given in the next section on

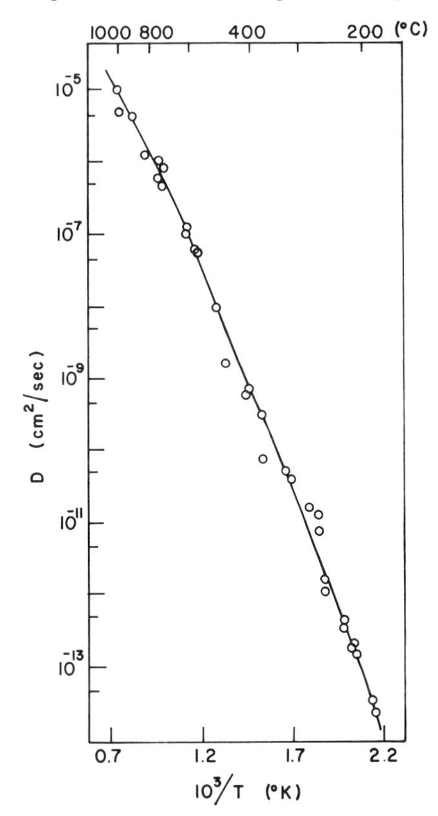

Fig. 2 Diffusion coefficient of sodium in vitreous silica as a function of reciprocal temperature, measured by Frischat.[31]

composition dependence. Frischat claimed that his diffusion data could be fit with three straight lines with break points at the temperatures of the α-β quartz (573°C) and the α-β cristobalite ($\sim$250°C) transformations. However, his data were not accurate enough to establish these breaks with certainty, and more precise measurements of electrical conductivity show that there are no breaks at the transition points, but only a gradual decrease in the activation energy as the temperature is increased.[5] The absence of a break at 250°C is shown in the resistance measurements in Fig. 3. Thus a continuous curve is drawn through the data of Frischat in Fig. 2.

In order to measure the electrical mobility of sodium ions in vitreous silica it was necessary to electrolyze the glass in fused salt to remove other mobile ions from the glass and replace them with sodium. This was done for the sample shown in Fig. 3; initially it contained mobile lithium ions as well as sodium ions. This sample was also used to test the Einstein equation, as

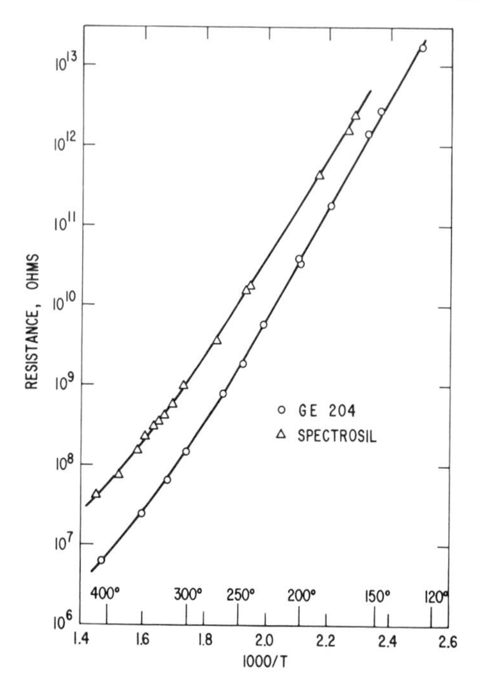

Fig. 3 Resistance of fused silicas containing sodium ions, as a function of $1/T$.

described in the preceding section, for Frischat's measurements. The concentration of sodium was calculated from the time course of electrolysis[5]; good agreement with the Einstein equation was found for the data in Fig. 3 and Frischat's results in Fig. 2.

At low temperatures the activation energy for sodium ion conduction in vitreous silica was constant, as shown in Fig. 3. Similar results were found for lithium and potassium ions. As the temperature was varied above a certain temperature, which was different for the different ions, the activation energy started to decrease and continued to decrease slowly as the temperature was raised. For sodium the activation energy for conduction at temperatures below about 280°C was 35 kcal/mole, whereas above about 600°C the activation energy was down to about 24 kcal/mole.[37] Veltri[38] measured the conductivity of vitreous silica from 1000 to 1800°C and found a constant activation energy of about 24 kcal/mole in this temperature range. Thus the activation energy for ionic transport is constant at high and low temperatures, but changes at intermediate temperatures. This change is particularly surprising in view of the unchanging structure of fused silica in this temperature range. Thus there is no apparent explanation for the change in activation energy for ionic transport in fused silica.

An equation of the form

$$\sigma = \sigma_0 \exp\left(\frac{-Q}{(T - T_0)}\right) \tag{14}$$

has been proposed[33,39] to account for the change in activation energy for conduction above the transition temperature. An empirical fit can be made to the data for fused silica at intermediate temperatures, since an additional adjustable parameter, T_0, is available, but the equation is not valid at lower temperatures, where it predicts a large curvature in log σ versus $1/T$; log σ is actually linear with $1/T$ at lower temperatures. A similar weakness of this equation for viscosity was discussed in Chapter 6.

Stress has long been known to effect the electrical conductivity of glasses. A quenched glass has a higher conductivity than an annealed one. The effects of compaction by hydrostatic stress and expansion by quenching on a soda-lime glass are shown in Fig. 4 from the work of Charles.[40] The figure

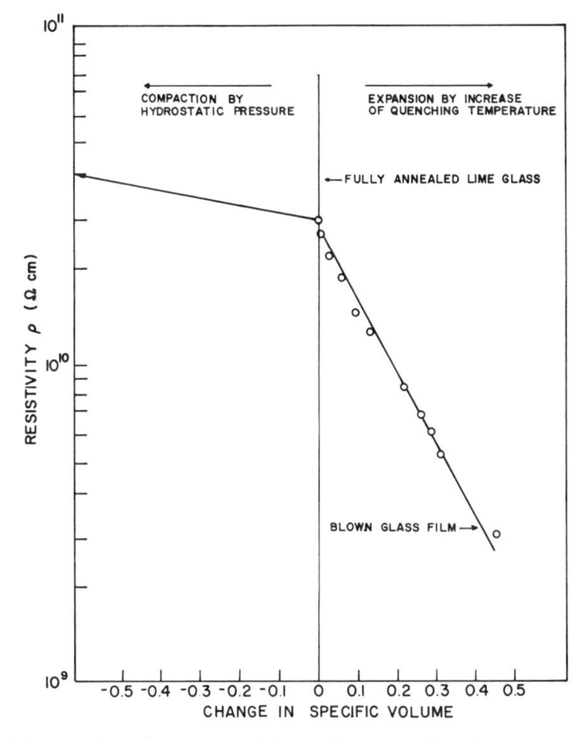

Fig. 4 Resistivity at 72°C of a commercial soda-lime glass (Corning No. 0080, see Chapter 6, Table 2) as a function of changes in its specific volume caused by hydrostatic pressure or by quenching from various temperatures.[40]

indicates a linear dependence of log conductivity on pressure, since for small volume changes this change is proportional to the pressure P. Thus

$$\sigma = \sigma_0 \exp\left(\frac{V^*P}{RT}\right) \tag{15}$$

where σ_0 and V^*, the "activation volume," are independent of pressure at constant temperature. Charles' results show that the effect of compressing the glass with hydrostatic pressure has a much smaller effect on resistivity of glass for unit volume change than do the structural changes brought about by quenching a sample to put it in tension internally. A V^* of about 3.7 cm^3/mole for hydrostatic compression is calculated from Fig. 4. Hamann[41] measured the effect of pressure on different alkali silicate glasses and found the following activation volumes in cm^3/mole: lithium, 1.0; sodium, 3.4; potassium, 6.0 These results can be compared with the molar volumes of the ions as calculated from the ionic radius: lithium, 0.55; sodium, 2.2; potassium, 5.9. In crystalline solids the activation volume is usually about half the molar volume, whereas in liquids it is closer to the molar volume. Thus diffusion in glasses is between these extremes and changes with the size of the diffusing ion.

The presence of a stress gradient in a glass gives an additional driving force for ionic transport. In the presence of a stress gradient $\partial P/\partial x$, Eq. 7 for the gradient in chemical potential becomes

$$\frac{\partial \mu}{\partial x} = \frac{RT}{c}\frac{\partial c}{\partial x}\frac{\partial \ln a}{\partial \ln c} + ZFE + V^* \frac{\partial P}{\partial x} \tag{16}$$

Weber and Goldstein[42] devised an ingenious experiment to measure V^* from this effect. They bent a glass plate to give a stress gradient across the thickness of the glass and measured the current J_p that flowed as a result of the applied stress ΔP. Then from this current and the electrical current I in a conventional measurement of electrical conductivity with an applied potential V they calculated V^* from the relation

$$V^* = \frac{J_p VF}{I\,\Delta P} \tag{17}$$

where F is the faraday. For a soda-lime glass they found $V^* = 1.3$ cm^3/mole, which is about one-third the value calculated from Charles' results. The reason for the difference is not clear.

EFFECT OF GLASS COMPOSITION ON IONIC TRANSPORT

The electrical conductivity is proportional to the product of concentration of the current-carrying ion and its mobility or diffusion coefficient, as shown by Eq. 11. Thus in a comparison of glasses with different concentrations of ions the diffusion coefficient is the more meaningful parameter. However, the activation energy for either property is not dependent directly on composition, and so it may be used to follow changes caused by, for example, changes in glass structure.

The tracer diffusion coefficient of sodium silicate glasses as a function of glass composition and temperature is shown in Fig. 5, from the measurements of Johnson.[25] The diffusion coefficient decreases as the concentration

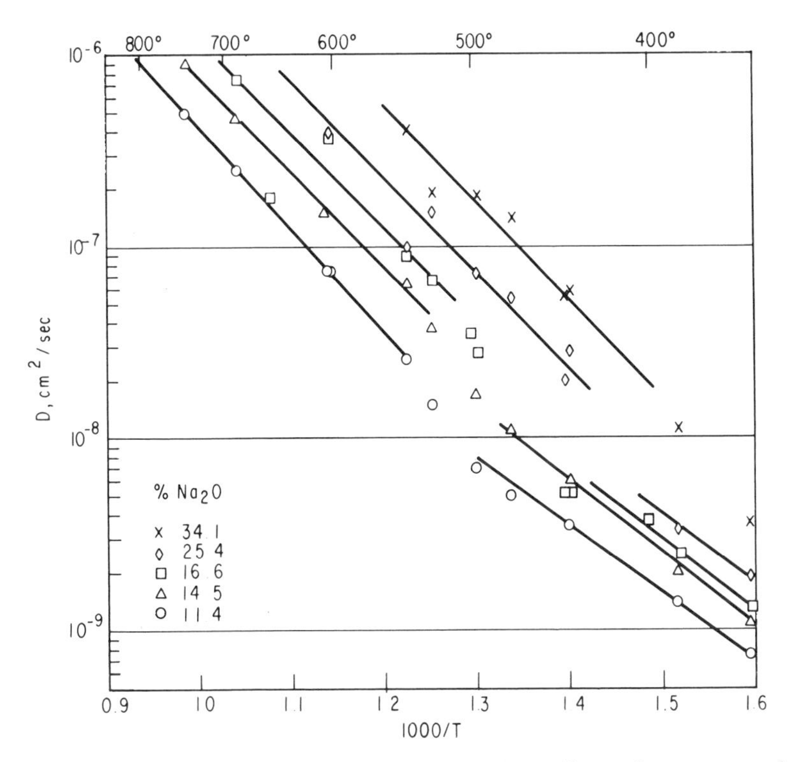

Fig. 5 Diffusion coefficients of sodium in binary sodium silicate glasses, measured by Johnson.[25]

of sodium ions decreases. Qualitatively the silica lattice is less broken-up the smaller the sodium concentration, making ionic transport more difficult. Similar trends are found for multicomponent silicate glasses, and in other alkali silicate glasses the diffusion coefficient of the alkali ion also decreases as its concentration decreases.

The activation energy for electrical conduction in binary alkali silicates is about 15 to 17 kcal/mole above an alkali concentration of about 20 mole %, and increases at lower alkali concentrations.[43] The data of different investigators do not agree very well,[43] particularly for sodium silicates.

Conduction in binary sodium and lithium silicates may be affected by phase separation in these glasses.

The addition of an oxide of a higher-valent metal ion to an alkali silicate leads to a decrease in the ionic mobility of the alkali ion. The decrease is related to the ionic size of the added ion for divalent ions; the largest change was found for barium ions, with a decreasing effect for lead, strontium, and calcium ions and a relatively small change for addition of magnesium, zinc, or beryllium oxides.[44,4] Addition of aluminum oxide increases the ionic mobility and decreases its activation energy,[45] as discussed further in the section on theory.

When a second alkali oxide is added to an alkali silicate glass the conductivities decrease sharply. This "mixed-alkali" effect is shown in Fig. 6

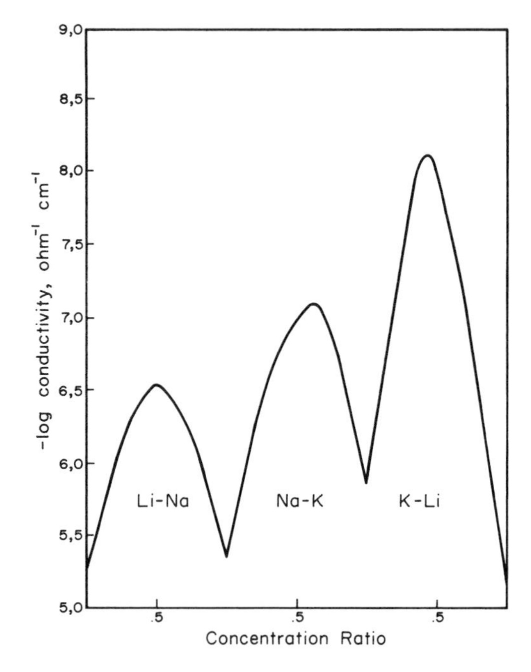

Fig. 6 The electrical conductivity of mixed binary alkali silicate glass containing 26 mole % alkali and 74% SiO_2 at 250°C, from Lengyel and Boksay.[46]

for mixed binary alkali silicate glasses.[46] Measurements of diffusion coefficients in such glasses show that the mobility of each ion is decreased by the addition of the other.[47–49] The decrease in conductivity results from these reductions in mobility, but the reason for the mobility reductions in terms of the mechanism of diffusion is uncertain.

The diffusion coefficient of sodium in vitreous silica is surprisingly high in view of its low concentration in this glass. Diffusion coefficients in different sodium silicate glasses at 386°C are shown in Fig. 7; in some soda-lime silicates the diffusion coefficient is less than in fused silica. This high value may result from a nonuniform distribution of aluminum ions in the silica. The alkali ions in fused silica are associated with alumina groups in the glass.[19] The aluminum ion substitutes for silicon in the silicon-oxygen network, and requires an associated positive ion for charge neutrality. The aluminum ions may be arranged in sheets or along "grain boundaries" that provide preferential paths for diffusion. There is no direct evidence for this speculative mechanism.

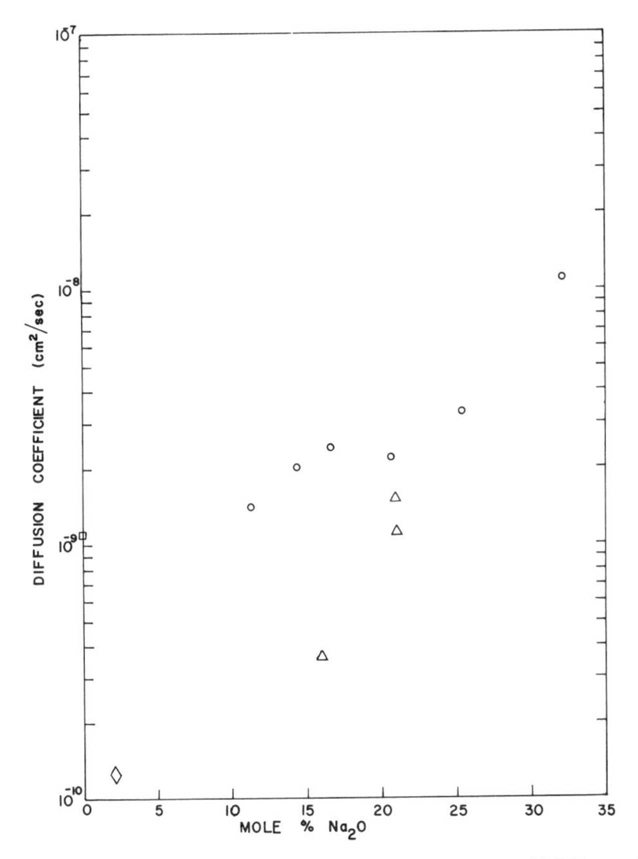

Fig. 7 Diffusion coefficients of sodium in sodium silicate glasses at 386°C as a function of sodium concentration. ○, binary sodium silicates[25]; △, sodium calcium silicates[25]; □, fused silica[31]; ◇, Pyrex borosilicate.[34]

INTERIONIC DIFFUSION AND ION SIZE EFFECTS

If glass is placed in contact with a medium containing monovalent cations, such as a fused salt or an aqueous solution, these cations can exchange with monovalent cations in the glass and interdiffuse with them into the glass. The ionic equilibrium between the glass and solution is considered in Chapter 14, as are the electrical potentials that result from this exchange and are responsible for the operation of glass electrodes. In this section the interdiffusion of like-charged ions is treated.

An alkali silicate glass can be considered as a matrix of immobile negative groups with associated mobile cations. An exchange cation normally has a different mobility from the original ion; therefore as interdiffusion proceeds one ion tends to outrun the other, and an electrical charge is built up. However, accompanying this charge is a gradient in electrical potential that slows down the fast ion and speeds up the slow one. To preserve electrical neutrality the fluxes of the two ions must be equal and opposite, and the electrical potential ensures this condition in spite of the difference in mobility of the two ions. The mobility u in an experiment with both ionic and electrical gradients is found from Eq. 7 and the definition of mobility as the velocity v per unit driving force; the flux J per unit time and area equals cv in the ionic concentration, so that

$$J = -u\left(RT\frac{\partial c}{\partial x}\frac{\partial \ln a}{\partial \ln c} + ZcFE\right) \tag{18}$$

An equation of this form applies to each mobile cation in the glass. These flux equations are usually called the Nernst-Planck equations, and were derived here from the assumption that the driving force is the gradient of electrochemical potential, and the definition of the mobility.

If two monovalent cations A and B interdiffuse, each has a flux equation (Eq. 18), with $Z = 1$. The conditions of electroneutrality require that $J_A = -J_B$ and $\partial C_A/\partial x = \partial C_B/\partial x$ because the number of negative groups is constant and immobile. From these four relations the gradient E in electrical potential is

$$E = \frac{RT}{F}\frac{u_B - u_A}{C_A u_A + C_B u_B}\frac{\partial c_A}{\partial x}\frac{\partial \ln a_A}{\partial \ln C_A} \tag{19}$$

To develop this electrical potential, small deviations in electrical neutrality must occur in the ionic system. However, the number of ions involved in these deviations is negligible compared to the total ionic concentrations.

When Eq. 19 is substituted into Eq. 18, the flux for ion A is

$$J_A = -u_A\left(\frac{C_A u_B + C_B u_B}{C_A u_A + C_A u_A}\right)\frac{\partial C_A}{\partial x}\frac{\partial \ln a_A}{\partial \ln C_A} \tag{20}$$

The coefficient of the concentration gradient is the same for ion B; if the activity coefficients are constant $\partial \ln a_A/\partial \ln C_A = 1$, and this interdiffusion coefficient $\tilde{D}$ is

$$\tilde{D} = \frac{D_A D_B}{N_A D_A + N_B D_B} \tag{21}$$

where $N_i = C_i/(C_A + C_B)$, and Eq. 8 is used to relate the mobilities to the tracer diffusion coefficient D. This latter substitution amounts to assuming that the mobilities in a tracer diffusion experiment, in which there are no gradients in ionic concentration, are equal to those in an interdiffusion experiment. Even if the Einstein equation (Eq. 11) is not obeyed, the ratio D_A/D_B still equals u_A/u_B if the deviations from this equation are the same for the two ions.

The diffusion equations with the concentration-dependent diffusion coefficient of Eq. 21 have been solved for various ratios of D_A/D_B and different geometries by Helfferich and Plesset,[50,51] with the assumption that the diffusion coefficients D_A and D_B were concentration independent. In Fig. 8 their curve is compared to experimental data for silver-sodium interdiffusion in a soda-lime glass.[24] The tracer diffusion coefficients of sodium and

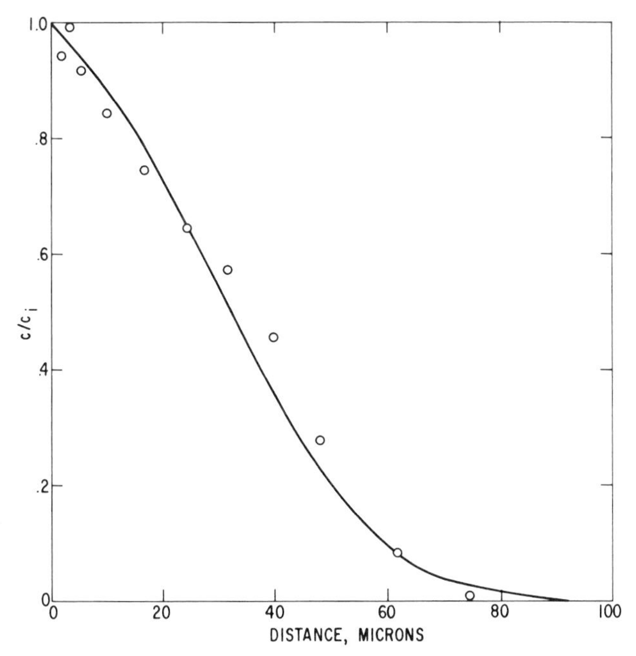

Fig. 8 Profile of silver that diffused from a silver nitrate melt into a soda-lime glass (Corning 0080, Chapter 6, Table 1) after 1097 min at 378°C. Line for $D_{Na}/D_{Ag} = 10$.

silver were measured, and the ratio D_{Na}/D_{Ag} and the absolute value of D calculated from these tracer diffusion coefficients agreed well with those calculated from the data in Fig. 8. Furthermore, measurements of the total amount of diffused material and the change in resistance with time of diffusion gave agreement with the interdiffusion coefficient of Eq. 21 in Eqs. 5 and 6, respectively. Thus the assumptions leading to Eq. 21 are confirmed: the validity of the Nernst-Planck equations, the equality of mobility ratio for tracer and interdiffusion experiments, and especially the independence of the (tracer) diffusion coefficients D_{Na} and D_{Ag} with changing ionic ratio in the glass. The equivalence of the mobilities in the two types of experiments is additional evidence that the Einstein equation (Eq. 11) is obeyed in this system, as was deduced from measurements of electrical conductivity and tracer diffusion of sodium in the initial glass.

In certain other glasses conditions are not so simple. For example, in Pyrex borosilicate glass, silver[52] and potassium[53] exchange for sodium are influenced by the two-phase structure of the glass. Silver ions are preferentially bound by one phase, so their effective diffusion coefficient is reduced at low silver ion concentrations.

A careful distinction should be made between interdiffusion experiments done well below the transition temperature of the glass and diffusion in silicate glasses melted with different or mixed alkali ions. The structure of the glass is determined by the melting conditions, and interdiffusion well below the transition temperature does not alter it. Thus the result that the tracer diffusion coefficient of an ion is constant with varying ionic ratio cannot be compared with the results on mixed alkali glasses in which a change in the ionic ratio changes the tracer diffusion coefficient; in this case, the structure of the glasses is different.

Interionic diffusion experiments have been carried out at temperatures not too far below the transition temperature.[54,55] In the sodium-potassium exchange of these experiments large stresses arise in the glass because of the difference in the sizes of the ions. This stress slowly relaxes at these temperatures, leading to a time-dependent change in the structure of the glass and also of the tracer diffusion coefficients of the ions. Thus interpretation of these experiments is difficult. When the stress is completely relaxed the diffusion coefficients have nearly the same dependence on ionic ratio as in the mixed alkali glasses melted with a a range of ionic ratios.[56]

Exchange of hydrogen ions from aqueous solutions with ions in silicate glasses shows special features. The first step in the dissolution of alkali silicate glass in water is exchange of the alkali ions with hydrogen ions from the water, as shown by Douglas and co-workers.[57–60] Interdiffusion of alkali and hydrogen ions also controls the rate of weathering of glass. At the glass surface alkali ions react with water to form alkali hydroxide on the surface and hydrogen ions that diffuse into the glass. The rate of interdiffusion of alkali and hydrogen ions controls the supply of alkali ions at the glass surface and therefore the rate of weathering. The addition of calcium to a binary sodium silicate glass reduces the diffusion coefficient of sodium in the glass, and this reduction is the reason that soda-lime glass is more resistant to weathering than binary sodium silicate glass. The reaction of water with glass is discussed further in Chapter 13; here interdiffusion is emphasized.

As the alkali-hydrogen ion exchange proceeds, the glass structure is stressed because of the much smaller size of the hydrogen ion; then further hydration by the reaction of water with the lattice leads to a gellike layer on the glass surface. Eisenman has shown that ionic diffusion in this layer is much more rapid than in unhydrated glass.[61] Thus the profiles of hydrogen ions diffusing into glass are complex. The profiles of sodium ions in various sodium alkaline-earth silicate glasses are shown in Fig. 9, from the

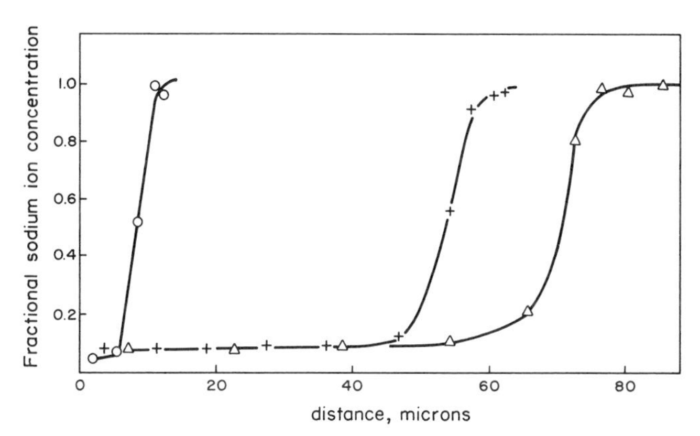

Fig. 9 Profile of sodium in glasses leached with water at 40°C. All glasses contain 68 mole % SiO_2 : △, 28% Na_2O, 4% BaO for 144 hr; +, 28% Na_2O, 4% SrO_2 for 144 hr; ○, 20% Na_2O, 12% SrO for 288 hr.

measurements of Boksay et al.[62] The leached layer in which the ionic diffusion coefficients are high leads to a low sodium concentration near the surface. In a potassium strontium silicate glass the effect of the leached layer is much less marked.[62]

In Table 2 the mobility ratios of diffusing ions in various rigid alkali silicate glasses are listed and can be compared with ionic radii in Table 1. In soda-lime glass the maximum mobility is found for sodium and lithium ions, with a large decrease for smaller (hydrogen) and larger (potassium and

Table 1 Ionic Radii

Ion	Radius(Å)	Ion	Radius(Å)
H^+	Small		
Li^+	0.68	Be^{2+}	0.33
Na^+	0.95	Mg^{2+}	0.67
Ag^+	1.26		
K^+	1.33	Ca^{2+}	0.99
Rb^+	1.48	Sr^{2+}	1.13
Cs^+	1.67	Ba^{2+}	1.35

Table 2 Ratios of Diffusion Coefficients of Monovalent Ions in Rigid Silicate Glasses

Glass type	Mole % M_2O	Temperature (°C)	Ions *A-B*	D_A/D_B	Refs.
Soda-lime	15.2 Na_2O	354	Na-Ag	4	21
			Na-Ag	12	24
			Na-K	500	63
	14.0 Na_2O	250	Ag-Li	2	64
	13.0 Na_2O	350	Na-H	1800	14
			Na-K	1400	14
			Na-Rb	$3.4(10)^5$	14
Potash-lime	12.5 K_2O	350	K-H	2	14
			K-Rb	190	14
Sodium binary	13 Na_2O	415	Na-K	25	47
	20 Na_2O	415	Na-K	39	47
	25 Na_2O	415	Na-K	48	47
Potassium binary	13 K_2O	415	Na-K	0.5	47
	20 K_2O	415	Na-K	0.5	47
	25 K_2O	415	Na-K	0.6	47
Sodium aluminosilicate	12.5 Na_2O	400	Na-K	~ 300	65
(hydrated)	27 Na_2O	25	Na-K	3; 7	61
Lithium aluminosilicate	5.2 Li_2O	415	Na-Li	1	66
Pyrex borosilicate	4 Na_2O	335	Na-Ag	1	52
		373	Na-K	~ 30	53
Vitreous silica	$1.1(10)^{-4}$ Na_2O	380	Na-Li	6.7	5
	$3.4(10)^{-4}$ Li_2O	380	Na-Ag	12	5
(G.E. 204)		380	Na-K	> 500	5
(Spectrosil)	?	380	Na-Li	33	5
		380	Na-K	1400	5
(Vitreosil)	~ $5(10)^{-4}$ Na_2O	1000	Na-H	~ 10^4	67

rubidium) ions. For fused silica the sequence is similar, except that lithium is less mobile than sodium. Silver is somewhat more mobile in lime glass than would be expected from its size, probably because it is much more polarizable than the alkali ions.

DIFFUSION OF ANIONS AND MULTIVALENT CATIONS IN GLASS

Anions and cations of higher valence diffuse much more slowly than monovalent cations in glass, and therefore have been much less studied. In two soda-lime glasses Frischat found that at 500°C the diffusion coefficient of calcium was four to five orders of magnitude lower than that for sodium.[68] For a glass of composition 15.8 wt. % Na_2O, 9.9% CaO, and 73.9% SiO_2 he found an activation energy for calcium diffusion of 53.7 kcal/mole from 490 to 600°C, compared to the value of 22.8 kcal/mole for sodium at temperatures below 500°C. Thus at lower temperatures the diffusion coefficient of calcium in lime glass is many orders of magnitude lower than that of sodium. Frischat[69] also measured calcium and aluminum diffusion in Infrasil fused silica at 1000°C, and found a value of $2.0(10)^{-8}$ cm^2/sec for calcium, about 400 times smaller than the diffusion coefficient of sodium at this temperature, and $1.3(10)^{-13}$ for aluminum, or 10^8 times as small as for sodium. From Eq. 13 and the viscosity of $4(10)^{16}$ P of fused silica at 1000°C, one would expect a diffusion coefficient of about 10^{-23} cm/sec for a diffusing entity with radius 2 Å. Thus aluminum must diffuse in silica by some sort of interstitial or defect mechanism, not by "flow."

Sah, Sello, and Tremere[70] calculated a diffusion coefficient of about $8(10)^{-14}$ cm^2/sec for phosphorus in silica at 1000°C from experiments on the reaction of P_2O_5 with SiO_2 layers on silicon. Their calculations were based on some uncertain assumptions; their calculation of an activation energy of 34 kcal/mole did not depend so critically on the assumptions and therefore may be more reliable. This value of activation energy seems low; however, a recent determination of the P_2O_5-SiO_2 phase diagram shows a liquid eutectic at about 830°C and 20% P_2O_5, so the P_2O_5-SiO_2 mixtures may be relatively fluid at 1000°C, giving a low activation energy for phosphorus diffusion.

Ghoshtagore[72] measured the diffusion coefficient of nickel in amorphous silica films and found a value of about 10^{-15} cm^2/sec at 1000°C, with an activation energy of 37 kcal/mole, about a factor of 100 below the diffusion coefficients of aluminum and phosphorous, but far below the value for calcium.

The diffusion of boron at 1000 to 1100°C in thin silica layers on silicon was measured by Horiuchi and Yamaguchi,[73] Brown and Kennicott,[74] and Schwenker.[75] An activation energy of about 88 kcal/mole was found at low boron concentrations, with a diffusion coefficient of about 10^{-17} cm^2/sec at 1050°C. This value is considerably lower than that for aluminum, but still higher than the value calculated from the Stokes-Einstein equation, using the viscosity of pure silica. However, it is likely that even a small amount of boron lowers the viscosity of silica considerably, so the boron may diffuse by "flow." Above a boron concentration of about 18%, diffusion was much more rapid, probably because of the decreased viscosity of the silica.

The diffusion coefficient of arsenic in silica films was found to be about $3(10)^{-16}$ cm^2/sec. at 1100°C and low arsenic concentrations, with an activation energy of about 110 kcal/mole.[75a,b] At higher arsenic concentrations the diffusion coefficient was higher and the activation energy lower. A number of experimental results on diffusion of boron, gallium, phosphorus, arsenic, and antimony in vitreous silica are reviewed by Ghezzo and Brown.[75b]

Interdiffusion in sodium-iron silicate[76] and potassium-strontium silicate[77] glasses has been studied at temperatures where the glasses are quite soft. These results are difficult to analyze because at least three components (alkali, alkaline earth, and silicon) are diffusing. Electrical fields set up by the faster-moving alkali ions are probably important.[77]

The diffusion coefficients of calcium,[78] aluminum,[79] and silicon[78] in a molten slag of composition 40% SiO_2, 39% CaO, and 21% Al_2O_3 had ratios of about 1:4:10, respectively. Approximate activation energies were in the range 60 to 70 kcal/mole. Results on diffusion of other ions in silicate melts are given by Winchell.[20,80]

THEORIES OF IONIC TRANSPORT

The diffusion coefficient D of Eq. 1 can be described in terms of a random walk of molecules in discrete steps. From the probability of finding a molecule at a certain place after a fixed number of steps

$$D = g\overline{\lambda^2}\Gamma \tag{22}$$

where g is a geometrical constant, $\overline{\lambda^2}$ is the average of the squares of the step lengths, and Γ is the average number of steps made in unit time. The jump rate Γ_m for ionic motion can be calculated from various models; its usual form is

$$\Gamma_m = \nu \exp\left(\frac{\Delta S_m/R - \Delta H_m}{RT}\right) \tag{23}$$

where ν is the vibration frequency of the diffusing atom, and ΔS_m and ΔH_m are the entropy and enthalpy of activation for the motion of an atom from one site to another. In some cases it is found that the preexponential frequency term depends on the temperature. In the preceding chapter results on diffusion of gases in glass showed such a dependence. However, the results on ionic diffusion have not been accurate enough or measured over wide enough temperature ranges to tell whether such a temperature factor is needed; therefore it will be ignored in this discussion.

The diffusion coefficients of monovalent cations in oxide glasses cannot be described by considering the glass to be a viscous liquid. Equation 13, the Stokes-Einstein equation, predicts much too low a diffusion coefficient, as described in the section on temperature effects. Furthermore, these cations diffuse much more rapidly than the lattice elements, such as silicon and oxygen, so the glass behaves like a rigid solid with mobile interstitial ions.

Ionic transport in rigid glasses is perhaps more like diffusion in ionic crystals, such as the alkali halides, or even in β-alumina, than in a viscous liquid. Diffusion in ionic crystals usually occurs by a defect mechanism. These defects can be vacant lattice sites or interstitial ions not in a normal site. In a glass interstitial ions are probably more mobile than vacancies, but in the following mathematical treatment the type of defects is unimportant.

The fraction N of defect sites per diffusing ion is given by an equation of the form

$$N = p \exp\left(\frac{-\Delta G_f}{RT}\right) \tag{24}$$

where p is a geometrical constant which equals the number of interstitial sites available to each diffusing ion for an interstitial model. Then the jump rate Γ in Eq. 22 is the product of Γ_m for motion with N. When these expressions are combined:

$$D = g\overline{\lambda^2}\nu\sqrt{p} \exp\left[\frac{\Delta S_f/2 + \Delta S_m}{R} - \frac{\Delta H_f/2 + \Delta H_m}{RT}\right] \tag{25}$$

This equation has the same form as the usual exponential dependence of D on $1/T$, with an activation energy of $\Delta H_f/2 + \Delta H_m$ and a D_0 independent of temperature. It is difficult to estimate the various quantities in Eq. 25 and none of them has been measured directly. Thus the applicability of a defect mechanism or Eq. 25 to ionic diffusion in glass is questionable.

In metals defects can be "quenched in" from higher temperatures, leading to higher rates of transport. This enhanced transport in a quenched material is evidence for the defect model, so I tried to quench defects into fused silica to test this model.[5] However, no increase in ionic mobility at 200°C in fused silica was found after quenching it from 900°C. Because of the low concentration of anionic sites and the low ionic mobility at 200°C, this test was considered to be optimum for quenching defects into glass. Since there is no direct evidence for a defect mechanism for ionic transport in glass, its applicability is uncertain.

The activation energy for molecular diffusion in glass is a strong function of the size of the diffusing molecule, as described in the preceding chapter. However, the activation energy for ionic diffusion in fused silica is about the same at low temperatures for the following ions[5,19,67]: hydrogen, lithium, sodium, and potasium. The activation energies for transport in various alkali silicate glasses shown in Fig. 5 increase with increasing size at lower ionic concentrations, but are all about the same at higher concentrations. Crudely one can think of two different processes being involved in ionic diffusion: a "squeezing" of the ions through the glass network, which should be harder the larger the ion, and a tearing away of the cation from the anionic site, which should be easier for the larger ions, since they are less tightly bound. Thus with these two competing effects one might expect a maximum in diffusion coefficient at some intermediate size, which is the result shown in Table 2 with a maximum rate of transport of sodium ion in various silicate glasses. However, this dependence apparently results mainly from preexponential factors, not the activation energy, because of the lack of size dependence of the latter in some glasses. It is difficult to understand which preexponential factors show this sort of size dependence.

The detailed ionic distribution in glass may be nonuniform, and may cause some of the uncertainties mentioned above. In Chapter 3, some evidence for a sheetlike structure of the alkali ions in alkali silicate glasses was described. Thus the diffusion in these sheets may be like diffusion in the layers of the β-alumina structure. A "free-ion" theory for the rapid diffusion of alkali ions in β-alumina has been proposed by Rice and Roth,[81] analogous to the free electron transport in metals. Diffusion in the randomly distributed and disjointed sheets in glass might be "free-ion" like and quite rapid, with difficulties in jumping to neighboring sheets. It was also suggested in the section on composition effects that a nonuniform distribution of aluminum ions in fused silica could lead to preferential paths for ionic diffusion. These mechanisms must be considered quite speculative and require testing with further experiments and calculations.

The effect of changing the structure of the anionic site of alkali ions in glass is shown by results on the electrical conductivity[45] and diffusion[28] in sodium aluminate glasses. The activation energy for sodium ion transport decreases as aluminum replaces silicon in the glass network. The aluminum-oxygen tetrahedron has a larger effective anionic radius than the silicon-oxygen group, since the negative charge of the aluminum group is effectively spread over the whole group. Thus the sodium ion is bound less tightly to this group, and "tearing" it away for a diffusive jump is easier.

CONDUCTION IN HIGH FIELDS

Results on the ionic transport in very high electrical fields may shed some light on the mechanism of ionic transport in glass, although the correct interpretation of these results is still not entirely clear. If a Boltzmann distribution of energy exists among ions, their jump rate is given by Eq. 23. When an electric field E is imposed on them, jumps in one direction become more probable than in the other, giving rise to the following equation for the conductivity:[82]

$$\sigma = \frac{2gcF\lambda\Gamma}{E} \sinh \frac{FE\lambda}{2RT} \tag{26}$$

For low fields $\sinh X = X$ and Eq. 26 becomes

$$\sigma = \frac{gcF^2\lambda^2\Gamma}{RT}$$

which is Eqs. 11 and 22 combined. When the fields are very high Eq. 26 becomes, in terms of the current $I = \sigma E$

$$I = 2gcF\lambda\Gamma \exp\left(\frac{F\lambda E}{2RT}\right) \tag{27}$$

so that the current depends exponentially on the field. This exponential dependence at high fields has been found by several different workers for sodium silicate glasses.[82–87] The hyperbolic sine law over the whole range of fields was found by Maurer[82] and Zagar and Papanikolau.[86] Maurer and Vermeer[85] found the expected temperature dependence of the preexponential and exponential factors of Eq. 27 from their low-field results on the same glasses. Maurer's jump distances calculated from Eq. 27 were high. However, when these λ values were corrected by using the effective local field E_L, calculated from the Lorentz-Lorentz relation

$$E_L = \left(\frac{\varepsilon + 2}{3}\right)E$$

where ε is the dc dielectric constant, instead of the external field E_0, reasonable values of λ (about 5 Å) were found. This correction has been questioned,[4] but it probably is valid. This distance of 5 Å is not too different from the average separation of the sodium ions in this glass. Zagar and Papanikolau[86] calculated values of the jump distance for several different sodium silicates, soda-lime silicate, and sodium borosilicate glasses, and found an average value of 28 Å for all glasses, with little difference between glasses and no effect of temperature between −30 and 100°C. Again the high value could be corrected by the Lorentz-Lorentz relation to a more reasonable value.

Barton and Leblond[87] criticized Eq. 27 because of the high-jump distances calculated from it and the differences between these distances and those calculated from the low-field conductivity. However, the latter are uncertain because of uncertainties in the factor g and correlation effects. Barton and Leblond suggested that a better equation would be that of Poole and Frenkel, in which the log current is proportional to the square root of the field at high fields. However, this mechanism would require a change from Eq. 27 to this different dependence at some intermediate field. There is no evidence for such a shift from Eq. 27 in the data of Maurer, and Zarar and Popanikolau, which agree with Eq. 27 at all fields. The jump distances calculated from the Poole-Frenkel equation are about 20 Å, and therefore are also higher than would be expected. Thus Eq. 27 still seems to be the best for representing conduction data at high fields.

REFERENCES

1. E. Warburg, *Ann. der Physik*, **21**, 622 (1884).
2. R. C. Burt, *J. Opt. Soc. Am.*, **11**, 87 (1925).
3. G. W. Morey, *The Properties of Glass*, 2nd ed. Reinhold, New York, 1954, p. 465.
4. A. E. Owen, "Electric Conduction and Dielectric Relaxation in Glass," in *Progress in Ceramic Science*, J. E. Burke, Ed., Macmillan, New York, 1963, p. 77.
5. R. H. Doremus, *Phys. Chem. Glasses*, **10**, 28 (1969).
6. K. Hughes, J. O. Isard, and G. C. Milnes, *Phys. Chem. Glasses*, **9**, 43 (1968).
7. K. K. Evstrop'ev and V. A. Khar'yuzov, *Dokl. Akad. Nauk USSR*, **136**, 140 (1961); Translation in *Proc. Acad. Sci. USSR*, **136**, 25 (1961).
8. M. Schwartz and J. D. Mackenzie, *J. Am. Ceram. Soc.*, **49**, 582 (1966).
9. W. G. Hagel and J. D. Mackenzie, *Phys. Chem. Glasses*, **5**, 113 (1964).
10. A. E. Owen, *Phys. Chem. Glasses*, **2**, 87, 152 (1961); **6**, 253 (1965).
11. H. A. Schaeffer and H. J. Oel, *Glastech. Ber.*, **42**, 493 (1969); *Z. Naturforsh*, **25a**, 59 (1970).
12. M. Schwartz, Thesis, Rensselaer Polytechnic Institute, Troy, N.Y., 1969.
13. E. Gough, J. O. Isard, and J. A. Topping, *Phys. Chem. Glasses*, **10**, 89 (1969).
14. P. Ehrmann, M. deBilly, and J. Zarzycki, *Verres Refrac.*, **18**, 164 (1964).
15. R. H. Doremus, "Ion Exchange in Glasses," in *Ion Exchange*, Vol. 2, J. A. Marinsky, Ed., M. Dekker, New York, 1969, p. 1.
16. J. E. Stanworth, *The Physical Properties of Glass*, Oxford University Press, London, 1950.
17. J. M. Stevels, "The Electrical Properties of Glass," in *Handbuck der Physik*, Vol 20, Springer-Verlag, Berlin, 1957, p. 350.
18. R. H. Doremus, "Diffusion in Non-crystalline Silicates," in *Modern Aspects of the Vitreous State*, Vol. 2, J. D. Mackenzie, Ed., Butterworths, London, 1962, p. 1.
19. R. H. Doremus, *J. Elect. Soc.*, **115**, 181 (1968).
20. P. Winchell, *High Temp. Sci.*, **1**, 200 (1969).
21. G. Shulze, *Ann. Phys. Lpz.*, **40**, 335 (1913).
22. J. Crank, *The Mathematics of Diffusion*, Oxford University Press, London, 1956, p. 26.
23. G. H. Frischat, *Phys. Chem. Glasses*, **11**, 25 (1970); *J. Am. Ceram. Soc.*, **53**, 285 (1970).

23a. G. L. McVay and E. H. Farnum, *J. Am. Ceram. Soc.*, **55**, 275 (1972).
24. R. H. Doremus, *J. Phys. Chem.*, **68**, 2212 (1964).
25. J. R. Johnson, R. H. Bristow, and H. H. Blau, *J. Am. Ceram. Soc.*, **34**, 135 (1951); see also J. R. Johnson, Thesis, Ohio State University, 1950.
26. E. Sedden, E. J. Tippet, and W. E. S. Turner, *J. Soc. Glass Tech.*, **16**, 450 (1932).
27. Y. Haven and Verkerk, *Phys. Chem. Glasses*, **6**, 38 (1965).
28. R. Terai, *Phys. Chem. Glasses*, **10**, 146 (1969).
29. J. V. Fitzgerald, *J. Chem. Phys.*, **20**, 922 (1952).
30. R. Terai, *J. Ceram. Ass. Jap.*, **72**, 817 (1964).
31. G. H. Frischat, *J. Am. Ceram. Soc.*, **51**, 528 (1968).
32. H. M. Garfinkel, *Phys. Chem. Glasses*, **11**, 151 (1970).
33. C. L. Babcock, *J. Am. Ceram. Soc.*, **17**, 329 (1934).
34. C. G. Wilson and A. C. Carter, *Phys. Chem. Glasses*, **5**, 111 (1964).
35. J. O. Bockris, J. A. Kirchener, S. Ignatowicz, and J. W. Tomlinson, *Trans. Far. Soc.*, **48**, 75 (1952).
36. J. T. Littleton, *Ind. Eng. Chem.*, **25**, 748 (1933).
37. A. E. Owen and R. W. Douglas, *J. Soc. Glass Tech.*, **43**, 159 (1959).
38. R. D. Veltri, *Phys. Chem. Glasses*, **4**, 221 (1964).
39. C. A. Angell, *J. Am. Ceram. Soc.*, **51**, 125 (1968).
40. R. J. Charles, *J. Am. Ceram. Soc.*, **45**, 105 (1962).
41. S. D. Hamann, *Austral. J. Chem.*, **18**, 1 (1965).
42. N. Weber and M. Goldstein, *J. Chem. Phys.*, **41**, 2898 (1964).
43. R. M. Hakim and D. R. Uhlmann, *Phys. Chem. Glasses*, **12**, 132 (1971).
44. O. V. Mazurin and E. S. Borisovski, *J. Tech. Phys. Moscow*, **27**, 275 (1957); *Sov. Phys. Tech. Phys.*, **2**, 243 (1957).
45. J. O. Isard, *J. Soc. Glass Tech.*, **43**, 113T (1959).
45a. E. L. Williams and R. W. Heckman, *Phys. Chem. Glasses*, **5**, 166 (1964).
46. B. Lengyel and Z. Boksay, *Z. Phys. Chem.*, **203**, 93 (1954); **204**, 157 (1955).
47. K. K. Evstrop'ev, in *The Structure of Glass*, Vol. II, Consultants Bureau, New York, 1960, p. 237.
48. G. L. McVay and D. E. Day, *J. Am. Ceram. Soc.*, **53**, 508 (1970).
49. J. P. Lacharme, *C.R. Acad. Sci. Paris*, **270C**, 1350 (1970); **275C**, 993 (1972).
49a. R, Terai, *J. Noncryst. Solids*, **6**, 121 (1971).
50. F. Helfferich and M. S. Plesset, *J. Chem. Phys.*, **28**, 418 (1958).
51. F. Helfferich, *J. Phys. Chem*, **66**, 39 (1962).
52. R. H. Doremus, *Phys. Chem. Glasses*, **9**, 128 (1968).
53. H. M. Garfinkel, *Phys. Chem. Glasses*, **11**, 151 (1970).
54. J. Tochon, *C.R. Acad. Sci. Paris*, **263C**, 829 (1966).
55. G. H. Frischat, *Glastech. Ber.*, **44**, 113 (1971).
56. G. H. Frischat, *J. Mat. Sci.*, **6**, 1229 (1971).
57. R. W. Douglas and J. O. Isard, *J. Soc. Glass Tech.*, **33**, 289T (1949).
58. M. A. Rana and R. W. Douglas, *Phys. Chem. Glasses*, **2**, 179 (1961).
59. R. W. Douglas and T. M. El-Shamy, *J. Am. Ceram. Soc.*, **50**, 1 (1967).
60. C. R. Das and R. W. Douglas, *Phys. Chem. Glasses*, **8**, 178 (1967).
61. G. Eisenman, "The Origin of the Glass Electrode Potential," in *Glass Electrodes for Hydrogen and Other Cations*, G. Eisenman and M. Dekker, Eds., New York, 1967, p. 133.
62. Z. Boksay, G. Bouquet, and S. Dobos, *Phys. Chem. Glasses*, **8**, 140 (1967).
63. R. H. Doremus, unpublished results.
64. C. M. Hollabaugh and F. M. Ernsberger, unpublished results; see Ref. 18, p. 33.
65. A. J. Burggraaf and J. Comelissen, *Phys. Chem. Glasses*, **5**, 123 (1964).
66. H. M. Garfinkel and C. B. King, *J. Am. Ceram. Soc.*, **53**, 686 (1970).

67. G. Hetherington, K. H. Jack, and M. W. Ramsay, *Phys. Chem. Glasses*, **6**, 6 (1965).
68. G. H. Frischat, *Glastech. Ber.*, **44**, 93 (1971).
69. G. H. Frischat, *J. Am. Ceram. Soc.*, **53**, 625 (1969).
70. C. T. Sah, H. Sello, and D. A. Tremere, *J. Phys. Chem. Solids*, **11**, 288 (1959).
71. J. Eldridge and P. Balk, *Trans. AIME*, **242**, 539 (1968).
72. R. N. Ghoshtagore, *J. Appl. Phys.*, **40**, 4374 (1969).
73. S. Horiuchi and J. Yamaguchi, *Jap. J. Appl. Phys.*, **1**, 314 (1962).
74. D. M. Brown and P. R. Kennicott, *J. Elect. Soc.*, **118**, 293 (1971).
75. R. O. Schwenker, *J. Elect. Soc.*, **118**, 313 (1971).
75a. J. Wong and M. Ghezzo, *J. Elect. Soc.*, **119**, 1413 (1972).
75b. M. Ghezzo and D. M. Brown, *J. Elect. Soc.*, **120**, 110, 146 (1973).
76. M. P. Borom and J. A. Pask, *Phys. Chem. Glasses*, **8**, 194 (1967).
77. A. K. Varshneya and A. R. Cooper, *J. Am. Ceram. Soc.*, **51**, 103 (1968); **55**, 220, 312 (1972).
78. H. Towers, M. Paris, and J. Chipman, *Trans. AIME*, **197**, 1455 (1953); **209**, 1 (1957).
79. H. Henderson, L. Yang, and G. Derge, *Trans. AIME*, **221**, 56 (1961).
80. P. Winchell, *J. Am. Ceram. Soc.*, **54**, 63 (1971).
81. M. J. Rice and W. L. Roth, *Solid State Chem.*, **4**, 294, (1971).
82. R. J. Maurer, *J. Chem. Phys.*, **9**, 79 (1940).
83. H. Schiller, *Ann. Physik*, **83M**, 137 (1927).
84. A. M. Venderovitch and V. J. Chenykh, *J. Tech. Phys. USSR*, **18**, 317 (1948).
85. J. Vermeer, *Physica*, **22**, 1257 (1956).
86. L. Zagar and E. Papanikolau, *Glastech. Ber.*, **42**, 37 (1969).
87. J. L. Barton and M. Lebland, *Verres Refrac.*, **24**, 225 (1970).

10

Electronic Conduction

"Glasses may replace silicon in transistors." "New electronic switches made of glass." With such headlines the applications of electronically conducting glasses have been heralded. These possible applications have led to great interest and activity in studying glasses that conduct electronically. However, these glasses have not yet been widely used in commercial devices, and a more sober analysis of their potential applications continues. A variety of experimental results on electronic conduction in glasses has been reported, but their interpretation and theoretical explanation are being debated.

Familiar semiconductors such as silicon and germanium have much higher conductivities than most electronically conducting glasses, but the term "amorphous semiconductors" is often used for the electronically conducting glasses, and in this discussion the two terms are used synonymously.

The papers from conferences on semiconducting glasses have been published in Volumes 2, 4, and 8 through 10 of the *Journal on Noncrystalline Solids*; individual papers from the first two volumes are referred to below. Adler has summarized the physics of electronically conducting glasses.[1] New work is appearing frequently in this active field of research.

In this chapter electronically conducting glasses are first classified; experimental results on electronic transport in the glasses and theories to explain these results are then summarized, and finally switching behavior and various explanations for it are described. In this entire discussion conduction phenomena through very thin amorphous films between metallic electrodes, such as silica films up to 1000 Å thick, are excluded because they depend as much on the properties of the electrodes as on those of the glassy films.

Different experimental methods show when electrons, rather than ions, are the conducting species in a glass.[1a] If continued electrolysis of a glass with metal electrodes does not change its electrical conductivity, one can be reasonably certain that the glass is an electronic conductor, because the motion of ions leads to space charges at the electrodes and discharge of ions there, both of which cause changes in conductivity. Appreciable voltages from cells with the glass as electrolyte, with appropriate electrodes, also are evidence for ionic conduction. A high conductivity at low temperatures and a low activation energy for conduction are often taken as evidence for electronic conductivity, but these are unreliable indicators in glasses.

TYPES OF ELECTRONICALLY CONDUCTING GLASSES

Electronic conductivity is found in certain glasses containing multivalent transition ions. The best known are the vanadate glasses, such as $BaO \cdot 2V_2O_5$, $0.6V_2O_5 \cdot 0.4Te_2O$, and various vanadium phosphate glasses, which were first investigated by Denton, Rawson, and Stanworth.[2] Many other glasses of this type have been prepared, using multivalent transition ions such as vanadium, iron, cobalt, and manganese in a matrix of phosphate, borate, or silicate glasses, often with additional components.[1a,3–5]

A second category of electronically conducting glasses has been intensely studied recently because of their potential applications, as mentioned in the Introduction. These are the "chalcogenide" glasses, those containing the Group VI (chalcogenide) elements, sulfur, selenium, and tellurium, alone or in combination with the Group V elements, phosphorus, arsenic, antimony, and bismuth, and often including other elements such as thallium and germanium. The glasses based on cadmium arsenide probably also belong in this category because of their similar behavior.

An apparently different category of electronically conducting glasses has been prepared by adding the platinum group metals, iridium, palladium, rhenium, and ruthenium, to thin films of oxide glasses.[6] However, there is some possibility that these films contain crystalline oxides.[7,8]

The common crystalline semiconductors, germanium, silicon, and indium antimonide, have not been prepared in the amorphous state by quenching from the melt, but can be made by deposition as thin films from the vapor. Also in this class are SiC, GeTe, and GeSi compounds or alloys (mixtures). Elemental tellurium is also in this class; although it is a chalcogenide, it has been prepared as glass only in film form. There is some question about the structure of these "amorphous" films; they may be crystalline on a very fine scale, as described in Chapter 3 on structure. Their electrical properties are different from those of the chalcogenide glasses, perhaps because of this difference in structure.

Glassy polymers can also show electronic conduction[9]; they are not discussed here.

ELECTRICAL PROPERTIES

Conductivity results are reported in terms of the equation

$$\sigma = \sigma_0 \exp\left(\frac{-Q}{RT}\right) \tag{1}$$

where σ_0 and Q are temperature-independent parameters, and Q is the activation energy.

The electronic conductivity of various vanadium phosphate[2,10–13] and iron phosphate[5] glasses was investigated as a function of glass composition. As the amount of transition metal oxide increased, the conductivity increased, as shown in Fig. 1. These were variations of conductivity with

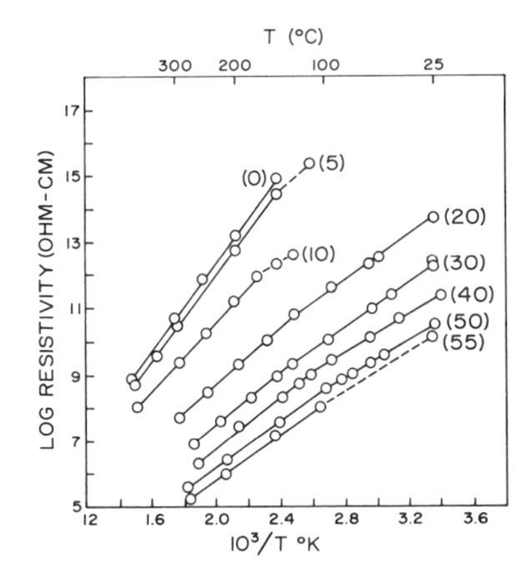

Fig. 1 The electrical resistivity of glasses with 55% (FeO + MgO) and 45% P_2O_5 as a function of $1/T$. Numbers on the graph are % FeO. From Ref. 5.

other constituents of the glasses, but they are probably related to changes in the valence state of the transition metal ions.[13] As the amount of transition metal ions is reduced, ionic conductivity can also become important at higher temperatures.

In vanadium phosphate glasses the conductivity is related to the relative amounts of penta and tetra valent (V^{5+} and V^{4+}) ions. The conductivity is a maximum when the molar ratio V^{4+}/V total is about 0.1 to 0.2, no matter what third components are present.[11] In iron phosphate glasses the conductivity is a maximum when the ratio Fe^{3+}/Fe total is about 0.4 to 0.6.[5] The thermoelectric power of these glasses is constant with temperature and is positive for most of the vanadium phosphate glasses, but changes from positive to negative in the iron phosphate glasses as the ratio Fe^{3+}/Fe total is increased.

The activation energy for conduction of a phosphate glass (80% V_2O_5, 20% P_2O_5) increased with increasing temperature.[14] Measurements of the electron-spin resonance in this glass showed that the V^{4+}/V^{5+} ratio did not change appreciably with temperature, indicating that the number of charge carriers was also constant with temperature.[15] Thus the increase in activation energy with temperature was attributed to a change in mobility of the charge carriers, rather than in their number.

The electrical conductivity of most chalcogenide glasses fits Eq. 1 and resembles that of intrinsic semiconductors, even down to quite low temperatures. The conductivity of selenium and arsenic selenide is typical,[16] as shown in Fig. 2.

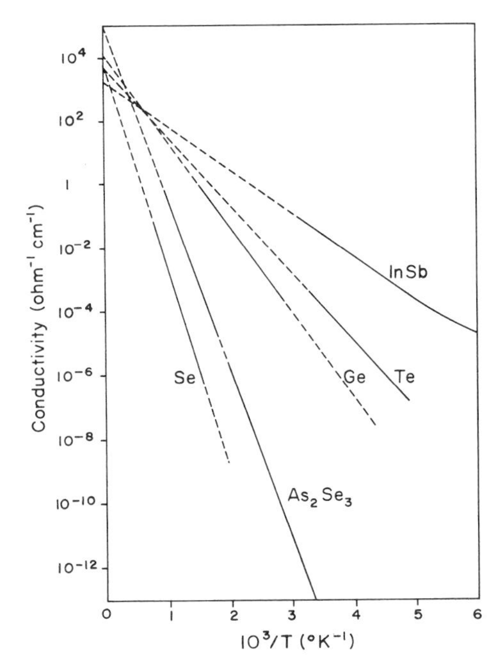

Fig. 2 Intrinsic electrical conductivity of various amorphous semiconductors as a function of $1/T$. Dashed lines are extrapolated. From Ref. 6, Fig. 17.

The "extrinsic" conductivity of the amorphous semiconductors deposited as thin films from the vapor is more pronounced than for the chalcogenides formed by quenching from the melt. As the films are annealed the conductivity in the extrinsic region is reduced, indicating that at least some of it results from dangling bonds created during evaporation. Annealing reduces the number of dangling bonds, as shown by spin-resonance measurements on silicon and germanium films.[17] The intrinsic conductivities of germanium, tellurium, and indium antimonide films are also shown, as a function of temperature, in Fig. 2.

The most striking characteristic of both calcogenide and film amorphous semiconductors is the insensitivity of their conductivity to impurities. In crystalline semiconductors tiny concentrations of foreign atoms cause large changes in conductivity and in the type of charge carriers, electrons, or electron holes. Impurity concentrations up to a few percent usually cause relatively minor changes in the conductivity of amorphous semiconductors. Some substantial changes in the conductivities of selenium[18] and As_2Se_3[19,20] by addition of impurities are attributed to changes in σ_0 and Q in Eq. 1, rather than to a change to extrinsic behavior.[21]

The values of σ_0 and Q in Eq. 1 are related, as shown by the extrapolation to nearly a common value of conductivity at infinite temperature in Fig. 2. The relation between room temperature conductivity and activation energy for a number of crystalline and amorphous semiconductors is shown in Fig.3. In most cases the intrinsic conductivity of the crystalline material is higher than for the amorphous.

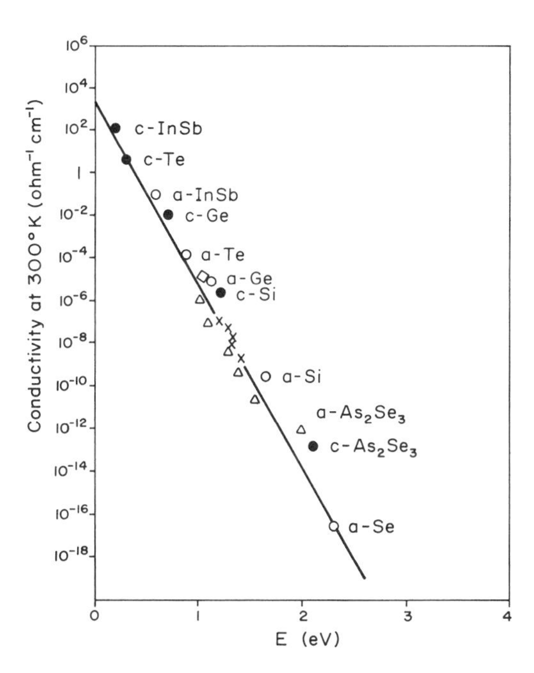

Fig. 3 Relation between electrical conductivity and activation energy E for conduction for various crystalline and amorphous semiconductors. From Ref. 16, Fig. 18.

Studies of the electrical conductivities of particular amorphous chalcogenides have been described in Volumes 2 and 4 of the *Journal of Noncrystalline Solids* and in later volumes of this journal, emphasizing the effects of sample history, composition, methods of preparation, and structure.

The question of the species that carries the charge in chalcogenide semiconductors has been investigated with various electrical properties. In several different systems Hall-effect measurements indicate *p*-type (electron holes) charge carriers, whereas thermoelectric power measurements point to *n*-type or electrons as carriers. Examples of Hall-effect measurements with the mobilities found are given in Table 1, as summarized by Roiler.[27] Little or no temperature dependence of the mobility was found. The only exception to *n*-type carriers in the table was the work of Nagels et al., on $Tl_2Te—As_2Te_3$ in which *p*-type carriers were found, in agreement with thermoelectric power measurements.[26]

Table 1 Hall-Effect Mobilities in Chalcogenide Glasses

System	Mobility (cm^2/V sec)	Temperature (°C)	Refs.
$Tl_2Se—As_2(Se, Te)_3$	≈ 0.03	Room	22
As—Te—I	0.08	20	23
	0.12	90	
As—Te—Br	0.01	20	23
	0.10	90	
$Tl_2Se—As_2Te_3$	0.09	−30–70	24
As—Te—Sc—Tl	0.1	20–500	25
$Tl_2Te—As_2Te_3$	0.14	Room	26
$As_2Se_3 \cdot 5\ As_2Te_3$	0.03	Room	27
$As_2Se_3 \cdot 3\ As_2Te_3$	0.05	Room	27

In a study of the photoconduction of amorphous As_2S_3, Ing et al. concluded that the photocurrent was carried by electron holes,[28] in agreement with the Hall-effect results on most other amorphous chalcogenides. These results and others, summarized by Fritzsche,[21] show that the drift mobility is limited by trapping. The slow decay of photoconductivity at liquid nitrogen temperatures is evidence for "deep traps."[29]

The ac conductivity of glassy chalcogenides at higher frequencies is proportional to some power n of the frequency, where n is usually between 0.7 and 1.0. The experimental results are summarized by Johnscher[30] and Pollak.[31] In As_2Se_3 and CS_2, n can equal 2 at frequencies higher than for

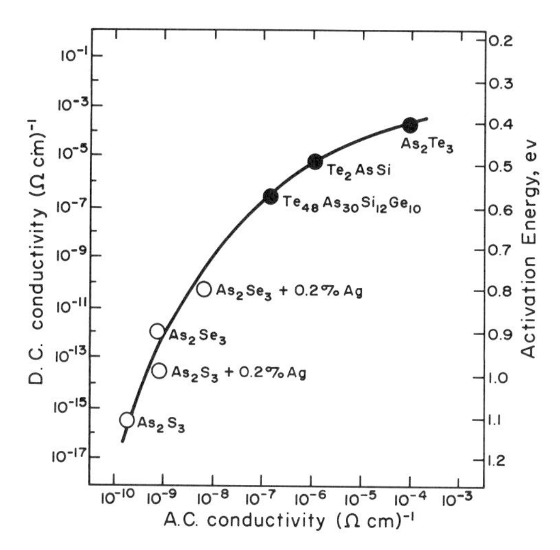

Fig. 4 Dc conductivity, its activation energy, and ac conductivity for various chalcogenide glasses, after Mott and Davis.[32]

the lower *n* values. The ac conductivity at higher frequencies is nearly independent of temperature. In Fig. 4 the relation between dc conductivity, ac conductivity, and activation energy for dc conductivity for various chalcogenide glasses is given. Ac losses in semiconducting glasses are also considered in Chapter 11.

THEORIES OF CONDUCTION

Great activity in devising theories for the electronic properties of disordered materials has led to understanding of some features of conduction in glassy semiconductors, but other aspects remain unexplained or at least debatable. In this section a qualitative summary of some theories of conduction is presented. New theories are appearing at such a rapid rate that this section will soon be out of date.

Conduction in the glasses containing transition metal oxides is generally considered to take place by "hopping" of carriers from one strongly localized state to another. The two states are the two valence states of the transition-metal ion. Experimentally a maximum in the conductivity is found for glasses containing mixtures of ions in the two valence states, in agreement with this model.

In the chalcogenide glasses conduction has been described as occurring in an energy "band" as in crystalline "broad-band" semiconductors. Cohen reviewed theories of this type of conduction,[33] which have been developed by several different authors, especially Mott and co-workers.[34] The density of electronic states in an energy band in the first models for disordered materials is shown in Fig. 5*a*. There is a band of states that extends throughout the material between two energies, and "tails" of localized states outside these energies that result from the disordered state of the material. The presence of structural defects modifies the tails, as shown in Fig. 5*b*. Mixtures of elements or compounds, called "alloys" by Cohen, give rise to compositional disorder and translational disorder enhanced by the requirement of satisfaction of local valences. The tails can overlap, as shown in Fig. 5*c*, giving a finite density of states at the Fermi energy E_F.

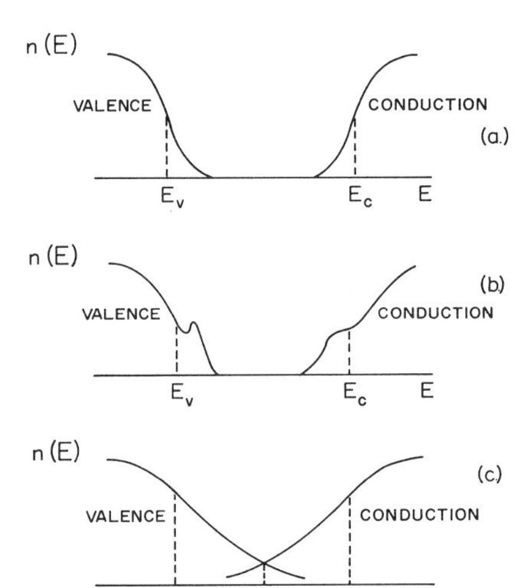

Fig. 5 The density of electronic states in energy band models for disordered materials. (*a*) The simplest model. (*b*) The effect of structural defects. (*c*) Possible situation in mixtures. From Ref. 33.

One theoretical problem with these models is to understand the validity of Eq. 1 for the conductivity when there are continuous densities of states. For crystals a band edge is required to give agreement with this equation. However, for glassy semiconductors Cohen[33] has suggested that this agreement can be explained if the mobility changes by several orders of magnitude near the critical energies E_v and E_c shown in Fig. 5. In this "mobility gap" model the transport mechanism changes near these energies from one characteristic of extended states, in which there is propagation with occasional scattering and a long mean-free path, to hopping between localized states.

Fritzche has discussed some problems with this picture of conduction in glassy semiconductors.[21] Electrical measurements show a large density of states, about 10^{19}/eV cm^3, at the Fermi energy in the middle of the "gap" shown in Fig. 5. Alternatively optical absorption measurements indicate a much smaller value, such as 10^{16}/eV cm^3, except near the critical energies. Furthermore, if the measurements of ac conductivity result from phonon-assisted hopping, then the density of states near the Fermi level becomes equal to or greater than 10^{19}/eV cm^3.

Fritzsche has proposed a heterogeneous model to account for these differences.[21] In it regions in which mostly holes are localized are different from regions in which electrons are localized. There are three kinds of electronic states: those localized in regions, channel states that extend through the material, but are excluded from certain regions, and extended states for which electrons have a finite probability of appearing anywhere in the material.

In this model the localized gap states are not found optically because there is a low probability for transitions between these states and the extended channel states. The latter are excluded from regions where the electrons are localized. Fritzsche discusses various other consequences for this model and compares it with some experimental data.[21] This model seems to account for some puzzling features of conduction in glassy semiconductors, but direct experimental evidence for the heterogeneous distribution of states is lacking. It could possibly be associated with phase separation in the glasses, which occurs easily.[35,36]

A "random-phase" model of conduction in glassy semiconductors has been proposed by Hindly.[37] The upper limit of the density of states, 10^{17}/eV cm^3, as calculated from this model is considerably lower than that calculated from the electrical measurements.

SWITCHING BEHAVIOR

Switching and "memory" effects in glasses have been reviewed by Pearson[35] and are discussed in many other papers, especially the *Journal of oncrystalline Solids*, Volumes 2, 4, and 8 through 10. These effects have been found in chalcogenide glasses, such as As—Te(I, Br, Se), AsTlSe, TiAs(Se, Te)$_2$,[40] and AsTeSiGe,[41] in vanadium phosphates,[39] in sodium borotitanate glass,[39] in amorphous silicon and germanium,[42] in organic semiconductors,[43] and probably occur in all semiconducting glasses.

The voltage-current characteristic of a glass containing 53 mole % As, 43% Te, and 4% I is shown in Fig. 6, as measured with a potentiometer circuit and a point-contact electrode.[35] The current was linear with voltage up to about 20 V, where it began to increase more rapidly. At 23 V the current suddenly jumped to a much higher value, and the voltage dropped to less than 2 V. In this highly conducting state the current increased very sharply with increased voltage. When the voltage was reduced to zero,

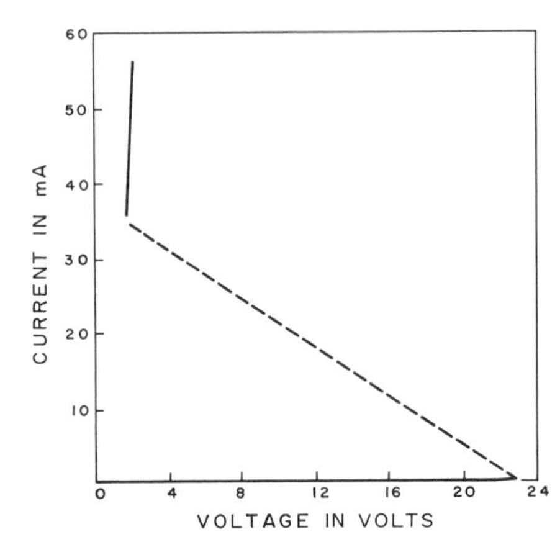

Fig. 6 The voltage-current characteristic of an AsTeI glass showing switching. From Ref. 35.

the glass returned to its highly resistive state. These changes did not depend on polarity and could be reproduced indefinitely, although the switching voltage was lower after the first cycle.

The mechanism of this switching behavior is being hotly debated. There is considerable evidence that the change in conductivity occurs only in certain small regions or filaments of the glass, which is usually studied as a thin film. In the high conductivity state the sample resistance does not depend on its area, whereas it shows the normal dependence in the low conductivity state.[44] Pearson and Miller reported microscopic evidence that a thin filament was associated with the high conductivity state,[45,35] and Stocker showed that switching occurs in localized regions by potential-probe measurements.[46] Some authors feel that the switching is accompanied by a phase change such as melting or crystallization.[35,44,45] Pearson calculated that the rate of a phase change, which could proceed at the speed of sound, would be sufficient to account for the high switching times. However, other authors have suggested that the switching occurs by some electronic process, such as electrode injection and increase in the number of carriers,[47] and energetic or " hot " electrons.[48] A " thermal avalanche " mechanism in which filaments of high temperature and conductivity has been proposed[49] and criticized.[47] Guntersdorfer showed that the melting of a $T_{48}As_{18}Ge_6Si_{28}$ glassy semiconductor accompanies switching, but could not decide whether the switching process was initiated thermally or electronically.[50] The experiments are not definitive as to which mechanism is the correct one, although I incline to the view that switching results from some sort of phase or structural transformation. Only further work can decide this point.

A " memory " state is also found in glassy semiconductors. It is established under " constant current " conditions, that is, when the increase in current is regulated in some way, for example, by having a high resistor in series with the glass. In this way the amount of current that flows in the sample is kept low compared to that in the switching experiment described above. The formation of the memory state was described by Pearson with the aid of Fig.7. With the initial application of voltage the current was proportional to voltage up to about 18 V, when it deviated from linear behavior and then increased erratically to the "forming current" of about 3 mA. After this step the device was in the memory state; it showed high conductivity even after long periods with no applied voltage. A short pulse of current somewhat higher than the forming current returned the device to its initial low conductivity.

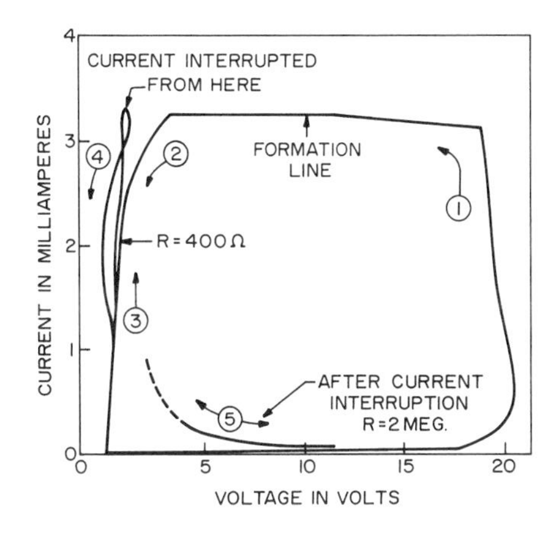

Fig. 7 Current-voltage characteristics of an AsTeI glass, measured under constant current conditions with a point-contact electrode, showing the formation and destruction of a memory state. From Ref. 35.

After this first cycle an increase of voltage under constant current conditions to about 12 V led to a reproducible "negative resistance" state shown in Fig. 7, in which an increase in the current gave a lower voltage. Change to the memory state would still occur at a higher current.

It is generally agreed that the formation of the memory state involves a structural transformation.[35,51–53] This transformation could be crystallization,[52] phase separation,[35] or some rearrangement of the structure at fairly short range.[53] The type of transformation could be different in different glasses. The agreement that a structural rearrangement is responsible for the memory state lends support to a similar mechanism for the switching process, since the two phenomena seem closely related. However, Bagley and Blair[52] found evidence for crystallization in three $As_2(Se, Te)_3$ glasses and an $As_{30}Te_{48}Si_{12}Ge_{10}$ glass that showed both switching and memory states, whereas a $2As_2Se_3 \cdot As_2Te_3$ glass showed no evidence of crystallization and only a switching state, but no memory state. More work is needed to establish with certainty the mechanism of switching and memory states in glassy semiconductors.

REFERENCES

1. D. Adler, *Crit. Rev. Solid State Sci.*, **2**, 317 (1971).

1a. J. D. Mackenzie, *J. Am. Ceram. Soc.*, **47**, 211 (1964).

2. E. P. Denton, H. Rawson, and J. E. Stanworth, *Nature*, **173**, 1030 (1954).
3. P. W. McMillan, in *Advances in Glass Technology*, Part 1, Plenum Press, New York, 1962, p. 333.
4. H. J. L. Trap and J. M. Stevels, *Verres Refract.*, **16**, 337 (1962).
5. K. W. Hansen, *J. Electrochem. Soc.*, **112**, 994 (1965).
6. C. C. Sartain, W. D. Ryden, and A. W. Lawson, *J. Noncryst. Solids*, **5**, 55 (1970).
7. P. R. Van Loan, *J. Noncryst. Solids*, **6**, 170 (1971).
8. C. C. Sartain, *J. Noncryst. Solids*, **6**, 172 (1971).
9. E. H. Martin and J. Hirsch, *J. Noncryst. Solids*, **4**, 133 (1970).
10. B. L. Boynton, H. Rawson, and J. E. Stanworth, *J. Electrochem. Soc.*, **104**, 237 (1957).
11. M. Munakata, *Solid-State Electron.*, **1**, 159 (1960).
12. L. A. Grechanik, N. V. Petrovykh, and Karpechenko, *Sov. Phys.-Solid State*, **2**, 1908 (1960).
13. V. A. Joffe, J. B. Patrina, and J. L. Peberovskaya, *Sov. Phys.-Solid State*, **2**, 609 (1960).
14. A. P. Schmid, *J. Appl. Phys.*, **39**, 3140 (1968).
15. G. T. Lynch, M. Soyer, L. L. Segel, and G. Rosenblatt, *J. Appl. Phys.*, **47**, 2587 (1971).
16. J. Stuke, *J. Noncryst. Solids*, **4**, 1 (1970).
17. M. H. Brodsky and R. L. Title, *Phys. Rev. Letters*, **23**, 581 (1969).
18. W. C. LaCourse, V. A. Twaddell, and J. D. Mackenzie, *J. Noncryst. Solids*, **3**, 234 (1970).
19. A. E. Owen, Glass Ind., **48**, 637, 695 (1967).
20. B. T. Kolomiets, Y. V. Rukhlyadev, and V. P. Shilo, *J. Noncryst. Solids*, **5**, 389, 402 (1971).
21. H. Fritzsche, *J. Noncryst. Solids*, **6**, 49 (1971).
22. B. T. Kolomiets and T. F. Nazarova, *Sov. Phys.-Solid State*, **2**, 369 (1960).
23. W. Y. Peck and J. F. Dewald, *J. Electrochem. Soc.*, **111**, 561 (1964).
24. E. B. Ivkin, B. T. Kolomiets, and E. A. Lebedev, *Bull. Acad. Sci. USSR, Phys. Ser.*, **28**, 1590 (1964).
25. J. C. Male, *Brit. J. Appl. Phys.*, **18**, 1543 (1967).
26. P. Nagels, R. Callaerts, M. Denager, and R. deConinck, *J. Noncryst. Solids*, **4**, 295 (1970).
27. M. Roiler, *J. Noncryst. Solids*, **6**, 5 (1971).
28. S. W. Ing, J. H. Newhart, and F. Schmidlin, *J. Appl. Phys.*, **42**, 696 (1971).
29. E. A. Fagen and N. Fritzsche, *J. Noncryst. Solids*, **2**, 180 (1970); **4**, 480 (1970).
30. A. K. Jonscher, *Thin Solid Films*, **1**, 2B (1967).
31. M. Pollak, *Phil. Mag.*, **23**, 519 (1971).
32. N. F. Mott and E. A. Davis, *Phil. Mag.*, **17**, 1269 (1968); *Electronic Processes in Non-crystalline Materials*, Oxford University Press, London, 1971.
33. M. Cohen, *J. Noncryst. Solids*, **4**, 391 (1970).
34. N. F. Mott, *Phil Mag.*, **19**, 835 (1969); **24**, 911, 935 (1971); *Contemp. Phys.*, **10**, 125 (1969); and many earlier articles by this author and his co-workers.
35. A. D. Pearson, *J. Noncryst. Solids*, **2**, 1 (1970).
36. G. V. Bunton, *J. Noncryst. Solids*, **6**, 72 (1971).
37. N. K. Hindly, *J. Noncryst. Solids*, **5**, 17 (1971).
38. G. B. Inglis and F. Williams, *J. Noncryst. Solids*, **5**, 313 (1971).

39. A. D. Pearson, J. F. Dewald, W. R. Northover, and W. F. Peck, in *Advances in Glass Technology*, Plenum Press, New York, 1962: Vol. 1, p. 357; Vol. 2, p. 145.
40. B. T. Kolomiets and E. A. Lebedev, *Radio Eng. Electron. USSR*, **8**, 1941 (1963).
41. S. R. Ovshinsky, *Phys. Rev. Letters*, **21**, 1450 (1968).
42. C. Feldman and K. Moorjami, *J. Noncryst. Solids*, **2**, 82 (1970).
43. A. Szymanski, D. C. Zarson, and M. M. Laber, *Appl. Phys. Letters*, **14**, 88 (1969).
44. L. A. Coward, *J. Noncryst. Solids*, **6**, 110 (1971).
45. A. D. Pearson and C. E. Miller, *Appl. Phys. Letters*, **14**, 280 (1970).
46. H. J. Stocker, *J. Noncryst. Solids*, **2**, 371 (1970).
47. H. Fritzsche and S. R. Ovshinsky, *J. Noncryst. Solids*, **2**, 393 (1970).
48. N. K. Hindly, *J. Noncryst. Solids*, **5**, 31 (1970).
49. F. M. Collins, *J. Noncryst. Solids*, **2**, 496 (1970).
50. M. Guntersdorfer, *J. Appl. Phys.*, **42**, 2566, 2577 (1971).
51. D. L. Eaton, *J. Am. Ceram. Soc.*, **47**, 554 (1964).
52. B. G. Bagley and H. E. Blair, *J. Noncryst. Solids*, **2**, 155 (1970).
53. E. J. Evans, J. N. Helbers and S. R. Ovshinsky, *J. Noncryst. Solids*, **2**, 3 (1970).
54. J. M. Lifshitz, *Adv. Phys.*, **13**, 483 (1964).

11

Dielectric and Mechanical Loss

In many substances mechanisms for dielectric and mechanical loss (internal friction) are entirely different. However, in nonmetallic solids there is often a loss of both kinds that is related to the electrical conductivity of the solid. Various mechanisms have been suggested for this loss, but none has been universally accepted. Therefore, in this discussion experiments on each kind of loss are discussed separately with no interpretation in terms of mechanism. For each type, experimental methods and effects of frequency, temperature, and glass composition on the loss are considered. In the final sections the relation between these two types of loss and theories for them are treated.

EXPERIMENTAL MEASUREMENT OF DIELECTRIC LOSS

The charge Q on a conductor is proportional to the potential difference V between this conductor and other conductors within the electric field of the charge, so that

$$Q = CV \tag{1}$$

where C is called the capacitance. The capacitance depends on the geometrical arrangement of the conductors and the material in between them called the dielectric and is measured in farads, which are coulombs/volt. The property of the material that determines the capacitance is called the dielectric constant ε. For a vacuum ε is called ε_0, the dielectric constant of free space, and equals $8.85(10)^{-12}$ F/m. If the conductors are flat plates of area A a distance d apart, the capacitance of this "condenser" or "capacitor" is given by

$$C = A\frac{\varepsilon}{d} \tag{2}$$

If an alternating voltage is impressed across the plates of a condenser with a vacuum dielectric, the current in the circuit lags behind the voltage by 90°. If the dielectric is not a vacuum, the current often is not exactly 90° out of phase with the voltage, giving rise to a power loss. This property can be described by considering the dielectric constant to be a complex number with a real and imaginary part:

$$\varepsilon^* = \varepsilon' - i\varepsilon'' \tag{3}$$

In this case the capacitance given by Eq. 2 can also be considered to be a complex quantity. The loss angle θ by which the current deviates from a 90° lag with the voltage is given by

$$\tan\theta = \frac{\varepsilon''}{\varepsilon'} \tag{4}$$

The complex dielectric constant is usually measured with an ac bridge, such as a Schering bridge.[1] A schematic diagram of such a bridge is given in

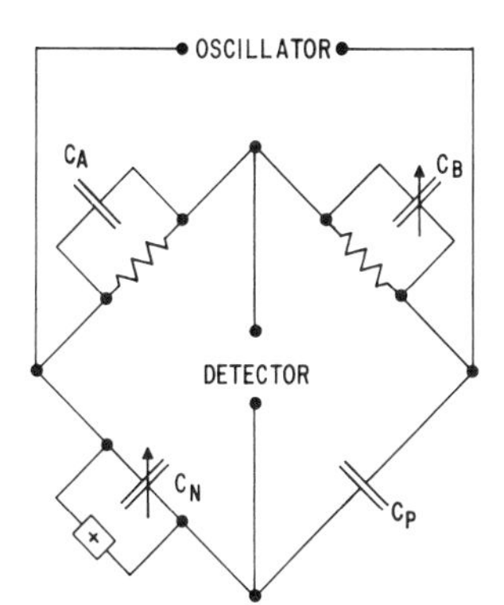

Fig. 1 Schematic diagram of a bridge circuit used to measure dielectric properties.

Fig.1. The variable capacitors C_N and C_B are alternatively adjusted until the detector shows a minimum current, first without the sample X and then with it. At low loss the real part of the capacitance C' of the sample is just the difference between the two readings of C_N, and $\tan\theta$ is proportional to the difference between the readings of C_B and C_N. For $\tan\theta$ greater than about 0.1, these readings must be corrected by adding squared terms.[1] If the sample is the dielectric of a capacitor with plates of area A a distance d apart, then ε' is given by $C' d/A$ from Eq. 2, and ε'' can be calculated from $\tan \theta$ and Eq. 4.

The type of electrodes used in measuring the dielectric loss of glass can substantially influence the results. The usual type of electrodes have been films of metal deposited on the glass. With this type of electrode it is necessary to make a correction for the contribution of the normal dc resistivity of the sample to the apparent value of ε''. This correction is inversely proportional to the frequency, so that at low frequency it is much larger than the other losses in the glass, and makes results at these frequencies quite unreliable. Another disadvantage of the film electrodes is that surface conductivity can contribute to the loss. This contribution can be avoided by a third "ring" electrode, but this electrode leads to additional complications in the measuring circuit. Silver electrodes can also react with the glass.

An alternative type of electrode is made of stainless steel disks, which are simply pressed against the flat glass sample.[2] The electrodes and sample can be held in vacuum for measurements at high temperatures. With this arrangement the glass is partially isolated from the electrodes, which are covered with a thin oxide layer. The dc resistivity of a sample between such electrodes was about three orders of magnitude lower than measured with silver electrodes painted on the same sample.

EFFECT OF FREQUENCY ON DIELECTRIC LOSS

At high frequencies (above about 10^6 cycles/sec at room temperature) the real part of the dielectric constant changes little with frequency, and the imaginary part is small. However, at lower frequencies there is a loss peak, and ε' decreases in the frequency range of the loss. Results at 253°C for an alkali borosilicate glass containing 7.2 wt. % Na_2O, 8.7% K_2O, and negligible Li_2O and CaO, are shown in Figs. 2 and 3. These measurements were made as described above, with stainless steel electrodes.[2] The increase in loss at low frequencies probably results from the dc resistivity of the sample, for which no correction was made. The values actually plotted are $(\varepsilon' - \varepsilon_\infty)/(\varepsilon_s - \varepsilon_\infty)$ and $\sqrt{2}\varepsilon''/(\varepsilon_s - \varepsilon_\infty)$, where ε_∞ is the high-frequency dielectric constant, and ε_s is the estimated value of the "static" dielectric constant, or the real part at zero frequency.

A number of earlier measurements of dielectric loss as a function of frequency have been made.[3–7] Taylor[3] found that his loss peaks for a number of soda-lime glasses and Pyrex borosilicate glass had about the same frequency dependence; an example is shown in Fig. 4. Charles[4] found a similar loss curve for a commercial soda-lime glass. The loss peaks found by

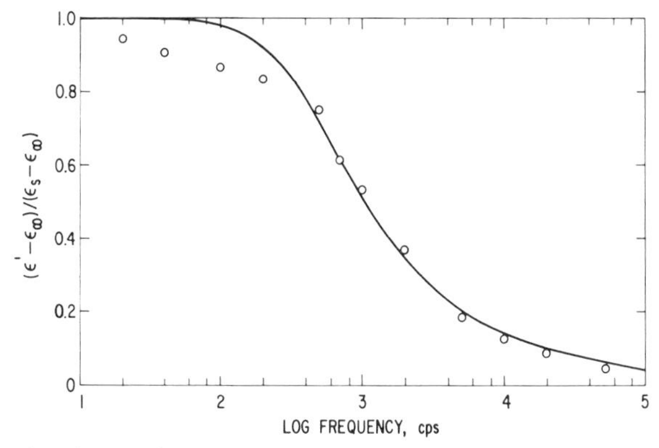

Fig. 2 Reduced real part of the dielectric constant for an alkali borosilicate glass. Points, experimental data; curve, theoretical.[2]

these investigators are somewhat broader than the one shown in Fig. 3. Taylor and Charles made their measurements on samples with painted silver electrodes. It is not certain whether the different experimental techniques are responsible for the differences in the shape of the loss curves. Isard[8] found some change in peak shape for $Na_2O{\cdot}4SiO_2$ glass as a function of temperature, using sodium amalgam electrodes, and he found an asymmetric peak, in agreement with Fig. 3. Asymmetric loss peaks have also been found for lithium[9] and cesium[10] silicate glasses.

Owen and Douglas measured some dielectric properties of fused silica.[7] They found that the dispersion curves of ε' had the same shape at different

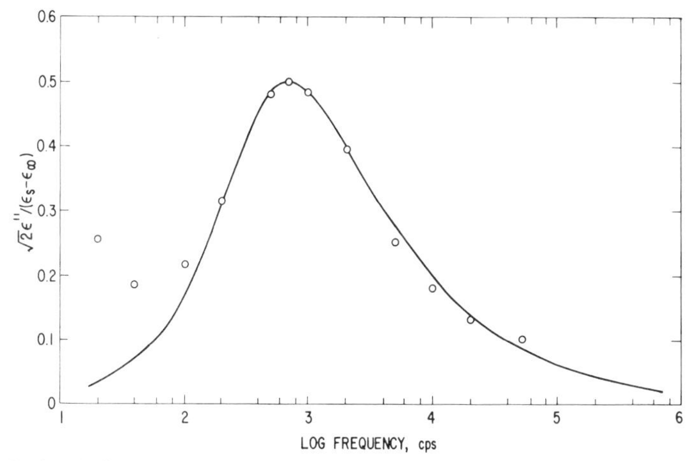

Fig. 3 Reduced dielectric loss as a function of frequency for an alkali borosilicate glass. Points, experimental data; curve, theoretical.[2]

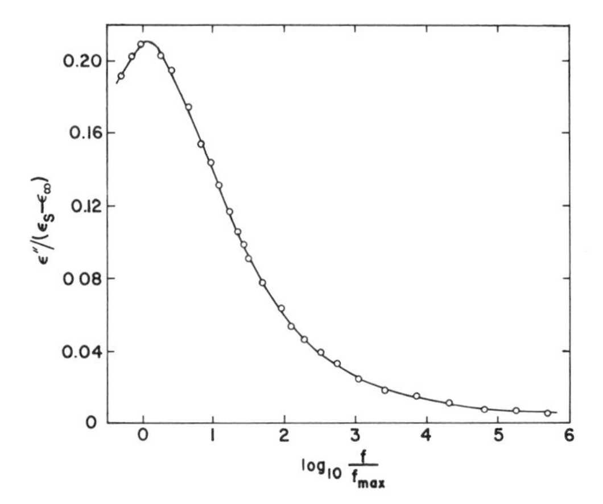

Fig. 4 Dielectric loss curve for a soda-lime glass, measured by Taylor.[3]

temperatures and for different fused silicas. Their loss curves were narrower than those measured by Taylor and Charles, but similar to the one shown in Fig. 2.

Topping and Isard[10a] measured the dielectric loss of sodium aluminate glasses at microwave frequencies. They found that the loss increased as a function of temperature because of increasing "migration" losses. The temperature-independent background was not associated with nonbridging oxygen atoms.

EFFECT OF TEMPERATURE ON DIELECTRIC LOSS

The loss peak and dispersion curve shift to higher frequencies as the temperature is increased. If the peak shape remains the same as the temperature is changed, the shift of the peak maximum gives a measure of the temperature dependence of the loss. For glasses, the frequency at which the loss is a maximum increases exponentially with reciprocal temperature following the relation

$$f_m = A \exp\left(\frac{-Q}{RT}\right) \tag{5}$$

where Q is the activation energy, R is the gas constant, T is the temperature, and A is independent of temperature.

The activation energies of relaxation and electrical conduction for some commercial and laboratory glasses are shown in Table 1. The activation energy for relaxation was close to that for conduction for every glass examined including many not listed in Table 1.

Table 1 Activation Energies for Dielectric Relaxation and Electrical Conductivity

Glass type	Composition (mole %)			Activation energy (kcal/mole)		Refs.
	SiO_2	Na_2O	CaO	Relaxation	Conductivity	
Soda-lime	15.3	10.5	74.2	20.6	19.8	3
Soda-lime	19.1	10.5	70.4	19.2	18.8	3
Soda-lime (commercial)	19.1	6.5	74.4	18.7	18.6	3
Pyrex borosilicate	80.0	4.1		23.9	22.2	3
Soda-lime (Corning 0080)	73.9	15.9	4.7	20.8	20.2	4
Fused silica (600–800°C)	100			23	22	7

Gough, Isard, and Topping found a dielectric loss at high temperatures in alkali-free borate glasses (alkaline earth or lead aluminoborates) that was related to the electrical conductivity of the glasses, even if the conductivity was electronic, as for bismuth borate glasses. Except for these latter glasses the conductivity is probably ionic and results either from the metal ions or residual hydrogen ions. In a review Isard also concludes that dielectric loss in glass is related to conductivity, whether it is electronic or ionic.[12] Mansingh et al. found a dielectric loss in an electronically conducting vanadium phosphate glass.[12a]

EFFECT OF COMPOSITION ON DIELECTRIC LOSS

High-frequency dielectric constants and power factors (power factor = $\cos\theta \approx \tan\theta$) at room temperature are summarized by Navias and Green[13] and Morey.[14] The high-frequency reduced dielectric constants $\varepsilon'/\varepsilon_0$ of some different glasses at 25°C are fused silica, 3.7; Pyrex borosilicate, 4.6; and commercial soda-lime (Corning 0080), 7.2. The dielectric constant usually increases as the amount of alkali in the glass increases. Factors for calculating the dielectric constant from the composition are given by Scholze,[15] as taken from Appen and Brecker.[16]

The power factor ($\tan\theta$) also increases as the amount of alkali increases. Many measurements of $\tan\theta$ for glasses, particularly for small $\tan\theta$ values, are not very accurate.

The dielectric loss at 10^3 cps of various fused silicas below room temperature down to 1°K was measured by Mahle and Cameron[17]; these authors also reviewed earlier work. This loss apparently is not related to the electrical conductivity of the silica. A loss peak at 240°K was found to be related to the aluminum content of the silica. Below about 100°K the background loss increased, the increase being greater the higher the hydroxyl (water) content of the silica. A peak in the loss at about 30°K has been attributed to structural relaxation of the silica network. A similar low-temperature mechanical loss has been found,[18] but it is not clear if water content can influence the mechanical loss,[17,18] or if the mechanism for dielectric and mechanical loss at these low temperatures is the same.

The dielectric loss in silicon dioxide layers on silicon is important in the stability and properties of electronic devices using these layers to insulate the semiconducting silicon. The presence of mobile conducting ions such as sodium and perhaps hydrogen in the silica layer leads to instabilities in these devices because of charge build-up at the silicon-silica interface in an electric field. A review of studies of these films is given in Ref. 19.

Dorda[20] found that the loss ($\tan \theta$) in a silica layer on silicon followed Eq. 5 and had an activation energy of about 7 kcal/mole at about 10^4 cps. The loss was affected by different gaseous atmospheres: hydrogen, nitrogen, and wet oxygen. At frequencies of 10^4 to 10^5 cps a dielectric loss peak at about 200°K in silicon dioxide on silicon was found that was related to the amount of water in the silica. This peak is apparently different from the low-temperature peaks found in bulk fused silica.[21]

Dielectric loss near the glass transition temperature in aliphatic alcohols was studied by Johari and Goldstein.[22] This relaxation probably results from relaxations of polyatomic groups, and is not related to processes at temperatures much lower than the glass transition. Reviews of earlier work on dielectric loss in a variety of silicate glasses are given in Refs. 21a and 6.

MEASUREMENT OF MECHANICAL LOSS

When a stress is applied to a material there is an instantaneous deformation and sometimes a further deformation that increases with time. When the stress is released there is an instantaneous elastic relaxation, but some of the deformation may relax more slowly.

If a periodic stress $S = S_0 \cos \omega t$ of frequency $\omega/2\pi$ is applied to the material, the deformation lags behind the stress and is proportional to $\cos(\omega t - \delta)$ because of these time-dependent relaxations. The tangent of the lag angle δ is called the internal friction. Two different experimental methods have been used to measure this mechanical loss or internal friction in glass: the torsion pendulum and the crystal oscillator.

In the torsion pendulum a fine glass rod is twisted and the damping of its oscillations observed. If there were no internal friction or external damping forces the pendulum would oscillate through the angle of the initial twist. However, internal friction causes this angle ϕ_1 to decrease to ϕ_2 in the subsequent cycle. The logarithmic decrement $\ln \phi_1/\phi_2$ is equal to one-half of the fractional energy loss $\Delta E/E$ per cycle. In turn $\Delta E/E$ is related to the internal friction $\tan \delta$:

$$\ln \frac{\phi_1}{\phi_2} = \frac{\Delta E}{2E} = \pi \tan \delta \tag{6}$$

Thus by measuring the deformation angles of a torsion pendulum after successive oscillations it is possible to calculate the internal friction. The frequency of the torsion pendulum usually is in the range from 0.05 to 10 vibrations/sec.

The internal friction of glass can be measured at frequencies in the kilocycle range by a piezoelectric driver-gage method.[23] One quartz bar drives a composite oscillator of which the sample is a part, and another quartz bar acts as the gage. The total energy loss in the composite oscillator is proportional to the ratio of the voltage V_d on the driver crystal to the voltage V_g from the gage crystal, and if the angle δ is small, so that $\tan \delta \approx \delta$, then

$$\delta = K \frac{V_d}{V_g m_t f_t^2} \tag{7}$$

where m is the total mass of the oscillator and f_t is the resonant frequency of the composite oscillator. The proportionality constant can be measured from the half-width of the resonance peak. To find the internal friction of glass, two samples differing in length by one-half of a wave length are used. Then

$$\delta = \frac{(m_L \delta_L - m_s \delta_s)}{m_L - m_s} \tag{8}$$

where the ms are the masses of the larger and smaller samples and the δs are the total losses calculated from Eq. 7, for the appropriate sample.

EFFECT OF FREQUENCY AND TEMPERATURE ON MECHANICAL LOSS

The internal friction of glass has been measured as a function of temperature at constant frequency because it is easier to change the temperature than the frequency in the techniques mentioned above. It is unfortunate that no measurements at constant temperature over a wide frequency range have been made, because many properties of glass change with temperature. Thus it is easier to compare theories of mechanical loss with measurements at constant temperature than with those at constant frequency.

The internal friction of a sodium silicate and a sodium-aluminum silicate glass over a wide range of temperatures is shown in Fig. 5. At high

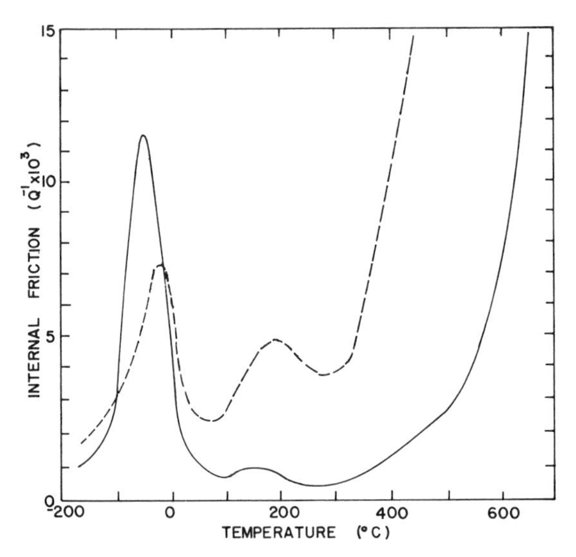

Fig. 5 Internal friction of a $Na_2O \cdot 3SiO_2$ glass (dashed line) and a $Na_2O \cdot Al_2O_3 \cdot 6SiO_2$ glass (solid line) at 0.4 cps as a function of temperature, measured with a torsion pendulum.[41]

temperatures the loss probably results from large-scale relaxations in the silicate network. The rate of these relaxations is related to viscous flow; they are discussed in more detail in Refs. 24 and 25. Viscous flow and the glass transition are discussed in Chapters 6 and 7, respectively.

At very low temperatures and high frequency the internal friction of glasses increases. Such an increase occurs in soda-lime, borosilicate, and fused silica glasses,[23,26,27] so it is apparently not related to ionic motion in the glass. A similar increase occurs in borate and germanate glasses.[24] Models for this increase usually involve bending of Si—O—Si bonds, but such models have not been directly confirmed, so the mechanism of the low-temperature loss remains uncertain.

At intermediate temperatures two internal friction peaks appear that are related to ionic motion in the glass. These peaks are shown in Fig. 5 for a sodium disilicate glass. The peak at low temperature is related to alkali ion movement in the glass, whereas the peak at higher temperatures is related to hydrogen ions in the glass. The reasons for these assignments are enlarged upon in the next paragraphs.

The temperature for maximum internal friction can be measured for different frequencies. At two different frequencies f_1 and f_2 the temperatures T_1 and T_2 of maximum tan δ are related by

$$\ln\frac{f_1}{f_2} = \frac{-Q}{R}\left(\frac{1}{T_1} - \frac{1}{T_2}\right) \tag{9}$$

where the activation energy Q is independent of temperature. The activation energies for the low-temperature peaks of a number of alkali silicate glasses are found to be close to those for electrical conductivity and alkali ion diffusion,[28–31] just as for the activation energies of the dielectric loss peaks. Furthermore, the height of the internal friction peak is related to the concentration of alkali ion in the glass, usually becoming smaller as this concentration becomes smaller, as described in the following section. Thus there appears to be a definite connection between alkali ion motion and the low-temperature peak. Mechanisms for this relaxation are discussed in the section on theories.

The internal friction peak at higher temperatures, sometimes called the "intermediate temperature" peak, has been attributed to the motion of nonbridging oxygen ions.[24,31–35] However, there are serious doubts about this attribution. The silicon-oxygen bond in fused silica has an energy of formation of over 100 kcal/mole, and in crystalline silicates the silicon-oxygen bond distance and bond angles are little affected by temperature or composition. Also the activation energy for viscous flow of silicates is high; for example, for soda-lime glasses it is about 100 kcal/mole at 500 to 600°C (see Chapter 6). These results show the rigidity of the silicon-oxygen bond and make doubtful the proposals that the internal friction peak in alkali silicate glasses at temperatures of 100 to 200°C results from the motion of nonbridging oxygen ions. From the figures above one would expect an activation energy of about 100 kcal/mole for transport of these ions, yet the activation energies for these peaks in a number of silicate glasses are much less, being as low as 32 kcal/mole. The diffusion coefficient of lattice oxygen in a soda-lime glass at 200°C (extrapolated) is 6×10^{-28} cm^2/sec;[36] it is unlikely that an ion with such a low diffusion coefficient could contribute to a relaxation process at a frequency of 1 cps. The average distance that an ion with this diffusion coefficient could move in 1 sec is less than 10^{-13} cm.

Several authors have claimed that the effect of glass composition on the second peak is consistent with the motion of nonbridging oxygen ions. However, several glasses containing many nonbridging oxygens associated with alkali ions show no intermediate peak,[30,35,37,38] and at least one glass with no alkali ions did show this peak.[39] Thus there is little support for the proposal that this internal friction peak is associated with the motion of nonbridging oxygen ions.

It is much more likely that this peak is related to the motions of hydrogen ions in the glass.[40] A number of experimental results point to the importance of hydrogen ions for this peak. First, there is a close correlation between the rate of weathering of a silicate glass and the occurrence of the peak. Glasses that weather rapidly show the peak, whereas those that are less reactive do not. Weathering results from the exchange of hydrogen ions from atmospheric water with alkali ions in the glass (see Chapters 13 and 14). The rate of weathering is controlled by the rate of interdiffusion of alkali and hydrogen ions in the glass. The addition of alkaline earth ions to the glass reduces the diffusion coefficient of alkali ions and consequently the rate of weathering. Aluminosilicates have a lower rate of weathering because the aluminosilicate groups in glass have a lower ratio of hydrogen-to-sodium ion affinity than do —SiO groups. Therefore, on the aluminosilicate glasses the fraction of surface alkali ions that exchange with hydrogen ions is lower than for other alkali silicate glasses. The internal friction peak at higher temperatures becomes smaller as more alkaline earth or aluminum oxide is added to an alkali silicate glass,[35,41] and is absent in some commercial glasses containing these oxides and known for good weatherability.[30,37,38]

Second, Vaugin et al. showed that when a fine glass fiber reacts with the atmosphere an internal friction peak slowly develops at higher temperatures than the alkali peak.[42] Their fibers were stored in a desiccator and measured in a vacuum, under which conditions this peak did not develop.

Third, DeWaal[43] studied the internal friction of a sodium disilicate glass before and after exchange with hydrogen ions from a melt of NH_4HSO_4. He treated the glass for $2\frac{1}{2}$ hr at 275°C. At this temperature the effective interdiffusion coefficient of sodium and hydrogen ions is about 1.3×10^{-11} cm^2/sec, so that the average distance of penetration of the hydrogen ions was about 5 μm. After this treatment DeWaal found a substantial increase in the height of the second internal friction peak at higher temperature, as would be expected if this peak resulted from motion of hydrogen ions.

Fourth, Abdel-Latif and Day[40a] were able to introduce a high-temperature peak in an $Li_2O \cdot Al_2O_3 \cdot 2SiO_2$ glass that should not contain nonbridging oxygen atoms. The original glass did not show this peak, but after hydrogen ions were introduced by ion exchange in molten ammonium acid sulfate for 21 hr at 366°C such a peak appeared. Therefore the experimental evidence is in favor of a connection between hydrogen ion motion in the glass and the appearance of the high-temperature peak.

EFFECT OF GLASS COMPOSITION ON MECHANICAL LOSS

As mentioned above the very low-temperature mechanical loss is not much affected by glass composition, and the losses at temperatures near the glass transition are related to the viscosity of the glass.

Many effects of composition on the "ionic" peaks were described in the preceding section. The effect of alkali ion concentration on peak height for number of different silicate glasses is shown in Fig. 6. There is considerable

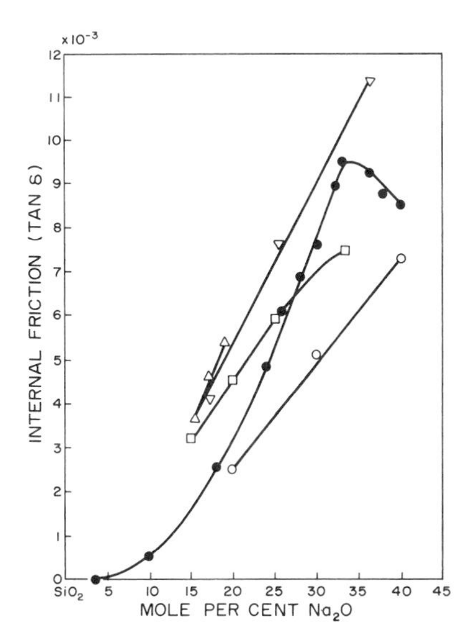

Fig. 6 Height of the internal friction peak in binary alkali silicate glasses as a function of alkali oxide concentration. ●, Coenen[28]; ○, Deeg[44]; □, Jagdt[45]; △, Mohyuddin and Douglas[31]; ▽, Forry.[33] From Ref. 24.

scatter in the results from one investigator to another for identical compositions, but for each investigator there is a consistent increase in peak height with alkali oxide concentration. These results show that the particular experimental technique used, and the history of the samples, can influence the background contributions to the peak height. Results on some glasses containing a third oxide are not consistent with those in Fig. 6; for example, in a glass containing 12.5 mole % Na_2O, 12.5% Al_2O_3, and 75% SiO_2 the peak height was about $12(10)^{-3}$,[41] and in Pyrex borosilicate glass (4% Na_2O) it was about 10^{-3}.[37] In fused silica with about 5 to 10 ppm by weight Na_2O, the peak height was measured to be about $3(10)^{-5}$.[23]

Internal friction in silicate glasses containing two alkali ions has been intensively studied.[34,41,46–48] As the concentration of a second alkali ion is increased in a binary alkali silicate glass, the alkali ion peak decreases in height and another peak with its maximum at a temperature between those

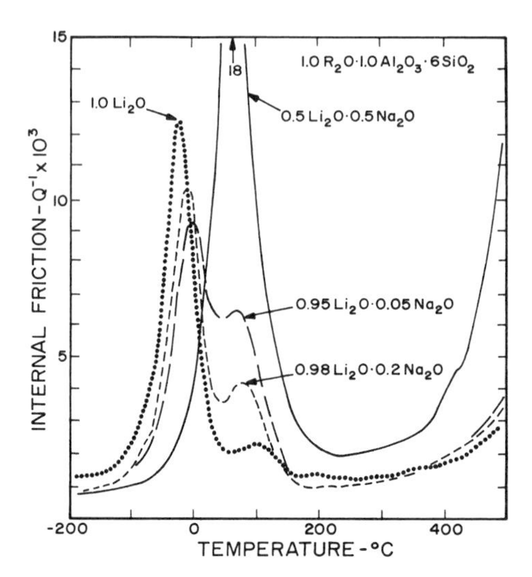

Fig. 7 Internal friction in lithium-sodium silicate glasses at a frequency of 0.4 cps.[41]

of the alkali and hydrogen ion peaks appears, as shown in Fig. 7. The height of this peak is greatest at some intermediate concentration ratio of the alkali ions, and the peak temperature is also lowest for some intermediate concentration ratio.[46,47,47a] The activation energy associated with this "mixed-ion" peak is correlated with the peak temperature, being higher for higher peak temperatures and lower for lower peak temperatures. For lithium-sodium silicate glasses the activation energies of the mixed peak were in the range of about 22 to 30 kcal/mole. The height of the mixed peak appears to be related to the tracer diffusion coefficient of the slowest diffusing species,[47,47a] but the reason for this relationship and the mechanism leading to the mixed peak are uncertain.

If a fiber of a binary alkali silicate glass is treated in a molten salt to exchange some of the alkali ions with different monovalent ions, the mixed-ion peak develops if the temperature of treatment is above about 200°C.[48,49,49a] If the exchange temperature is lower, there is little change in the internal friction peaks.[48] Thus the exchange of ions without structural rearrangement in the glass apparently does not lead to a mixed-ion peak; to develop this peak some structural modification of the binary silicate glass is necessary. This modification occurs at temperatures well below the glass transition temperature, so it probably involves only short-range changes in the glass structure. Only a very thin layer of exchanged glass (15 μm in a -mm fiber) gives rise to a considerable reduction and shift of the single alkali ion peak and to introduction of the mixed-ion peak.[48]

In sodium borate glasses with Na_2O concentrations greater than about 15 mole %, the alkali ion peak behaves much the same as in sodium silicates.[28,50] The temperature of maximum loss decreases as the sodium concentration increases, as does the peak height, paralleling an increase in conductivity in the glass. Below about 15% Na_2O the peak is obscured by background damping, since the temperatures approach the glass transition range.

In sodium phosphate glasses containing from 35 to 50% sodium, two "ionic" peaks are found.[24,51] As the sodium ion concentration increases, the height of the lower temperature peak decreases and that of the higher temperature peak increases, and the temperatures of maximum loss become farther apart. These two peaks may result from sodium and hydrogen ion motion, as in silicate glasses, but more work is needed to establish this assignment with certainty.

The internal friction of sodium germanate glasses is quite similar to that of sodium silicate glasses.[52,53] The alkali ion peak has an activation energy close to that for electrical conductivity of the glasses and for sodium ion diffusion in them, and the peak height is related to the sodium ion concentration.[53] A higher temperature peak is similar to the same peak in sodium silicate glasses, and is also probably related to hydrogen ions in the glasses.

A mechanical loss peak has been found in an iron phosphate glass; this peak has the same activation energy as the conductivity for the glass.[54] This result is curious because the conductivity in this glass is electronic rather than ionic.

Day and Rindone found that when a lithium silicate glass was partially crystallized, the internal friction decreased as the volume fraction of crystals increased, and both peaks gradually disappeared.[55] The electrical conductivity of the crystals was much lower than of the glass matrix, so the glass behaved as if its effective volume for mechanical loss was decreased by the formation of crystals. In a study of partially crystallized lithium-potassium aluminosilicate glasses, Li Chia-Chih et al. found different results.[56] Crystallization of their glasses led to intensification of an internal friction peak at intermediate temperatures, perhaps the mixed-ion peak, possibly because of changes in the ratios of the two alkali ions in the residual glass.

Fraser measured the acoustic loss of various fused silicas over a wide temperature range.[27] He found a minimum in $\tan \delta$ at a frequency of $1.6(10)^6$ cps and temperatures between 200 and 500°C at a level below 10^{-6}, which is much smaller than the background values of about 10^{-4} found in the torsion experiments on alkali silicate glasses. Marx and Sivertsen found similar low levels in their loss measurements on fused silica at $3.7(10)^4$ cps. Fraser found a correlation between the minimum background loss and the state of oxidation of the silica; the more reduced the silica the greater was the loss. Apparently a stoichiometric silica would show the lowest background loss. This relation with the state of oxidation implies either that reduced groups, such as an oxygen ion vacancy, contribute to the loss, or that reduction leads to easier motions in the silica lattice.

THEORIES OF DIELECTRIC AND MECHANICAL LOSS IN GLASS

Most theories of the alkali ion peaks in dielectric and mechanical loss in glass have considered these peaks to result from short-range rearrangements of the alkali ions in the glass. In dielectric loss this rearrangement leads to an effective dipole moment, and in mechanical loss it leads to a time-dependent change in modulus. In none of these theories has it been possible to predict the peak temperature.

The shapes of the loss peaks can be calculated by assuming an exponential decay of current or deformation. For dielectric loss this assumption leads to a complex dielectric constant of the form

$$\varepsilon^* = \varepsilon' - i\varepsilon'' = \varepsilon_\infty + \frac{\varepsilon_0 - \varepsilon_\infty}{1 - i\omega\tau} \tag{10}$$

or

$$\varepsilon' = \varepsilon_\infty + \frac{\varepsilon_0 - \varepsilon_\infty}{1 + \omega^2\tau^2} \tag{11}$$

and

$$\varepsilon'' = \frac{(\varepsilon_0 - \varepsilon_\infty)\omega\tau}{1 + \omega^2\tau^2} \tag{12}$$

Here ε_0 is the static (dc) dielectric constant, ε_∞ is the dielectric constant at high frequency, ω is 2π times the frequency, and τ is a relaxation time.

For the internal friction tan δ can be found, based on similar equations, to be[57]

$$\frac{\tan\delta}{(\tan\delta)_{\max}} = \frac{2\omega\tau}{1 + \omega^2\tau^2} \tag{13}$$

Equations 12 and 13 give symmetrical loss peaks that are considerably narrower than those measured for dielectric and mechanical loss in glass. To explain the broader peaks it is necessary to assume that there is a spectrum of relaxation times τ, rather than a single time. It has not been possible to relate such spectra of relaxation times to any other properties of glasses.

In order to overcome some of the limitations of previous treatments, I developed a new theory for mechanical loss in ionic conductors.[58] From this theory it is possible to calculate peak temperatures and shapes that are in good agreement with some measured values; however, the theory has some serious deficiencies and must be considered as preliminary. At the least it gives empirical formulas for peak shape and temperature. The reader is referred to Ref. 58 for a complete description of this theory; here it is briefly discussed.

When a stress gradient is applied to a solid ionic conductor, the ions in it move to reduce the resulting strain. Weber and Goldstein have measured this motion in a sodium silicate glass.[59] If the moving ions are blocked at the sample surface, an electric field is set up as a result of their motion. The excess of the ions at the surface is often called a "space charge." This nonuniform distribution of ions in the glass leads to a change in the modulus of the glass that occurs as the ions move, and so is time-dependent. The internal friction as given by this treatment is

$$\tan\delta = \frac{M_R}{M_B} + B\left(\frac{(v^2+1)^{1/2}-1}{(v^2+1)}\right)^{1/2} \tag{14}$$

where B is a frequency-independent parameter and

$$v = \frac{\omega\varepsilon_\infty}{\sigma} \tag{15}$$

where ω is the angular frequency $2\pi f$, ε_∞ is the high-frequency dielectric constant, and σ the electrical conductivity of the glass.

The first term M_R/M_B results from a frequency-independent background in the internal friction.

The shape of the internal friction peak as a function of frequency in this theory is given by Eq. 14 without any adjustable parameters, as shown in Fig.8. The peak is broader than for a single relation time (Eq. 13), and is asymmetric. There are apparently no measurements of internal friction as a function of frequency at constant temperature to make a reliable comparison with the theoretical curve.

The internal friction is a maximum when $v^2 = 3$, or at a frequency f_m:

$$f_m = \frac{\sqrt{3}\sigma}{2\pi\varepsilon_\infty} \tag{16}$$

If the internal friction is measured at a constant frequency and the temperature is varied, the maximum internal friction occurs at the temperature where σ is given by Eq. 16, with f_m equal to the measuring frequency. A plot of the peak shape as a function of temperature with an activation energy for conduction of 20 kcal/mole is given in Fig. 9. Again the peak is broader than for a single relaxation time, and is asymmetric. The measured peak shape is often broader than that calculated by the theory. However, Vaugin

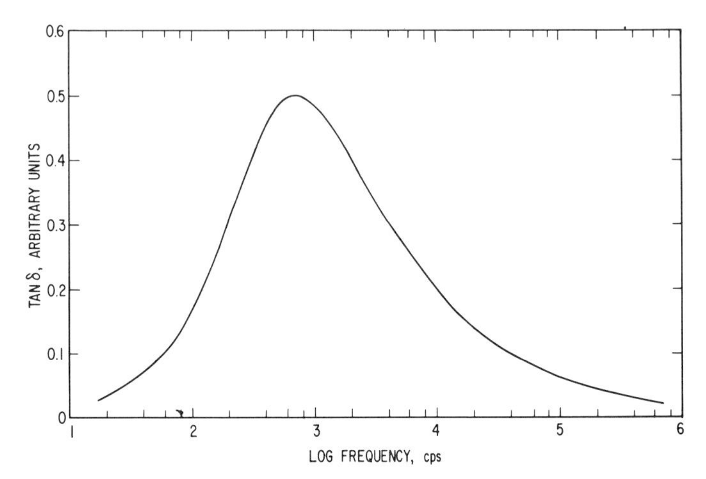

Fig. 8 Internal friction as a function of frequency at constant temperature, as calculated from Eq. 14 with no background contribution.

et al.[42] measured peak widths at half-maximum for $Na_2O \cdot 4SiO_2$ glass rods as low as 50°, as did Coenen[28] and Day and Steinkamp[41] on other glass compositions. These widths are close to the calculated one of 46° for an activation energy of 15 kcal/mole. One possible contribution to the broader peaks found by other workers is the exchange of surface sodium ions with hydrogen ions from atmospheric water. Vaugin et al. and Coenen used vacuum or dry air for their measurements, and aluminosilicate glasses of Day and Steinkamp have lower affinity for hydrogen ions, so that the influence of these ions was reduced in the experiments with narrower peaks.

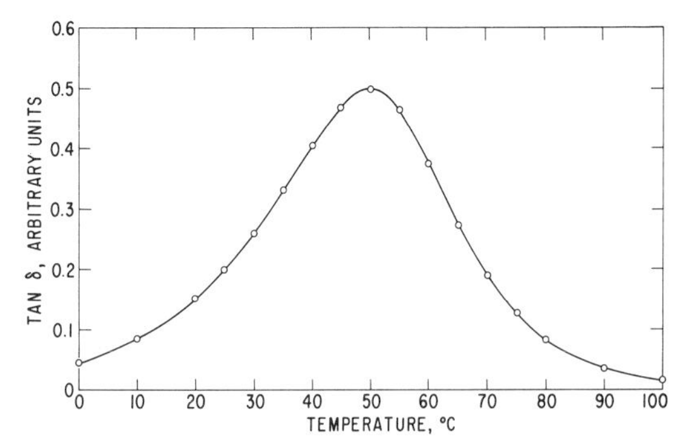

Fig. 9 Internal friction as a function of temperature at a constant frequency, as calculated from Eq. 14 with no background contribution.

The theoretical peak temperature can be calculated from Eq. 16 and experimental results on electrical conductivity and dielectric loss. The conductivity should be measured on the same sample for which an internal friction measurement was made, since these conductivities are very sensitive to the thermal history of the sample. Such conductivity measurements are not usually available, and there is some uncertainty in using measurements by other investigators for comparison. In Table 2 measured and calculated

Table 2 Measured and Calculated Temperatures of Maximum Internal Friction for Some Silicate Glasses.

Glass type	Measurement frequency (cps)	High frequency dielectric constant	Temperature for maximum $\tan\delta$ (°C) Measured	Calculated[a]	References Internal friction	Resistivity
Commercial soda-lime, chilled	0.224	7.0	48	35	30	30
Annealed	0.224	7.0	62	60	30	30
Pyrex borosilicate	1.12	4.6	106	111	37	3
	9.06	4.6	134	140	37	3
	37,000	4.6	318	299	23	3
12.5% Na_2O 12.5% Al_2O_3 75% SiO_2	0.4	7.0	−50	−53	41	60
Fused silica	37,000	3.5	646	657	23	7
19.9% Na_2O 81.1% SiO_2	0.383	7.0	−7	15	31	61
33% Na_2O 67% SiO_2	5	7.0	−33	−23	43	61

[a] See Eq. 16.

peak temperatures are compared. The results are close except for the binary alkali silicate glasses. Thus Eq. 16 seems to be valid for most glasses.

The theory predicts proportionality of the peak height to the square root of the ionic concentration. Because of the variations in peak height in different glass compositions and different studies, as discussed in the preceding section, it is difficult to test this prediction from these data. Possibly, a more reliable test can be made by comparing results on alkali silicate glass with those for fused silica. For a silicate glass with 33% Na_2O ($= 0.027$ mole/cm^3) the maximum $\tan\delta$ is about 0.01, and in fused silica with about $3(10)^{-7}$ mole Na_2O/cm^3 it is about $3(10)^{-5}$, from the results of Marx and Sivertsen.[23] Thus the ratio between $\tan\delta$s is about 300, which is close to the square root of the concentration ratio of 10^5.

One difficulty with this theory of internal friction in glass is its prediction that the height of the internal friction peak should be inversely proportional to the sample size, whereas experimentally one finds no size dependence. Thus further work is needed to test the basic mechanism proposed to see if it is valid and if the theory can be modified to meet this difficulty. However, the peak shape, peak temperature, and concentration dependence of peak height seem to be given reliably for most glasses, so the theory is worthy of consideration.

REFERENCES

1. F. A. Laws, *Electrical Measurements*, McGraw-Hill, New York, 1938, p. 417.
2. R. H. Doremus, General Electric Research and Development Center Report 69-C-231, September 1969.
3. H. E. Taylor, *J. Soc. Glass Tech.*, **41**, 350 (1957); **43**, 124 (1959). These articles contain a review of earlier work on dielectric relaxation in glass.
4. R. J. Charles, *J. Appl. Phys.*, **32**, 1115 (1961); *J. Am. Ceram. Soc.*, **45**, 105 (1962).
5. J. Volger, J. M. Stevels, and C. van Amerongen, *Philips Res. Rep.*, **8**, 452 (1953).
6. J. M. Stevels, in *Handbuch der Physik*, Vol. 20, Springer-Verlag, Berlin, 1957, p. 350.
7. A. E. Owen and R. W. Douglas, *J. Soc. Glass Tech.*, **43**, 159 (1959).
8. J. O. Isard, *Proc. IEEE (London)*, **109B**, 22, 440 (1961).
9. L. Heroux, *J. Appl. Phys.*, **29**, 1639 (1958).
10. D. R. Uhlman and R. M. Hakim, *J. Phys. Chem. Solids*, **32**, 2652 (1971).

10a. J. A. Topping and J. O. Isard, *Phys. Chem. Glasses*, **12**, 145 (1971).

11. E. Gough, J. O. Isard, and J. A. Topping, *Phys. Chem. Glasses*, **10**, 89 (1969).
12. J. O. Isard, *J. Noncryst. Solids*, **4**, 357 (1970).

12a. A. Mansingh, J. M. Reyes, and M. Sager, *J. Noncryst. Solids*, **7**, 12 (1972).

13. L. Navias and R. L. Green, *J. Am. Ceram. Soc.*, **29**, 267 (1946).
14. G. W. Morey, *The Properties of Glass*, Reinhold, New York, 1954, p. 502.
15. H. Scholze, *Glas*, Vieweg and Sohn, Braunschweig, Germany, 1965, p. 192.
16. A. A. Appen and R. J. Bresker, *J. Tech. Physics USSR*, **22**, 946 (1952).
17. S. H. Mahle and R. D. Cameron, *Phys. Chem. Glasses*, **10**, 222 (1969).
18. J. T. Krause and C. R. Kurkjian, *J. Am. Ceram. Soc.*, **51**, 226 (1968), and earlier studies referred to in this article.
19. E. Kooi, *The Surface Properties of Oxidized Silicon*, Springer-Verlag, New York, 1967.
20. G. Dorda, *Surface Sci.*, **15**, 14 (1968).
21. R. Nannoni and M. J. Musselin, *Thin Solid Films*, **6**, 397 (1970).

21a. A. E. Owen, in *Progress in Ceramic Science*, J. E. Burke, Ed., MacMillan, New York, 1963, p. 77.

22. G. P. Johari and M. Goldstein, *J. Chem. Phys.*, **55**, 4245 (1971).
23. J. W. Marx and J. M. Sivertsen, *J. Appl. Phys.*, **24**, 81 (1953).
24. J. L. Hopkins and C. R. Kurkjian, in *Physical Acoustics*, Vol. II, Part B, Academic, New York, 1965, p. 91. This paper contains a review of internal friction measurements on glass.
25. R. W. Douglas, P. J. Duke, and O. V. Mazurin, *Phys. Chem. Glasses*, **9**, 169 (1968).
26. A. H. Meitzler and A. H. Fitch, *J. Appl. Phys.*, **40**, 1614 (1969).
27. D. B. Fraser, *J. Appl. Phys.*, **41**, 6 (1970).
28. M. Coenen, *Z. Elektrochem.*, **65**, 903 (1961).
29. H. Rötger, *Glastech. Ber.*, **19**, 192 (1941).

30. J. V. Fitzgerald, *J. Am. Ceram. Soc.*, **34**, 314, 339, 399 (1951).
31. J. Mohyuddin and R. W. Douglas, *Phys. Chem. Glasses*, **1**, 71 (1960).
32. J. V. Fitzgerald, K. M. Laing, and G. S. Bachman, *J. Soc. Glass Tech.*, **36**, 90 (1952).
33. K. E. Forry, *J. Am. Ceram. Soc.*, **40**, 90 (1957).
34. H. Rötger, *Glastech. Ber.*, **31**, 54 (1958).
35. R. J. Ryder and G. E. Rindone, *J. Am. Ceram. Soc.*, **43**, 662 (1960).
36. W. D. Kingery and J. A. Lecron, *Phys. Chem. Glasses*, **1**, 87 (1960).
37. P. L. Kirby, *J. Soc. Glass Tech.*, **39**, 385 (1955).
38. G. J. Copley and Oakley, *Phys. Chem. Glasses*, **9**, 141 (1968).
39. P. W. L. Graham and G. E. Rindone, *J. Am. Ceram. Soc.*, **50**, 336 (1967).
40. R. H. Doremus, *J. Noncryst. Solids*, **3**, 369 (1970).
40a. A. I. A. Abdel-Latif and D. E. Day, *J. Am. Ceram. Soc.*, **55**, 254 (1972).
41. D. E. Day and W. E. Steinkamp, *J. Am. Ceram. Soc.*, **52**, 571 (1969).
42. L. Vaugin, J. C. Breton, P. Gobin, *Verres Refrac.*, **23**, 174 (1969).
43. H. deWaal, *J. Am. Ceram. Soc.*, **52**, 165 (1969).
44. E. Deeg, *Glastech. Ber.*, **31**, 1, 85, 124, 229 (1958).
45. R. Jagdt, *Glastech. Ber.*, **33**, 10 (1960).
46. J. E. Shelby and D. E. Day, *J. Am. Ceram. Soc.*, **52**, 169 (1969).
47. G. L. McVay and D. E. Day, *J. Am. Ceram. Soc.*, **53**, 508 (1970).
47a. J. W. Fleming and D. E. Day, *J. Am. Ceram. Soc.*, **55**, 186 (1972).
48. H. deWaal, *Phys. Chem. Glasses*, **10**, 108 (1969).
49. J. D. Taylor and G. E. Rindone, *J. Am. Ceram. Soc.*, **51**, 289 (1968).
49a. A. I. A. Abdel-Latif, *J. Am. Ceram. Soc.*, **55**, 279 (1972).
50. K. H. Karsch and E. Jenckel, *Glastech. Ber.*, **34**, 397 (1961).
51. R. L. Myerson, M.Sc. Thesis, Massachusetts Institute of Technology, Cambridge, 1961.
52. C. R. Kurkjian and J. T. Krause, *J. Am. Ceram. Soc.*, **49**, 134 (1966).
53. J. E. Shelby and D. E. Day, *Phys. Chem. Glasses*, **11**, 224 (1970).
54. R. A. Miller and K. W. Hansen, *J. Elect. Soc.*, **116**, 254 (1969).
55. D. E. Day and G. E. Rindone, *J. Am. Ceram. Soc.*, **44**, 161 (1961).
56. Li Chia-chih, Lo-Chie-yüeh, and Cheng Hu-ming, in *Structure of Glass*, Vol. 7, Consultants Bureau, New York, 1966, p. 193.
57. C. Zener, *Elasticity and Anelasticity of Metals*, University of Chicago Press, 1948, pp. 6 ff.
58. R. H. Doremus, *J. Appl. Phys.*, **41**, 3366 (1970).
59. N. Webber and M. Goldstein, *J. Chem. Phys.*, **41**, 2898 (1964).
60. V. A. Tsekhomskii, O. V. Mazurin, and K. K. Evstrop'ev, *Sov. Phys. Solid State*, **5**, 426 (1963).
61. E. Seddon, E. J. Tippet, and W. E. L. Turner, *J. Soc. Glass Technol.*, **16**, 450 (1932).

Part Three

CHEMICAL AND SURFACE PROPERTIES

The next three chapters deal with chemical reactions in glass, many of which go on at or near the glass surface. These reactions are important in a number of properties of glass, such as strength, chemical durability, weathering, adhesion and sealing, surface conductivity, and potentials in glass electrodes, and in glass manufacture in melting and fining (removal of bubbles). The division of material between the three chapters is roughly based on how far beneath the glass surface a process goes on. If only the first few molecular layers, up to a depth of about 100 Å, are involved, the phenomenon is discussed in Chapter 12, whereas phenomena involving more of the glass, even if still a thin layer from a bulk viewpoint, are included in Chapter 13 or 14. Ion exchange and potentials of glass electrodes are discussed separately in Chapter 14, since these potentials depend on ion exchange, and this exchange is a rather special kind of chemical reaction. Surface defects affecting glass strength and surface treatments to improve strength are discussed in later chapters.

12

Surface Properties

The properties and processes going on at a glass surface are affected by its structure, so the first section of this chapter is devoted to a discussion of the molecular groups found on a glass. The overwhelming majority of work has been carried out on silica surfaces, which are described in detail. In subsequent sections physical adsorption, chemisorption and surface chemical reactions, adsorption from solution, and surface conductivity are discussed. Some surface properties of glass are not discussed because they have been adequately covered in previous publications, and not much recent work has been done on them. Among these properties are surface tension of molten glasses (Ref. 1, p. 191; Ref. 2, pp. 212 ff.; Ref. 3, pp. 1 ff.); adhesion and wetting (Ref. 3, pp. 348 ff., 394 ff.; Ref. 4, pp. 181 ff.); and cleaning (Ref. 3, pp. 290 ff.). A review on properties of glass surfaces was written by Ernsberger.[4a]

SURFACE STRUCTURE

The surface structure of oxide glasses depends on the reactions of "dangling" oxide bonds. Thus in a silicate glass a surface leads to Si—O— and Si— bonds that are unsatisfied. These bonds react rapidly with atmospheric water to form SiOH groups. Therefore the surface of an oxide glass is normally composed of metal hydroxyl groups. The thickness and structural arrangement of the hydrated surface layer depend on the composition of the glass, its thermal history, humidity, and surface treatment after melting and cooling.

Extensive studies of silica (SiO_2) surfaces have been made because of the widespread use of silica gel as a catalyst and adsorbent. Results of these studies are discussed here because they give a basis for the understanding of the surface structure of silicate glasses.

Hydroxyl groups on silica surfaces have been examined in detail by infrared spectroscopy.[5–8a] In these studies silica particles less than about 1 μ in diameter were pressed into thin porous disks for observation in the spectrophotometer. Spectra of such disks after different heat treatments are shown in Figs. 1 and 2. The broad band at about 3450 cm^{-1} and another at

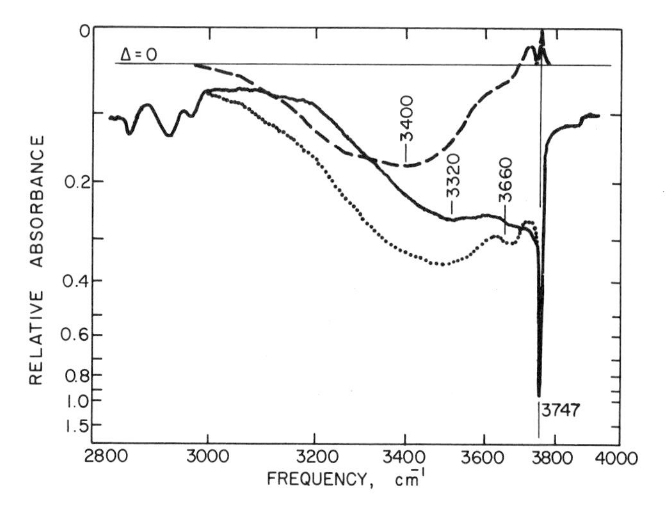

Fig. 1 Infrared spectra of SiOH groups on a cabosil silica. Dotted line, in air at room temperature; solid line, after 3 hr in vacuum at 30°C; dashed line, difference between solid and dotted line. From MacDonald.[5]

about 1250 cm^{-1} result from the OH vibrations in molecular water. Therefore the presence of these bands indicates physically adsorbed molecular water. Consistent with this interpretation is the result that these bands disappear after pumping at room temperature or heating to 150°C for a short time. These treatments reveal another band at 3660 cm^{-1}, in addition to the sharp band at 3747 cm^{-1}. As the temperature is raised above 150°C the band at 3660 cm^{-1} slowly disappears, until at high temperatures only the sharp band at 3747 cm^{-1} remains. The latter band is attributed to isolated SiOH groups, whereas the one at 3660 cm^{-1} is considered to result from surface hydroxyl groups close enough together to be hydrogen-bonded. Schematic diagrams of these and other surface hydroxyl groups are given in Fig. 3.

Another type of silanol group exists near a silica surface, namely, internal SiOH groups.[6–7,9–11] The presence of this group was originally inferred from deuterium exchange reactions on glass surfaces.[7] At room temperature

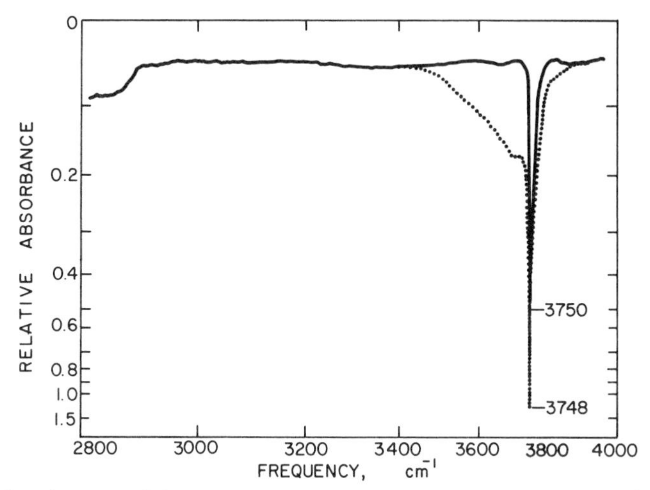

Fig. 2 Infrared spectra of SiOH groups on a cabosil silica. Dotted line, after 30 min in vacuum at 500°C; solid line, after 8.5 hr in vacuum at 940°C. From MacDonald.[5]

only part of the hydroxyl groups reacted with D_2O vapor to form SiOD; the other groups were presumed to be somewhat beneath the surface and so inaccessible to the D_2O. This internal SiOH group results from the diffusion of water molecules into silica and their subsequent reaction with the silica lattice to form two SiOH groups, as described in Chapter 8. This process becomes important above about 100°C, as shown by an interpretation[9] of results of water adsorption on Vycor surfaces.[12] At room temperature these internal silica groups do not form[11] because the diffusion coefficient of water in bulk silica is too low at this temperature.[9]

The density of isolated SiOH groups on a silica surface has been calculated as 1.4 groups/100 Å, and that of hydrogen-bonded groups as 3.2 groups/100 Å, or 1.6 pairs.[13]

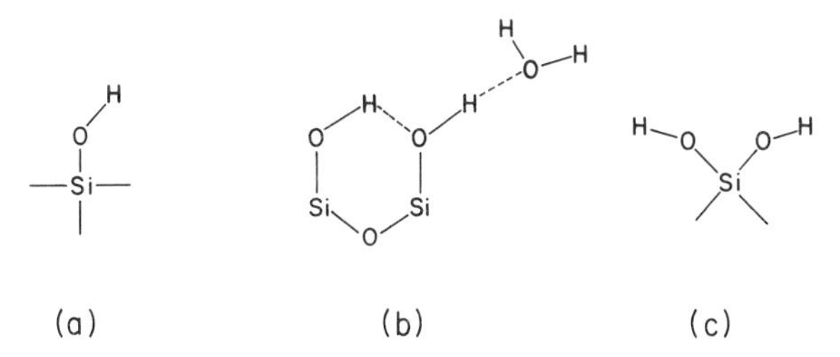

Fig. 3 Schematic diagrams of hydroxyl groups on a silica surface. (*a*) Isolated group. (*b*) Hydrogen-bonded groups with an adsorbed water molecule. (*c*) Two hydroxyls on one silicon atom.

A number of different models for the hydroxylated silica surface, as a particular plane of a crystalline modification of silica, have been proposed.[13–16] However, Armistead et al. concluded that the surface corresponded to an array of different crystal planes, some of which contain widely separated hydroxyl groups and others with closer spacing that leads to hydrogen bonding.[13]

There is evidence[15,17–18] that some of the isolated surface hydroxyls groups on silica occur as pairs on the same silicon atom (Fig. 3*c*). Hair and Hertl[17] and Bermudez[18] reached this conclusion from studies of reactions of chlorosilanes and boron trichloride with surface hydroxyls. They claimed that groups such as

```
O     O     R        O     O
 \   / \   /          \   / \
  Si    Si             S     B—Cl
 /   \ /   \          /   \ /
O     O     R        O     O
```

are formed in reactions of these compounds with two hydroxyl groups on one silicon atom. However, such a four-membered ring is not often observed in silicate or borosilicate structures, and should be highly strained. Furthermore, Hair and Hertl found that all isolated hydroxyls were equally reactive; it would be surprising if the paired hydroxyls had the same reactivity as the single ones. These authors concluded that the reaction order of 1.5 for multichlorosilanes, as compared to 1.0 for monochlorosilane, showed the presence of paired groups. However, this different order for multichlorosilanes could also be caused by other reactions and mechanisms, such as the additional reaction of a chlorine on the bound silane to form a SiCl group. It is also possible to interpret the results of Bermudez without involving geminal hydroxyl groups (Ref. 18, footnote 41), so that more evidence for such groups is needed before their existence on a silica surface can be accepted.

To summarize, the following types of hydroxyl groups exist on silica surfaces: isolated SiOH groups, hydrogen-bonded SiOH groups, internal SiOH groups, and molecularly adsorbed water. The relative amounts of these different groups on silica depends on the thermal and atmospheric history of the glass, and the temperature and humidity at which it is being observed.

The surfaces of silicate glasses with additional components probably have the same types of groups as pure silica, modified by the following considerations. If other glass formers are in the glass network, they will also provide sites for hydroxyl groups. Thus AlOH, BOH, and POH groups are likely. Monovalent cations R^+ in a silicate glass can exchange with water by the following reaction:

$$H_2O + SiO\,R^+ = SiOH + ROH \qquad (1)$$

This reaction gives rise to SiOH groups at the cation sites. Since the hydrogen ion is smaller than other monovalent cations, this exchange leads to a tensile stress at the glass surface, which can cause enhanced reactivity and further hydration, or in some cases even cracking. Ion exchange reactions are discussed further in Chapters 13 and 14. Internal hydroxyl groups are less likely in multicomponent silicate glasses because the diffusion of water molecules in the more dense structure of these glasses is much retarded. However, a hydrated layer can form more easily because of the ion exchange reaction (reaction 1) and subsequent enhanced reactivity.

Ion exchange reactions of water with cations of higher valence or anions of a glass surface have not been experimentally verified, so they presumably do not lead to major alterations in the surface structure of silicate glasses, at least as far as hydroxyl groups are concerned.

The surface of a silicate glass can become depleted of volatile components during glass melting and cooling to room temperature. The alkali oxides are particularly volatile. Optical measurements of the surface refractive index sometimes show a lower index in a thin layer near the surface, indicating either a loss of alkali by volatilization or a hydrated layer.

From these considerations one can conclude that the surface of a silicate glass containing only silicon as a glass former should be quite similar to that of pure silica. Dangling surface Si— or SiO— groups are hydroxylated, and any surface SiO alkali groups are also hydroxylated from ion exchange with water in the atmosphere. SiO groups bonded to higher valent cations may not exchange with hydrogen ions, but these cations are tightly bound to two or more SiO— groups, and so should not play a major role in surface reactions or adsorption. If other glass formers are present, different hydroxyl groups will occur on the glass surface. It is possible that one glass former or another will appear preferentially at the surface, giving a distribution of surface hydroxyl groups different from what would be expected from the bulk composition.

Various other experimental techniques have been used to examine glass surfaces, and are reviewed in Refs. 3 and 4. With electron microscopy Navez and Sella[19] showed changes in the morphology of polished and ion-bombarded surfaces of soda-lime and borosilicate glasses. The electron microprobe[19] and electron diffraction[20] showed that surfaces of polished soda-lime glass were deficient in sodium, probably because of ion exchange with water (reaction 1).

A favorite material for adsorption studies is "thirsty glass," a porous glass made by leaching a phase-separated borosilicate glass.[8,21] Thirsty glass is an intermediate in the manufacture of 96% Vycor or silica glass, as described in

Chapter 4. The surface of thirsty glass is heterogeneous and contains a high proportion of borate groups (B—OH), much more than would be expected from the bulk composition of 4% B_2O_3.[21] Thus results of adsorption studies with this material are not directly comparable with those on silica, although there are many similarities in the surface behavior of the two materials.[8]

PHYSICAL ADSORPTION

The nature of physical or chemical adsorption on glass is strongly dependent on the surface structure discussed in the preceding section. The surface hydroxyl groups are the most important sites for adsorption and reaction, so the thermal history of a glass is important in its adsorption behavior. Unfortunately only recently has detailed knowledge of the silica surface been available, so that most older adsorption studies were carried out on glass surfaces of unknown hydroxyl configuration. On the other hand, adsorption studies on pure silica can be related to the hydroxyl structure if their thermal history is known, so that adsorption studies on silica will be emphasized here.

McDonald found that the 3749-cm^{-1} band on Aerosil 2491 silica was shifted by physical adsorption of a number of gases.[22] Since this band results from isolated SiOH groups on the silica surface, these shifts indicate that the gases are adsorbed on or near these hydroxyl groups. At low coverages at 83°K the shifts for argon, krypton, xenon, nitrogen, oxygen, methane, and perfluoromethane were small, the largest being 32 cm^{-1} for methane. There was no direct relation between the band shifts and the polarizabilities of the adsorbed molecules. At higher gas pressures the bands shifted more, indicating interaction of more than one adsorbed molecule with each group. In a detailed study of argon, oxygen, and nitrogen adsorption, McDonald found that much more of the latter gas was adsorbed at the same gas pressure than the other two gases. For nitrogen the original 3749-cm^{-1} hydroxyl band disappeared at a pressure several times lower than that for coverage with a "BET monolayer," indicating that the nitrogen molecules were preferentially adsorbed at the free SiOH groups, and that a monolayer on silica actually involves multiple adsorption at such a group.

Water, methanol, and benzene adsorbed on silica in appreciable amounts at room temperature, causing much greater band shifts.[22] Adsorption of these gases with hydrogen bonding is intermediate between physical and chemical sorption.[22b] Cyclohexane also adsorbed at temperature, but caused a shift about the same as for methane. Benzene adsorption on porous glass was studied in detail by Cusumano and Low,[22a] who also reviewed extensive earlier studies of adsorption of aromatics on silica and porous glass.

Physically adsorbed molecules are apparently bonded to the hydrogen atom on an SiOH group. Larger band shifts for water and methanol, indicating stronger hydrogen bonds with these molecules, are expected from their hydrogen bonding in the liquid state. A study of diethylamine adsorption on silica showed that it strongly preferred the isolated silanol groups to those groups hydrogen-bonded in pairs,[23] whereas water appears to prefer the adjacent (hydrogen-bonded) silanol groups.[8,24]

A theoretical calculation was made of the band shift expected from nitrogen adsorption, considering the adsorbed molecule to be a quadrupole with its axis lying along the direction of the OH group, and rotating about an axis perpendicular to this group.[25] Reasonable agreement with the experimental shift was achieved.

Basila found a relation between band shifts for several substituted benzenes physically adsorbed on silica and their ionization potentials.[23] He argued that since the hydrogen bonding reaction is a sort of charge-transfer reaction, this relation would be expected. A relation between band shifts and ionization potentials was also found for adsorption of halogenated methanes on silica.[23]

The band shifts should be correlated with heats of adsorption if they are related to the strength of the hydrogen bond with the adsorbed molecule. Kislev and his colleagues measured the heats of adsorption and band shifts for a number of organic molecules adsorbed on silica and found a good correlation between these quantities.[26]

The studies above show that physical adsorption on silica takes place on particular sites, depending on the nature of the adsorbing molecule. The sites are hydroxyl groups, either isolated or hydrogen-bonded in pairs. Since neither of these types of sites fills the surface completely, the concept of a monolayer of adsorbed molecules loses significance. Thus absolute surface areas determined by the BET method are suspect because the density of isolated hydroxyl groups is about 1.4 groups/100 Å^2. If, for example, the area of an adsorbed nitrogen molecule is 16.2 Å^2,[27] then when one molecule of nitrogen is adsorbed per group, only about one-sixth of the surface is actually covered. For silica the BET method often gives the same effective surface areas for a number of different physically adsorbed molecules,[28] but these areas may all be inaccurate.

The variation of the amount of adsorbed gas as a function of the gas pressure is called an adsorption isotherm. A number of equations have been proposed for these isotherms,[27] both on an empirical and theoretical basis, but none is entirely satisfactory. Hydrated silica surfaces have been extensively studied as adsorbants, and over narrow pressure ranges a variety of equations fit the experimental isotherms. Thus such a fit is not conclusive evidence for the validity of the assumptions from which the isotherm is derived. The experimental results are summarized in many sources, for example, Refs. 27, 29, and 30.

Adsorption studies on the surfaces of other silicate glasses have been less extensive than for silica. Langmuir studied adsorption on microscope coverslides, presumably of soda-lime glass.[31] He found that the amount of gas adsorbed at 90°K increased in the order of oxygen, argon, nitrogen, carbon monoxide, and methane. These gases did not adsorb appreciably at room temperature, although carbon dioxide did. Langmuir found good agreement with his data on oxygen and argon and his adsorption isotherm in the form

$$Q = \frac{Q_s bP}{1 + bP} \tag{2}$$

where Q is the amount of gas adsorbed, Q_s is the amount adsorbed when the surface is covered with a monolayer, P is the gas pressure, and b is a coefficient dependent only on temperature. Q_s was actually treated as an adjustable parameter, and its relation to the surface area is questionable, as discussed above. Data on carbon monoxide did not fit this simple equation. Zeise[32] measured adsorption of hydrogen, oxygen, nitrogen, and methane on coverslides and found better agreement if Q was squared in Eq. 2.

Hobson measured the physical adsorption of nitrogen,[33,34] helium,[33] and argon[35] on Pyrex borosilicate glass over a wide pressure range. His data for nitrogen near liquid nitrogen temperatures are shown in Fig. 4. These data fit none of the simple equations for adsorption isotherms. Hobson found a reasonable fit with the Dubinin-Radushkevich equation:

$$\log \frac{Q}{Qs} = k\left(\log \frac{P}{Po}\right)^2 \tag{3}$$

where Po is the equilibrium vapor pressure of the adsorbing gas, and k is a coefficient dependent only on temperature. Again Q_s has been considered to be the amount adsorbed in a monolayer, but for glasses probably should be considered only as an empirical coefficient related to this amount.

The decrease in the amount of adsorption at low pressures, shown in Fig. 4, and the agreement with Eq. 3, might be interpreted as resulting from heterogeneous adsorption sites on the Pyrex surface. These sites could possibly be SiOH and BOH groups. However, a careful study of the adsorption with infrared techniques, similar to that of McDonald on silica, seems necessary to confirm this possibility.

Hobson did not find agreement with Henry's law:

$$Q = kP \tag{4}$$

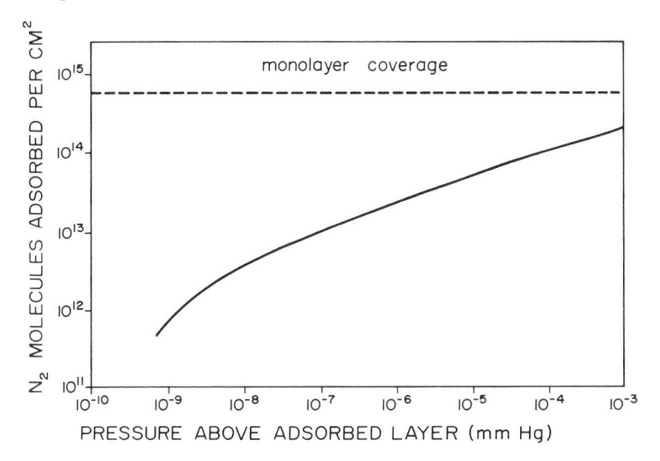

Fig. 4 Adsorption of nitrogen on Pyrex borosilicate glass. From Hobson.[34]

where k is a coefficient independent of pressure, even at the lowest pressures. He suggested that such agreement should occur at pressures below those of his measurements. It is also possible that the presence of different types of adsorption sites could explain the lack of agreement with Henry's law (see discussion in Ref. 35). Haul and Gottwald also concluded that Pyrex borosilicate glass had different types of adsorption sites from adsorption measurements by a molecular flow technique.[36]

Hobson studied the adsorption of helium on Pyrex at 4.2°K.[33] He found proportionality between the logarithm of the amount of gas adsorbed and the logarithm of the pressure over the range where the data were reliable. The effect of pressure on adsorption was very small, since the amount of gas adsorbed changed only by a factor of about 1.4 as the pressure changed from 10^{-9} to 10^{-4} mm Hg.

Hobson and Armstrong[35] calculated the following values for energies of adsorption on Pyrex, in kcal/mole: nitrogen, 1.7; argon, 1.2; and helium, 0.16. These absolute values must be considered as quite uncertain, but are about the right relative value. A number of authors found adsorption energies of about 6 to 10 kcal/mole for adsorption of water on silicate glasses.[4] The spread of values may represent heterogeneous adsorption sites or uncertain techniques for estimating the energies.

The adsorption of argon and krypton on Pyrex glass was measured by Kindle et al.[37] They found agreement with Eq. 3 for adsorption of krypton, but deviations from this equation for argon adsorption. They explained this deviation as resulting from changes in the surface structure of the glass during annealing, but in view of the empirical nature of Eq. 3 and uncertainty about the surface structure of the glass, such a possibility remains unconfirmed. Similar results for krypton adsorption on Pyrex were found by others.[38,39]

A number of other studies of physical adsorption of gases on glasses are summarized in Ref. 4.

CHEMISORPTION AND SURFACE CHEMICAL REACTIONS

Various gases react with the SiOH groups on a glass surface. Ammonia becomes hydrogen-bonded to the isolated SiOH groups, as shown by the strong perturbations of the SiOH vibrations after treatment of silica with ammonia, but does not react with the hydrogen-bonded groups.[40–42] Both physically and chemically adsorbed carbon monoxide are found on silica surfaces at 78°K.[43] Other chemisorption reactions on silica surfaces are reviewed in Refs. 8 and 8a.

Surface hydroxyl groups can be replaced by halogen atoms, which change some surface properties substantially. A partial substitution of fluoride for hydroxyl groups is achieved by treatment with ammonium fluoride or hydrofluoric acid solutions. The latter, and probably the former, etch away the hydrated surface and also substitute fluoride ions for some of the hydroxyls. Heating this surface to 700°C causes complete removal of hydroxyl groups, and the resultant surface is hydrophobic.[24] On the partially fluorinated surface the hydrogen ions on the hydroxyl groups are more easily ionized, apparently because of neighboring SiF groups.[42,44] Gaseous hydrofluoric acid also reacts with surface hydroxyls on silica, replacing all of them with fluoride ions at about 700°C. If the surface had been drastically dehydrated by heating to a higher temperature, reaction of surface Si-O groups to form Si-F was slower. A further discussion on the reactions of fluorides with glass is given in the next chapter.

A silica surface can be chlorinated by reaction with sulfuryl chloride (SO_2Cl_2) or silicon tetrachloride,[45] or by heating in chlorine at 700 to 950°C or carbon tetrachloride[40,43,46] at 350 to 600°C. Morrow and Devi have referred to many other studies of chemisorption of metallic halides on silica.[47] Other reactions are esterification, ammination, carboxylation, etherification, and reactions with Grignard reagents, organosilicon compounds,[48] diborane,[8] and $Al(CH_3)_3$.[49]

Morterra and Low reported the formation of some reactive sites on silica surfaces after degassing a methylated silica at temperatures above 600°C, and investigated the reaction products of these sites with a number of gases.[50]

Ashmore et al. studied the recombination of chlorine atoms on Pyrex and silica surfaces and found that the recombination coefficient was about 10^{-5}

for silica at room temperature and increased slowly as a function of temperature.[51]

Glass surfaces can act as catalysts for various biochemical reactions,[52] for example, for rapid clotting of blood. They are also sometimes used as substrates for culturing cell growth.

The acid ionization constants of various surface hydroxyl groups on silica were found by Hair and Hertl from the frequency shifts of the OH band maxima when different molecules were adsorbed on them.[53] BOH and POH groups were formed on the silica surface by reaction with BCl_3 and PCl_3, respectively. Acid ionization constants pk_a for ionization reaction

$$XOH = XO + H^+$$

were found to be SiOH, 7.1; BOH, 8.8; and POH, −0.4. They can be compared with values in aqueous solution of SiOH, 9.7 and POH, 2.0.

Reactions of carbon dioxide[54,55] and oxygen[56] with freshly formed silicate surfaces were studied in ultrahigh vacuum by Antonini and Hochstrasser. Fresh surfaces were obtained by fracturing or abrading glass samples inside the vacuum system. On a fresh silica surface the energy of adsorption of carbon dioxide was about 11 kcal/mole, as compared to the more usual heats of physical adsorption of about 1 to 2 kcal/mole. The desorption of gas was followed as the surface was heated; the increase of temperature was linear with time. A single desorption peak was found for fused silica and a calcium borosilicate glass, whereas for an alkali borosilicate and a sodium silicate (75% SiO_2, 25% Na_2O) glass two desorption peaks were found. The two peaks were taken to be evidence for two different types of adsorption sites, perhaps one being associated with unperturbed silicon atoms and the other with silicon perturbed by the presence of modifying ions, probably alkalis. This interpretation seems reasonable, but needs confirmation with other techniques.

These reactions were reversible as long as the sample was kept in the vacuum system at room temperature. Heating the samples to higher temperatures progressively destroyed the active reaction sites, perhaps because of diffusion and reaction of atoms from the interior of the glass.

Antonini et al. found that oxygen reacted with the unpaired electrons (dangling bonds) at silicon atoms on a fresh silica surface to form an $Si^+O_2^-$ complex.[56] The oxygen gas did not dissociate but retained its molecular form, although it was tightly bound with an adsorption energy of about 75 kcal/mole. The surface mobility of the adsorbed oxygen was immeasurably low. For an alkali borosilicate, a calcium borosilicate, and a sodium silicate glass the surface coverage was about the same as for silica, indicating a similar density of silicon dangling bonds for these glasses as for silica. The adsorption energy for the alkali borosilicate was somewhat lower than for silica.

Examination of the fresh surfaces with electron spin resonance[55] showed the presence of dangling silicon bonds, which could be described as unpaired electrons. This type of defect is similar to the E'_2 center of Weeks, which results from gamma or neutron irradiation of fused silica. These centers progressively disappeared as they reacted with carbon dioxide or oxygen. Reaction led to a transfer of charge from the silicon atom to the adsorbed gas, giving it an effective negative charge (CO_2^- and O_2^-), and desorption reversed the charge transfer.

ADSORPTION FROM SOLUTION

A number of studies of adsorption of trace amounts of radioactive ions on glass in aqueous solution have been tabulated by Kepak.[57] In water the surface silanol groups behave like very weak acids. Thus in alkaline solution the glass surface should carry a negative charge, and cations should adsorb on the ionized SiO^- groups. However, the adsorption is apparently complicated by a variety of effects. Adsorbed amounts of iron[58] and mercury[59] on a borosilicate glass (75% SiO_2, 7% B_2O_3, 6% Al_2O_3, 7% Na_2O, 4% BaO, 1% CaO) did not respond to changes in pH and concentrations of other ions as one would expect from simple adsorption on a weak-acid anionic site. The authors concluded that traces of ions, such as chloride, in solution could influence the complexing of the trace quantities of radioactive ions in solution. Colloidal particles of iron hydroxide were considered to be the adsorbing species at pH from 5 to 8.[58] Neutral mercury molecules were suggested as the adsorbing species in mercury solutions. However, more mercury was adsorbed on a soda-lime glass at the same pH than on the borosilicate, which might result from a larger number of silanol groups on the surface of the soda-lime glass, resulting from exchange of the sodium ions in the glass with hydrogen ions by reaction 1. Also different types of sites are present on the glasses, for example, BOH and more AlOH on the borosilicate. Thus the adsorption process is complicated by many factors, and its complete definition requires more carefully controlled systems.

In dilute alkali ion solutions and neutral pH the ions are adsorbed on silica[60,61] in the sequence $Cs^+ > Rb^+ > K^+ > Na^+ > Li^+ > (C_2H_5)N^+$. This decreasing adsorption as the size of the hydrated ion increases would be expected for ionic interaction with a directed anionic site of small effective radius, such as SiO^- (see Chapter 14). At higher electrolyte concentrations and higher pH the adsorption affinity changes[62] to $Li^+ > Na^+ > Cs^+ > K^+$. This changed sequence probably results from partial dehydration of the ions as they exchange more deeply into the hydrated layer and even the dry glass.[61]

I made some preliminary observations of adsorption of silver ions from a fused sodium nitrate bath that show the importance of surface treatment on adsorption from solution. A fused silica tube was cleaned in hot nitric acid for a short time, presumably hydrating the surface to a depth of several molecular layers. Radioactive silver ions adsorbed strongly onto this surface from a trace solution in the molten sodium nitrate at 350°C. The total amount of silver ions was roughly estimated to be three or four times the number of surface silanol groups, showing that adsorption occurred for several molecular layers into the glass. Then the glass was etched in 5% hydrofluoric acid for about 30 sec. This treatment removed the adsorbed silver and presumably the hydrated layer, and replaced some (perhaps half) of the surface hydroxyl groups with fluoride ions, giving SiF groups (see discussion of HF adsorption in the preceding section). After this treatment no appreciable amount of silver was adsorbed from the sodium nitrate bath containing radioactive silver, showing that the adsorption sites were removed or blocked by the HF treatment.

More careful study of adsorption from solution seems warranted, in view of its importance in container problems.

SURFACE CONDUCTIVITY

The surface resistivity of clean glass in dry air is very high: 10^{14} ohms/square or higher at room temperature.[63] Geddes found no appreciable surface leakage of freshly broken glass surfaces in clean dry air.[64] Since these results are at the limit of reliable resistivity measurements, the inherent resistivity of a dry glass surface is uncertain.

The surface resistivities of most glasses decrease rapidly as the relative humidity at room temperature increases above about 40%.[63–65] (See also Ref. 30, pp. 461 ff., and Ref. 8.) However, the surface resistivity of fused silica, cleaned in chromic acid and then in distilled water, remains high even at quite high relative humidity.[63,65]

These experimental observations can be explained in the following way. The inherent surface resistivity of a glass is very high. If there is any ionic contamination on the surface it can conduct current in humid air, since physically adsorbed water reacts with these contaminants to provide mobile ions. Glasses containing monovalent cations can react with water by ion exchange (reaction 1) forming metallic hydroxides on their surfaces. These hydroxides react further with water, forming mobile ions on the surface. In some cases the hydroxides absorb so much water that a liquid solution is formed on the glass surface, giving a highly conductive surface film. This possibility was first recognized by Faraday in his remarkable study of the electrical properties of glass.[66]

If these explanations of surface conductivity of glass are correct, this conductivity either results from chance contamination or from ion exchange in the glass. The rate of the latter process is controlled by diffusion of the exchanging ions in the glass, so the surface conductivity of glass should be a function of the time the glass is exposed to water. This was found to be the case in the experiments of Pike and Hubbard.[67] Furthermore, the more rapid the exchange proceeds the greater the increase in surface conductivity should be. This exchange process of reaction 1 is also responsible for weathering and for dissolution of glasses in water, as described in the following chapter. Thus one would expect a relationship between surface conductivities and chemical durability of glasses, which was also found by Pike and Hubbard.

The surface conductivity with alternating current behaves very similarly to the dc conductivities.[68]

Iizima found that the surface conductivity of electronically conducting arsenic-tellurium-germanium glass increased from $5(10)^{-6}$ to 10^{-3} Ω/square after 90 min in boiling water, and to $3(10)^{-4}$ Ω/square in wet air.[69] He claimed that the conducting species were not ions, but in view of the results above this conclusion is questionable.

In view of the mechanisms of surface conductivity on glass described above, reliable quantitative results of this phenomenon should be difficult to obtain. Chance contamination, time dependence, and removal of reaction products from the surface are all difficult to avoid.

REFERENCES

1. G. W. Morey, *The Properties of Glass*, Reinhold, New York, 1954.
2. H. Scholze, *Glas*, Vieweg and Sohn, Braunschweig, Germany, 1965.
3. L. Holland, *The Properties of Glass Surfaces*, Chapman and Hall, London, 1964.
4. M. L. Deribere-Desgardes and M. M. Bre, "Connaissances sur L'etat de Surface des Verres," in *La Surface du Verre et ses Traitments Modernes*, Union Scientifique Continentale du Verre, Charleroi, Belgium, 1967, pp. F5 and E.5 (French and English translations).

4a. F. M. Ernsberger in *Annual Review of Materials Science*, **2**, 529 (1972).

5. R. S. McDonald, *J. Phys. Chem.*, **62**, 1168 (1958).
6. G. J. Young, *J. Colloid Sci.*, **13**, 67 (1958).
7. V. Ya. Davydov, A. V. Kiselev, and L. T. Zhuravlev, *Trans. Far. Soc.*, **60**, 2254 (1964), and earlier articles by these authors referred to in this paper.
8. M. L. Hair, *Infrared Spectroscopy in Surface Chemistry*, Marcel Dekker, New York, 1967.

8a. L. H. Little, *Infrared Spectra of Adsorbed Species*, Academic, London, 1966.

9. R. H. Doremus, *J. Phys. Chem.*, **75**, 3147 (1971).
10. F. H. Hambleton, J. A. Hockey, and J. A. G. Taylor, *Trans. Far. Soc.*, **62**, 801 (1966).
11. A. J. Tyler, F. H. Hambleton, and J. A. Hockey, *J. Catal.* **13**, 35 (1969).

12. V. R. Deitz and N. H. Turner, *J. Phys. Chem.*, **74**, 3823 (1970).
13. C. G. Armistead, A. J. Tyler, F. H. Hambleton, S. A. Mitchell, and J. A. Hockey, *J. Phys. Chem.*, **73**, 3947 (1969).
14. J. B. Peri and A. L. Hensley, *J. Phys. Chem.*, **72**, 2926 (1968).
15. J. A. Hockey and B. A. Pethica, *Trans. Far. Soc.*, **57**, 2247 (1961).
16. J. A. DeBoer, M. E. A. Hermans, and J. M. Vleeskens, *Proc. Koninikl, Ned. Akad. Wetenschap.*, **B60**, 44 (1957).
17. M. L. Hair and W. Hertl, *J. Phys. Chem.*, **73**, 2372 (1969).
18. V. M. Bermudez, *J. Phys. Chem.*, **75**, 3249 (1971).
19. M. Navez and C. Sella, *C.R.*, **250**, 4325 (1960); **251**, 529 (1960); **254**, 240 (1962); **257**, 4183 (1963).
20. J. J. Antal and A. H. Weber, *Phys. Rev.*, **89**, 900 (1953).
21. M. J. D. Low and N. Ramasubramanin, *J. Phys. Chem.*, **70**, 2740 (1966); **71**, 730, 3077 (1967).
22. R. S. McDonald, *J. Am. Chem. Soc.*, **79**, 850 (1957).
22a. J. A. Cusumano and M. J. D. Low, *J. Phys. Chem.*, **74**, 792, 1950 (1970).
22b. P. A. Sewell and A. M. Morgan, *J. Am. Ceram. Soc.*, **52**, 136 (1969).
23. M. R. Basilo, *J. Chem. Phys.*, **35**, 1151 (1961).
24. T. H. Elmer, J. D. Chapman, and M. E. Nordberg, *J. Phys. Chem.*, **67**, 2219 (1963).
25. G. J. C. Frohnsdorff and G. L. Kinton, *Trans. Far. Soc.*, **55**, 1173 (1959).
26. A. V. Kiselev, *Surface Sci.*, **3**, 292 (1965), and earlier articles referred to in this paper.
27. E. A. Flood, Ed., *The Solid-Gas Interface*, M. Dekker, New York, 1966.
28. D. L. Kanta, S. Brunauer, and L. E. Copland, in Ref. 27, pp. 413 ff.
29. D. M. Young and A. D. Crowell, *Physical Adsorption of Gases*, Butterworths, London, 1962.
30. A. Adamson, *The Physical Chemistry of Surfaces*, Wiley-Interscience, New York, 1967.
31. I. Langmuir, *J. Am. Chem. Soc.*, **40**, 1361 (1918).
32. H. Zeise, *Z. Phys. Chem.*, **136**, 385 (1928).
33. J. P. Hobson, *Can. J. Phys.*, **37**, 300, 1105 (1959).
34. J. P. Hobson, *J. Chem. Phys.*, **34**, 1850 (1961).
35. J. P. Hobson and R. A. Armstrong, *J. Phys. Chem.*, **67**, 2000 (1963).
36. R. Haul and B. A. Gottwald, *Surface Sci.*, **4**, 321, 334 (1966).
37. B. Kindl, E. Negri, and G. F. Cerofolini, *Surface Sci.*, **23**, 299 (1970).
38. F. Ricca, R. Medona, and A. Bellardo, *Z. Phys. Chem. NF*, **52**, 291 (1967).
39. N. Endow and R. A. Pasternak, *J. Vac. Sci. Tech.*, **3**, 196 (1966).
40. J. B. Peri, *J. Phys. Chem.*, **70**, 2937 (1966).
41. N. W. Cant and L. H. Little, *Can. J. Chem.*, **43**, 1252 (1965).
42. J. D. Chapman and M. L. Hair, *J. Catal.*, **2**, 145 (1963); *Trans. Far. Soc.*, **61**, 1507 (1965).
43. A. W. Smith and J. M. Quets, *J. Catal.*, **4**, 163 (1963).
44. S. S. Jorgensen and A. T. Jensen, *J. Phys. Chem.*, **71**, 745 (1967).
45. M. Folman, *Trans. Far. Soc.*, **57**, 2000 (1961).
46. M. Shimiza and M. J. D. Low, *J. Am. Ceram. Soc.*, **54**, 271 (1971).
47. B. A. Morrow and A. Devi, *Trans. Far. Soc. I*, **68**, 403 (1972).
48. W. Hertl and M. L. Hair, *J. Phys. Chem.*, **75**, 2181 (1971).
49. R. J. Peglar, F. H. Hambleton, and J. A. Hockey, *J. Catal.*, **20**, 309 (1971).
50. C. Morterra and M. J. D. Low, *J. Phys. Chem.*, **73**, 321, 327 (1969); **74**, 1297 (1970).
51. P. G. Ashmore, A. J. Parker, and P. E. Stearne, *Trans. Far. Soc.*, **67**, 3081 (1971).
52. P. B. Adams, *New Scientist*, **41**, 25 (1969).
53. M. L. Hair and W. Hertl, *J. Phys. Chem.*, **74**, 91 (1970).
54. J. F. Antonini, in Ref. 4, p. 69.
55. G. Hochstrasser, in Ref. 4, p. 79.

56. J. F. Antonini, G. Hochstrasser, and P. Acloque, *Verres Refrac.*, **23**, 169 (1969).
57. F. Kepak, *Chem. Rev.*, **71**, 357 (1971).
58. P. Benes, J. Smetana, and V. Majer, *Collect. Czech. Chem. Commun.*, **33**, 3410 (1968).
59. P. Benes, *Collect. Czech. Chem. Commun.*, **35**, 1349 (1970).
60. H. T. Tien, *J. Phys. Chem.*, **69**, 350 (1965).
61. T. F. Tadros and J. Lyklema, *J. Electroanal. Chem.*, **17**, 267 (1968).
62. R. W. Dalton, J. L. McClanahan, and R. W. Mautmun, *J. Colloid Sci.*, **17**, 207 (1962).
63. H. L. Curtis, *Bull. Bur. Stand.*, **11**, 359 (1915).
64. S. Geddes, *J. Roy. Tech. Coll.* (*Glasgow*) **3**, 551 (1936).
65. A. Ya. Kouznetzov, *J. Chem. Phys. USSR*, **27**, 657 (1953).
66. M. Faraday, *Phil. Trans.*, Part 1, 49 (1830); *Experimental Researches in Electricity*, J. M. Dent, London, 1914, pp. 38 ff.
67. R. G. Pike and D. Hubbard, *J. Res. Nat. Bur. Stand.*, **59**, 127 (1957).
68. W. A. Yager and S. O. Morgan, *J. Phys. Chem.*, **35**, 2026, 2040 (1931).
69. S. Iizima, *Solid State Comm.*, **9**, 795 (1971).

13

Chemical Reactions

The reactions considered in this chapter involve the bulk of the glass, rather than just surface molecules, although some of the processes, such as dissolution, do take place in a thin layer near the glass surface. These reactions are of great importance in glass manufacture, for example, in dissolution of component oxides and in fining, as well as in properties such as chemical durability, oxidation state, and sealing.

In this chapter the following subjects are discussed: reactions with gases, fining, reactions in the melt, sealing, reactions in aqueous solution, and weathering.

REACTIONS WITH GASES

Certain gases such as water and hydrogen can react with the silicon-oxygen network of silicate glasses, breaking it up and consequently modifying such properties as strength and viscosity. Hydrogen and oxygen in glass determine its oxidation state and can reduce or oxidize ions in the glass. All these gases dissolve and diffuse in silicate glasses as molecules, as described in Chapter 8, and the rates of their reaction with the glass are determined by their diffusion coefficients in it.

Water

Water reacts with the silicon-oxygen bond as follows:[1,2]

$$\begin{array}{c} | \quad\;\; | \\ -\mathrm{Si}-\mathrm{O}-\mathrm{Si}- \\ | \quad\;\; | \end{array} + H_2O = \begin{array}{c} | \\ -\mathrm{Si}-\mathrm{OH} \\ | \end{array} \;\; \begin{array}{c} | \\ \mathrm{HO}-\mathrm{Si}- \\ | \end{array} \tag{1}$$

forming pairs of adjacent silanol groups. These groups and the hydroxyl ions so formed are very immobile, even at temperatures as high as 1000°C. (See

Chapter 8 and Refs. 3 and 4.) Thus water molecules must diffuse in and out of the silicate lattice to form or remove these hydroxyl groups.

The solubility of water in fused silica (Vitreosil and O.G. grades, Thermal Syndicate, Walland, England) was determined by Roberts and co-workers from the height of the infrared absorption peak resulting from Si-OH groups[5] and from tracer diffusion studies.[6] This solubility is proportional to the number of silanol groups formed by reaction per unit volume when the glass is in equilibrium with water vapor of a fixed pressure. Below about 1000°C the solubility depended on the thermal history of the glass.[7] For a sample heated at 1100°C the solubility was nearly constant from 1200° to 700°C[7] at about $3(10)^{-3}$ SiOH groups formed per SiO_2 group in the glass, with a water pressure of 700 mm Hg. The solubility is not a function of impurity content in the fused silica, since Heterington and Jack found the same solubility for a very pure silica (Spectrosil) and one containing more impurities, for example, about 60 ppm aluminum and 4 ppm sodium.[8] The solubility is proportional to the square root of the water vapor pressure,[5,8,9] as would be expected from reaction 1.

Equilibrium between the SiOH groups and dissolved water molecules gives the above solubilities. The insensitivity to impurities and the square root dependence on pressure suggest that there are a number of equivalent sites for reaction by reaction 1 in the silica; in fact, probably most of the Si—O—Si bonds in the glass are about equally susceptible to reaction 1.

The solubility of water in molten alkali silicates increases linearly with the alkali concentration,[9] and also in the order Li < Na < K silicates.[10] There is a slight increase in solubility with increasing temperature.[10,10a] In molten B_2O_3 there is a decrease in solubility of water with increasing temperature.[11]

In fused silica the infrared spectrum of the silanol groups formed by reaction with water shows no indication of hydrogen bonding,[1] but in the alkali silicates the number of hydrogen-bonded OH groups increases as the amount of alkali increases and also in the order Li < Na < K.[9] The number of hydrogen-bonded hydroxyls decreases as aluminum oxide is added to the glass melt.[9] Apparently the OH groups can form a hydrogen bond to nonbridging oxygen anions that have an associated alkali cation. As the bonding of the oxygen ion to the alkali ion becomes stronger (the smaller the alkali ion the stronger the bond) the tendency to hydrogen bonding is reduced. Addition of aluminum removes nonbridging oxygen ions, since each aluminum oxide tetrahedron has an associated alkali ion. The increased hydrogen bonding is perhaps at least partly responsible for the increased solubility in alkali silicates, and for the order of solubility among the various alkali silicates.

An additional factor in the water solubility is probably the reactivity of the silicon-oxygen bond. As alkali oxide is added to the silica structure this bond is weakened, as shown by the greatly decreased viscosity resulting from such addition. The activation energy for viscous flow in fused silica in the temperature range 1100 to 1400°C is about 170 kcal/mole, whereas for binary sodium silicate glasses it is about 100 kcal/mole in the transition region (450 to 600°C). Thus the greater reactivity of the bond pushes reaction 1 to the right, giving greater water solubility as the alkali ion concentration increases. Other structural interpretations of these results on water solubility have been given.[9,12]

Hydrogen

Hydrogen also reacts with silica to form hydroxyl groups.[13,14] In fused silica containing 50 to 100 ppm of aluminum the reaction becomes perceptible above about 700°C.[14] In purer silica it does not occur appreciably at these temperatures,[14] but at higher temperatures or pressures hydrogen forms hydroxyl groups from Si—O—Si bonds. The reaction can be written as follows:

$$-\overset{|}{\underset{|}{\mathrm{Si}}}-\mathrm{O}-\overset{|}{\underset{|}{\mathrm{Si}}}- + \mathrm{H_2} = -\overset{|}{\underset{|}{\mathrm{Si}}}\mathrm{OH} + \mathrm{H}-\overset{|}{\underset{|}{\mathrm{Si}}}-$$

The presence of the SiH groups has been established from the growth of the SiH stretching band at 2250 cm^{-1} (4.45 μ) in vitreous silica containing hydrogen.[15–17] In this case the rate of introduction or removal of OH groups is controlled by the diffusion of hydrogen molecules,[13] which is much faster than the diffusion of the larger water molecules (see Chapter 8).

The rate of exchange of deuterium with SiOH groups in glass was studied by Lee.[14] He found a first-order reaction for silica with low impurities in which all the hydroxyl groups were attached to silicon atoms. In a fused silica with about 100 ppm of aluminum, he found that the first-order reaction coefficient decreased with time, probably because both AlOH and SiOH groups were present in this glass, and the exchange rate was different for these two groups.

These results show that there can be at least three different types of hydroxyl groups in a silicate glass.[25]

1. An aluminum hydroxyl group can be thought of as having the structure

$$\begin{array}{c} -\mathrm{O}\ \mathrm{R}^+ \\ | \\ \mathrm{O}-\mathrm{Al}-\mathrm{OH} \\ | \\ \mathrm{O} \end{array}$$

where R^+ is either an alkali ion or another hydrogen ion, more or less associated with one of the oxygens bound to the aluminum. These groups result from the reaction of hydrogen with the Al—O—Si bond and are called "loosely bound" by Lee.[14]

2. Hydroxyl groups in pairs are formed during melting or annealing of the glass by reaction with water:

$$\begin{array}{cc} | & | \\ -\text{SiOH} & \text{HOSi}- \\ | & | \end{array}$$

These groups are introduced or removed by diffusion of water, which is slow compared to hydrogen diffusion.

3. Hydrogen ions are introduced by ion exchange with alkali ions, either by interion diffusion or by electrolysis. These groups cannot be removed by gaseous diffusion because they are needed for electroneutrality.

Hydrogen can also reduce ions in various oxide glasses to the atomic state, for example, gold,[18] silver,[19,20,23] lead,[20,21,22] bismuth,[20,21,22] and antimony.[20,21,22] At lower temperatures these reactions are often limited to layers of the glass near its surface because of the slow diffusion of hydrogen in the glass at these temperatures. In addition, hydrogen can reduce the valence of an ion in glass without reducing it completely to the atomic state. In this way dissolved hydrogen affects the oxidation state of the glass. A study of fluorescence of terbium and europium in glass demonstrated the reduction of these ions from 3+ to 2+ valence;[24] the rate of the reduction was controlled by diffusion of molecular hydrogen.

Oxygen

Oxygen dissolved in glass determines the "state of oxidation" of the glass if one or no ions with different valence states are present. As an example the reaction of oxygen with iron ions in a silicate glass can be written as

$$2\{2(-\overset{|}{\underset{|}{\text{Si}}}-\text{O}^-)\,\text{Fe}^{2+}\} + -\overset{|}{\underset{|}{\text{Si}}}-\text{O}-\overset{|}{\underset{|}{\text{Si}}}- + \tfrac{1}{2}\text{O}_2 = 2\{3(-\overset{|}{\underset{|}{\text{Si}}}-\text{O}^-)\,\text{Fe}^{3+}\} \quad (2)$$

The reaction is written in this rather cumbersome way to emphasize that the oxidation results in the formation of an additional anionic $Si—O^-$ or nonbridging oxygen group. The actual oxygen coordination number of iron is probably four or six, so more bridging oxygen ions could be included on either side of the reaction. The equilibrium constant for reaction 2 can be written as

$$K = \frac{a_3\, a_n^{1/2}}{a_2\, f_0^{1/4} a_b} \quad (3)$$

where a_3 and a_2 are thermodynamic activities of iron in the 3+ and 2+ state, respectively, a_n and a_b are the activities of nonbridging and bridging oxygen ions, and f_0 is the fugacity of gaseous oxygen in equilibrium with the glass. In most cases activities are hard to calculate, and they are replaced by concentrations, with the result that K may depend on concentration.

A number of authors have written reaction 2 in terms of free oxygen ions in the glass rather than in terms of bridging and nonbridging oxygen ions bonded to silicon ions.[10,26] The choice of one method or the other is a matter of taste; I have chosen reaction 2 because the actual concentration of oxygen ions in silicate glasses with more than 67% SiO_2 is very small and cannot be measured directly. It is always possible to express equilibrium constants in terms of minor components; Eq. 3 has the advantage of being expressed in terms of groups present at substantial concentrations.

If the gas is ideal, the equilibrium constant should be related to a power of the partial pressure of oxygen, this power being $n/4$, where n is the difference between valence states. Thus the Cr^{3+}—Cr^{6+} oxidation-reduction reaction was related to the 3/4 power of p_{O_2},[28] and the Fe^{2+}—Fe^{3+} and Ce^{3+}—Ce^{4+} equilibria depended upon $p_{O_2}{}^{1/4}$.[29,30]

Douglas and co-workers studied the Cr^{3+}—Cr^{6+}, Fe^{2+}—Fe^{3+}, and Ce^{3+}—Ce^{4+} equilibria in binary lithium, sodium, and potassium silicate glasses and reviewed earlier work on these systems in various glasses.[31,32] They found that oxidation was more favorable as the alkali content of the glass increased, and as the ionic size increased in the order of lithium, sodium, and potassium.

From reaction 2 these composition effects can be interpreted as resulting at least in part from changes in the reactivity of the silicon-oxygen bonds, just as for water solubility as described above. As alkali is added to the glass the silicon-oxygen bonds are weakened, increasing the tendency to oxidation.

The order of ability to oxidize with size of alkali ion can perhaps be interpreted in terms of the strength of bonding between the ion and the nonbridging oxygen ions. As the alkali ion size increases the oxygen-alkali bond decreases in strength, so the nonbridging oxygens are more "available" for coordination with the multivalent ions, and thus the ion with the higher charge is favored. This speculative explanation needs further support for confident acceptance. Douglas et al.[27] and Franz[41] have interpreted results on oxidation of ions in glass in terms of the oxygen ion activities.

In discussing the oxidation state of a glass and other chemical properties the concept of acidic or basic constituents is often used. The glass-forming oxides such as silica and boric oxide are considered to be acidic, and the alkali and alkaline earth oxides basic. This designation can be justified in the "Lewis acid" sense from the negative charge on the glass-forming groups (SiO^-) that results from the reaction with a metal oxide, making the silica an electron acceptor and the metallic cation an electron donor. Also when the glass-forming oxides are dissolved in water they form very weak acids, whereas the alkali and alkaline-earth oxides form strong bases. For oxides with higher valent cations, such as those of transition and refractory metals (TiO_2, Ta_2O_5, Nb_2O_5) the acid-base designation is not so clear. In acid-base terms the increase in ability to oxidize with increasing alkali concentration is considered to result from the increased basicity of the glass. A reaction such as

$$2Na_2O + Cr_2O_3 + \tfrac{3}{2}O_2 = 2Na_2CrO_4$$

can be considered to result in the formation of chromate ion in the glass, and is enhanced by increasing the alkali concentration. It is not clear how a group such as Na_2CrO_4 is incorporated into the network structure of the glass.

When more than one ion with two valence states is present in appreciable quantity in the glass, its oxidation state is determined by the equilibrium reaction between these two ions.[33] The relative ability to oxidize various ions in glass was calculated by Tress from thermodynamic data of the free oxides.[34] He found the order given in Table 1. Thus, of the pairs shown, chromium is the most easily reduced, and tin is the most easily oxidized. An ion *A* above ion *B* tends to oxidize *B*, being reduced itself.

Table 1 Oxidizing Tendency of Ions in Glass

Reaction
$CrO_3 = Cr_2O_3$
$Mn_2O_3 = MnO$
$CeO_2 = Ce_2O_3$
$As_2O_5 = As_2O_3$
$Sb_2O_5 = Sb_2O_3$
$Fe_2O_3 = FeO$
$SnO_2 = SnO$

Paul and Douglas studied the equilibrium amounts of various ion pairs in different states of oxidation and found ratios consistent with the reaction proposed.[33] For example, for chromium and arsenic

$$2Cr^{6+} + 2As^{3+} = 2Cr^{3+} + 3As^{5+} \qquad (4)$$

so that

$$K = \left(\frac{Cr^{6+}}{Cr^{3+}}\right)^2 \left(\frac{As^{3+}}{As^{5+}}\right)^3 \qquad (5)$$

This relation was confirmed experimentally. For reaction 4 to occur the ions must transfer electrons between themselves. The diffusion coefficients of multivalent ions in glass are probably in the range 10^{-7} to 10^{-6} cm^2/sec at 1400°C, the temperature of these experiments.[35,36] The samples were held at this temperature for many hours to attain equilibrium, so there should have been enough "collisions" of the ions to reach equilibrium according to reaction 4. At lower temperatures the ionic diffusion coefficients decrease sharply, whereas the diffusion of oxygen molecules should change more slowly. Thus it is possible that at lower temperatures the electron transfer takes place via oxygen diffusion rather than directly.

Nitrogen

Nitrogen can react with glass melted under reducing conditions, forming three Si—N bonds.[37,38] In the presence of water or ammonia, N—H bonds are formed, as demonstrated by the 3-μ infrared absorption band. The solubility of nitrogen increases with temperature, but decreases as the amount of alkali in the glass increases.[39,40] From an oxidation-reduction point of view this result is consistent with the finding of more reducing glasses as the silica content increases, and perhaps weakens the above explanations in terms of silicon-oxygen bond strengths.

Carbon Dioxide

The solubility of carbon dioxide in sodium silicate glasses was found to increase rapidly below about 1100 or 1200°C, depending on the composition.[42,43] Above these temperatures the solubility did not change as much with temperature, and for several glasses was about 10^{-4} wt. %/atm, which is roughly the amount of physical solubility expected. The logarithm of the solubility in wt. % was linear with reciprocal temperature, so the enthalpy of solution ΔH can be calculated from the solubilities S at two different temperatures T:

$$\Delta H = -\frac{R \ln (S_1/S_2)}{(1/T_1 - 1/T_2)} \tag{6}$$

where R is the gas constant. From the data of Pearce on binary alkali silicates ΔH is about -56 kcal/mole.

One possible way to write the reaction of carbon dioxide with a sodium silicate glass is

$$CO_2 + 2(Na^+ \ \bar{O}\overset{|}{\underset{|}{Si}}-) = -\overset{|}{\underset{|}{Si}}-O-\overset{|}{\underset{|}{Si}}- + Na_2CO_3 \tag{7}$$

The entity formed in the glass is probably not a simple sodium carbonate molecule, but some more complex combination of these ions with the silicate lattice. Strnad[44] found that the solubility of carbon dioxide in a 30.3 mole % Na_2O, 69.7% SiO_2 melt was proportional to pressure below atmospheric pressure and at 1000°C was consistent with reaction 7. Both Pearce and Strnad found a large increase in solubility of carbon dioxide as the amount of Na_2O in the melt increased above 25%. The increase for melts above 35% Na_2O was much greater than proportional to the square of the Na_2O composition, as would be expected from reaction 7. Apparently in melts of higher Na_2O, the Na_2O activity increases sharply. This increase can also be deduced from the sharp increase in the volatility of soda in these melts.[45]

Pearce also measured the reaction of carbon dioxide with sodium borate melts and found a similar increase in solubility as the temperature decreased; however, the enthalpy of solution was considerably lower than for the silicates.[46]

Sulfur Dioxide

Sulfur dioxide also dissolves in silicate melts. Its reaction to an alkali sulfate is complicated by the need for oxygen to form the sulfate from sulfur dioxide. Thus its solubility in alkali silicates increases as the alkali concentration increases,[42,47–49] and decreases as the temperature increases; however, under reducing conditions the solubility is much reduced. The reaction with alkali silicate can be written as

$$SO_2 + \tfrac{1}{2}O_2 + 2(Na^+\ \bar{O}\overset{|}{\underset{|}{Si}}-) = -\overset{|}{\underset{|}{Si}}-O-\overset{|}{\underset{|}{Si}}- + Na_2SO_4 \qquad (8)$$

It has long been known that furnace gases are beneficial for the weathering properties of glass. These effects arise from the reaction of sulfur dioxide in the gases with the glass surface. Water and oxygen, as well as sulfur dioxide, must be present, and the overall reaction can be written as

$$2Na^+\ (\text{glass}) + SO_2 + \tfrac{1}{2}O_2 + H_2O = 2H^+\ (\text{glass}) + Na_2SO_4$$

This "dealkalization" reaction is an ion exchange that is made possible by the removal of the sodium ion as sodium sulfate, as confirmed by Douglas and Isard.[50] They found that the rate of formation of sodium sulfate was proportional to the square root of time, and was controlled by the interdiffusion of hydrogen and sodium ions in the glass. The rate of weathering of a glass with a dealkalized layer is less because weathering requires ion exchange with sodium ions.

FINING

Fining, or the removal of bubbles from a glass melt, is one of the major technological problems in glass melting. It is usually solved by holding the glass for some time at a temperature somewhat below the highest melting temperature, and by adding certain minor constituents to the original glass batch. The mechanisms by which these additions aid bubble removal are still somewhat uncertain, although much progress in understanding has been made in recent years.

Bubbles can be removed from a melt by either of two ways. They can rise to the surface or the gas in them can dissolve in the glass. The rate of rise is given by the following equation:

$$\frac{dh}{dt} = \frac{2\rho g R^2}{9\eta} \tag{9}$$

where ρ is the density of the glass, g is the gravitational constant, R is the bubble radius, and η is the viscosity of the glass. For a viscosity of 100 P, typical for melting temperatures, the rate of rise of bubbles 0.1 mm in diameter is about 10 cm/day, which is too small to eliminate them from a normal glass furnace. Thus small bubbles can be removed from glass melts only by dissolution of their gas into the glass melt, although larger bubbles can rise to the surface.

Arsenic oxide is a common fining agent added to glass to help remove bubbles. For many years it was thought that the arsenic released oxygen at glass melting temperatures, which "swept out" the bubbles in the glass. However, the calculation above shows that such a mechanism would not eliminate small bubbles, and the elegant experiments of Greene and co-workers showed that arsenic enhances dissolution of oxygen bubbles in glass.[51–53] Thus the importance of arsenic and antimony oxide additions to the glass is to aid in removal of fine bubbles, rather than to generate more gas.

The mechanism by which these agents enhance dissolution of gas from fine bubbles is still being debated. I suggested that the role of arsenic is to reduce the oxygen concentration in the glass.[54] At low temperatures arsenic oxide contains mostly pentavalent arsenic, but at higher temperature the As_2O_5 decomposes to As_2O_3 and oxygen. Thus a considerable portion of arsenic ion in glass is in the trivalent state, as has been confirmed by chemical analysis.[55,56] This trivalent arsenic can react with oxygen physically dissolved in the glass, reducing its concentration and increasing the rate of oxygen diffusion into the glass. This hypothesis is confirmed by an analysis of the results of Greene and co-workers on the rate of shrinking of oxygen bubbles in different glasses. As the amount of arsenic in the glass increased, the amount of oxygen initially dissolved in the glass was reduced, and the rate of shrinking of the bubbles increased.[54] Thus in the case of oxygen bubbles the role of arsenic and antimony fining agents is to reduce the amount of oxygen dissolved in the glass.

In actual glass-melting practice carbonates are the usual source of sodium and calcium oxides. Thus one would expect to find carbon dioxide in bubbles in glass as well as oxygen and nitrogen from air. A number of analyses of bubbles in glasses made from carbonates showed that these three are indeed the main gases found in the bubbles.[57,58] The value of arsenic for dissolving oxygen from the bubbles is clear from the remarks above, but the elimination of the other two gases requires further consideration.

Cable et al. showed that the addition of arsenic to a glass batch increases the rate at which carbon dioxide diffuses out of bubbles at 1200°C in a soda-lime (73.5 wt. % SiO_2, 16.5% Na_2O, 10.0% CaO) glass.[59] The solubility of carbon dioxide in this base glass is probably small, judging from results cited in the preceding section that show a decrease in solubility as the soda concentration decreases. Since it was shown by Greene's work that the arsenic in the glass lowered the concentration of oxygen in the glass, it is reasonable to relate this lowered oxidation state to the enhanced solubility of carbon dioxide. The decomposition of carbon dioxide to oxygen and carbon monoxide can be written as

$$CO_2 = O_2 + CO \tag{10}$$

A decrease in the oxygen concentration or oxidation state of the glass would favor this decomposition. The smaller carbon monoxide molecule should diffuse more rapidly in the glass than carbon dioxide, so decomposing dissolved carbon dioxide should increase its rate of absorption from bubbles. This hypothesis for the increase in rate of absorption of carbon dioxide by arsenic is speculative and needs confirmation with further experiments.

Fining by arsenic and antimony is improved if some of the alkali carbonates in the glass batch are replaced by nitrates. To explain this result it has been assumed that trivalent arsenic or antimony oxides react with the nitrate to form pentavalent oxides, which then release their oxygen at the glass-melting temperature to form larger bubbles. As mentioned above this mechanism is unlikely; furthermore, this reaction of these oxides with nitrates under the conditions of glass manufacture is improbable. Potassium and sodium nitrates decompose completely to the oxides between about 400 and 800°C, whereas in this same temperature range the arsenic and antimony oxides are in the pentavalent states as the result of reaction with oxygen.[60,61] It is only above 900°C that the pentavalent compounds begin to decompose. Thus at no temperature above 500°C are the nitrates and trivalent oxides in the glass batch together.

The role of the added nitrates is probably to increase the amount of oxygen in the bubbles at the expense of nitrogen. The nitrates decompose before the glass becomes molten,

$$NaNO_3 \rightarrow NO + \tfrac{1}{2}O_2 + Na_2O$$

increasing the proportion of oxygen in the mixed batch materials and consequently in the bubbles of entrapped gas. Therefore it is suggested that arsenic and antimony increase the rate of fining in glass by decreasing the concentration of oxygen dissolved in the glass, which leads to increased rates of absorption of oxygen and carbon dioxide. Nitrates in the glass batch improve fining rates by replacing nitrogen in entrapped gas by oxygen.

Sulfates are often used as fining agents. Lyle showed that in soda-lime glasses sulfate additions could either speed or retard fining, depending on the composition of the glass.[62] He found that when the ratio of soda to silica was greater than $0.45 + 0.20/W_s$, where W_s is the weight fraction of silica, fining was retarded, whereas for lower ratios it was enhanced, at least for silica concentrations from 69 to 76%. A very similar effect of soda-to-silica ratio was found on the color produced by carbon and sulfur in soda-lime silicate glasses. For production of a stable amber color the soda-to-silica ratio had to be less than $0.5 + 0.22/W_s$; with a greater ratio the amber color was not stable and the glasses became blue-green.[63] Furthermore, it was found that bubbles in stable amber glasses were removed quickly, whereas the blue-green glasses foamed for a long time at 1475°C. These effects of composition are mainly related to the oxidation state of the glass.

Effects of composition on the oxidation state of glasses containing sulfur have been studied in detail in connection with the stability of the amber color.[64] This color is stable in equilibrium with a partial pressure of oxygen between about 10^{-10} and 10^{-8} atm.[65] The coloring group is ferric iron in tetrahedral coordination with one of the surrounding oxygen ions replaced by sulfur.[66] As the concentration of soda in a soda-lime glass was increased, the glass became more oxidizing, and the optical absorption by the ferric sulfide group decreased, indicating a decreased concentration of these groups.[64]

The presence of sulfide in the more reducing glasses means it can react with oxygen and carbon dioxide from bubbles and therefore enhance their removal from the bubbles.[66a] Harding and Ryder found that an increase in the sulfate level in a soda-lime glass reduced the retention of nitrogen.[67] However, when the sulfate was replaced by sulfide (more reducing conditions) nitrogen retention increased, again showing the effect of state of oxidation of the glass on reaction of nitrogen with it, and indicating that sulfide also helps to remove nitrogen from bubbles.

The published reports of a symposium on fining describe some interesting studies.[68] Swarts has given a brief review of fining, including some discussion of the solubility of nitrogen from bubbles.[40]

REACTION IN THE LIQUID PHASE

Potts carried out a series of experiments on the rate of "melting" in glass batches, and found that for soda-lime glasses the rate of melting increased as the silica content decreased, soda being more effective than lime in increasing the melting rate.[69] These results are just what one would expect from the effects of these constituents on the viscosities of these glasses.

The rate of dissolution of silica in alkali silicate melts probably controls the overall rate of glass melting, and dissolution is important in the attack of glass melts on refractories. The rate of dissolution of fused silica in sodium silicate,[70–72] potassium silicate,[70] sodium calcium silicate, and sodium carbonate[73] was shown to be controlled by diffusion in the melt. The rate of dissolution in the alkali silicates increased as the amount of silica in the melt decreased, as would be expected from the accompanying decrease in viscosity. The activation energy for dissolution of fused silica in molten sodium silicate was about 30 kcal/mole between 900 and 1250°C and 39 kcal/mole between 1200 and 1400°C in soda-lime silicate melts. These values are lower than the activation energies for viscous flow in these melts. Truchlarova and Vesprek concluded that the dissolution of fused silica proceeded in two stages: first, a transformation to cristobalite and then dissolution of the cristobalite.

Cooper and co-workers also found that the rate of dissolution of alumina, anorthite, mullite, and fused silica in a calcium aluminum silicate melt was controlled by transport in the melt,[74,75] as was solution of quartz in sodium silicate.[76]

SEALING

The two most important requirements for obtaining a good seal are that the materials being sealed form a chemical bond and that high stresses in the seal are avoided. These two requirements are discussed in turn. In *Glass-to-Metal Seals* Partridge[77] discusses the technology of these seals; the choice of materials, their preparation, different types of seals, equipment for making seals, and strain in seals. Andrews treated sealing of porcelain enamels to iron and steel,[78] and Pask reviewed his work on bonding at glass-metal interfaces.[79]

In sealing one glass to another glass the requirement of chemical bonding is met by heating the glasses until their viscosities are low enough so that they flow together and mutually dissolve. These two processes ensure a large surface area of contact and good chemical bonding. If sealing at a lower temperature is desired it is sometimes possible to use a low-melting glass to effect a union between two high-melting ones.

Experimentally it has been found that a metal must have an oxide coating before it can be sealed to a glass. The amount of oxide needed depends on the metal. If the glass dissolves all the oxide the seal becomes weak, but if the oxide coating is too thick it may be weak because of its poor adherence to the metal.

The mechanism of bonding between a glass and a metal has been discussed extensively.[77–79] My viewpoint is that a continuous series of covalent bonds between metal and glass are needed for a good bond. Since the solubility of metals in glass is very low, this requirement cannot be met by a bare metal in contact with glass. The metal must have an oxide coating which is strongly bonded to the metal. This oxide then partially dissolves in the glass, giving a region of continuously decreasing oxide concentration from the oxide of the metal being sealed into the glass.[80]

If a glass is saturated with the oxide of a metal it can be readily sealed to that metal[81] because the glass cannot dissolve more oxide. When the glass is liquid the glass-oxide surfaces react to form covalent bonds and the oxide remains tightly bonded to the metal surface, so a continuous series of covalent bonds from metal to glass is established.

"Adherence oxides" are often added to porcelain enamels to aid in their adherence to iron.[82] These oxides contain metal ions such as cobalt or nickel which have a lower oxidation potential than iron. I postulate that these adherence oxides aid adherence by oxidizing the iron, thus providing sufficient oxide for making a continuous bond. The reduced cobalt in some of these oxides forms metallic dendrites in the enamel near the enamel-glass interface. Ions with two valence states, such as manganese, can also act as adherence agents, again presumably by oxidizing the iron and being reduced to their lower valence state.

Although not normally considered to have oxidized surfaces, both gold and platinum can adhere to glass if treated at high temperatures in oxygen.[79,83,84] Apparently the gold and platinum oxides are stabilized as silicates. Gold shows some solubility in glass at temperatures above about 1000°C,[85] probably as an aurate.

Pask and his co-workers have observed a correlation between the contact angle between glass and a metal and their adherence.[79,86] For low-contact angles (below 40 to 50°) the glass adheres to the metal, whereas for contact angles above these values the materials do not adhere.

The second requirement for a good seal is that it not contain high stresses, which can cause failure. Transient stresses can arise from temperature gradients resulting from sharp temperature changes of the seal. The low thermal conductivity of glass enhances the possibility of these stresses, but they can be removed by careful annealing and slow cooling of the seal. The annealing relieves the stress by allowing the glass to flow.

Permanent stresses arise because of differences in the thermal expansivity of the metal and glass. The calculation and measurement of permanent stresses in cylindrical glass-wire seals was discussed by Hull and Berger.[87] These stresses are minimized by matching the expansion coefficients of the glass and metal as closely as possible. A rough guide is that the expansion coefficients should differ by no more than 10%, or that the expansion curves differ by no more than $1.5(10)^{-4}$ at the set point of the glass. For the best fit the entire contraction curve of the glass, from its set point to room temperature, must match the contraction curve of the metal as closely as possible. The set point of a glass is normally considered to be the temperature about midway between the annealing temperature and the strain temperature (see Chapter 6).

A new method of sealing glass to a metal has been developed by Wallis and Pomeranz at P. R. Mallory and Company.[88–91] This method involves the application of an electrical field to a glass-metal junction. Satisfactory sealing can be accomplished at temperatures much below those usually necessary; for example, a sodium borosilicate glass, Corning No. 7052, was sealed to an iron-nickel-cobalt alloy at 480°C, whereas the usual sealing temperature for this combination is about 1100°C.[91] In this method the glass and metal must fit together closely, since at 480°C the glass maintains its gross configuration and does not flow appreciably. If the metal surface is heavily oxidized, the fit can be somewhat poorer. Matching of expansion coefficients is important in making a good seal, as in other techniques.

The mechanism of this field-assisted sealing is uncertain. One proposal is that it involves electrostatic forces, which hold the glass and metal tightly together.[90] It seems more likely that the field enhances chemical reactions at the glass-metal interface.[91a] Ionic transport in the glass requires oxidation and reduction reactions at the glass surface, and these reactions would enhance sealing. More work is needed to elucidate conclusively the sealing mechanism.

REACTION WITH WATER AND AQUEOUS SOLUTIONS

The rate of reaction of a glass with water is an important element in its chemical durability, which is a crucial property in determining the applicability of a glass for such uses as containers of liquids and structural or optical elements in a humid atmosphere. The usual tests for chemical durability involve exposure of the glass to water, an aqueous solution, or a humid atmosphere for a certain time and temperature. Since the results depend critically on these conditions, which have not been standardized, the chemical durability of glass has remained a qualitative concept without precise definition. Various tests for chemical durability of glasses are reviewed and discussed in Refs. 92 to 94. A bibliography on chemical durability is also available.[94a]

The dissolution of glass in water involves breaking silicon-oxygen bonds in the glass network by reaction with water. The simplest glass for understanding reactions with water is fused silica, which is therefore considered first; then more common glass types are discussed.

The solubility of amorphous silica in water at 25°C was found to be about 0.012 wt. % by Alexander et al.,[95] in close agreement with previous work cited in their article. The dissolved species was found to be unionized monomeric silicic acid, $Si(OH)_4$. The solubility was about constant at pH values from 1 to 8. Above pH 8 the solubility rose sharply because of the ionization of the silicic acid:

$$Si(OH)_4 + OH^- = \overline{SiO_4H_3} + H_2O \tag{11}$$

However, the amount of unionized $Si(OH)_4$ in solution remained constant, even though the total silica in solution was increased by the formation of the SiO_4H_3 ions. Between 25 and 200°C the solubility of amorphous silica in water increased linearly with temperature, being about 0.83 wt. % at 200°C.

Jorgensen measured the rate of dissolution of silica in 1 *M* sodium chlorate solution at 50°C and found a continuously decreasing rate as a function of time.[96] He fitted his data to a third-order reaction, but over the first 40% or so of the process it could be equally well considered a diffusion-controlled process, with the amount of dissolved silica being proportional to the square root of time. The data of O'Connor and Greenberg[97] on the rate of solution of silica in water at 50°C show a similar time dependence, although these authors treated their data in terms of a surface reaction involving removal and addition of SiO_2 groups. Stober has discussed the rates of dissolution of amorphous and crystalline silicas in terms of silicic acid species adsorbed on the silica surface.[98]

The mechanism of silica dissolution must involve breaking silicon-oxygen bonds by reaction 1. This could happen right at the silica surface, but then one would expect a constant dissolution rate controlled by the surface chemical reaction. Furthermore, it is found that larger molecular units apparently dissolve first and hydrolyze only slowly to the equilibrium monomolecular $Si(OH)_4$ form.[95,96] A possible mechanism is that the dissolution is controlled by diffusion of water molecules a few molecular distances into the silica and their reaction with internal silicon-oxygen bonds. This mechanism explains the diffusion-controlled kinetics and the initial dissolution of larger molecular units, but needs further confirmation for confident acceptance.

The solubilities of cristobalite and quartz in water are considerably lower than that of amorphous silica,[99] being about 2 to $3(10)^{-3}$ wt. % at room temperature.[100] This result is rather surprising because one might expect about the same solubility for fused silica and the crystalline silicas with their similar chemical structure.

The initial attack of water on an alkali silicate glass involves exchange between alkali ions in the glass and hydrogen ions in the water:

$$Na^+ \text{ (glass)} + H_2O = H^+ \text{ (glass)} + NaOH \tag{12}$$

The sodium hydroxide formed leads to increasingly rapid attack of the silicate network by reaction 11, so it must be removed or kept to a negligible level when the mechanism of water attack is being studied. The amount of alkali removed is initially proportional to the square root of time, indicating a diffusion-controlled process. This time variation was confirmed by Douglas and co-workers[101,102]; Rana and Douglas reviewed earlier work in which this functional variation was found for reaction of water with a variety of silicate glasses.[101]

The rate of reaction 12 is apparently controlled by interdiffusion of alkali and hydrogen ions in the glass. The ionic diffusion coefficients in the hydrated surface layer are much higher than in the dry glass, particularly for sodium silicate glasses. This result has been confirmed by measurements of ionic diffusion coefficients in the hydrated layer with radioactive tracers,[103] by comparison of rates of removal of alkali with the electrical conductivity,[50,102] and from the profile of sodium ions in the glass after treatment with water.[104]

This increase in ionic diffusion coefficient could result from either or both of two causes. First, the replacement of sodium ions by smaller hydrogen ions allows more rapid diffusion through the glass network because it becomes more "open." One argument against this mechanism is the result of Ehrmann et al.[105] They found that the replacement of sodium by hydrogen ions at 350°C does not seem to lead to an increase in the ionic diffusion coefficients.

A second cause could be the formation of a hydrated layer by diffusion of water molecules into the glass and their subsequent reaction with the glass network by reaction 1. This mechanism was postulated for the dissolution of silica above. This reaction leads to a weakened network through which ions can diffuse more easily. In this case one would expect that the larger ions would be speeded more than smaller ones by hydration. The results of Eisenman are consistent with this requirement, since he found that in the hydrated layer sodium ions diffused only about five to ten times faster than potassium ions,[103] whereas in the dry glass the sodium is about 1000 times more mobile than potassium.[105] The relative importance of these two factors can only be assessed by further work.

Since the initial rate of attack of water on a glass involves ionic diffusion, reducing the rate of this diffusion should increase the chemical durability of a glass. Thus the increase in durability of a sodium silicate glass as alkaline-earth oxides are added probably results from reduction of the diffusion coefficient of sodium ions by the alkaline-earth ions, as described in Chapter 9. Many effects of glass composition on chemical durability, as reviewed by Morey,[92] can be understood from the changes in the diffusion coefficients of alkali and hydrogen ions in the glass as the composition changes.

The addition of alumina to an alkali silicate glass increases its chemical durability, even though alumina additions actually increase the ionic diffusion coefficients, as described in Chapter 9. The alumina increases durability by decreasing the number of surface sodium ions that exchange with hydrogen ions from solution, as will be discussed in more detail in the following chapter, not by its affect on alkali diffusion.

The excellent durability of certain borosilicate glasses, such as Corning 7740 (Pyrex), apparently does not result from the lower sodium diffusion coefficient in them since it is about the same as in certain soda-lime glasses, but rather from the special phase-separated structure of these glasses. As shown in Chapter 4 these glasses have a silica matrix with a very fine (20 to 30 Å diameter) second phase of sodium borosilicate. The reactivity of the glass to aqueous solutions is therefore close to that of silica except in highly acidic solutions. Even if the borosilicate phase were continuous it could not be etched out because of the high pressure needed to force liquid water through capillaries 30 Å in diameter.

At the same time that the alkali ions are exchanging with hydrogen ions, silica dissolves in the water. Initially the amount of silica dissolved is also proportional to the square root of time for a number of sodium silicate glasses.[102] The rate of silica dissolution is closely related to that of alkali removal[102]; thus glasses that give up their alkali rapidly also dissolve silica more quickly. This result is apparently not related to an increase in the alkalinity of the attacking solution, which is the main reason for accelerated dissolution of glasses in practical applications, because the experimenters took care to renew the attacking solution so as to minimize the concentration of alkali in it. Therefore the similarity of silica dissolution rate to that of alkali comes from changes in the structure of the glass as a result of the exchange of hydrogen ions for alkali ions.

These results can be understood if it is postulated that the rate of silica dissolution is controlled by diffusion of water into the glass just as for pure amorphous silica. As the ion exchange proceeds the surface layer becomes more open, and the water molecules can diffuse in it more rapidly, thus increasing the rate of reaction of water with the silica lattice. Also the silicon-oxygen bonds in the surface layer may be more reactive after ion exchange because of the strain resulting from the replacement of a large ion with a smaller one.

During later stages of the dissolution of glasses in water the rate of removal of both silica and alkali increases and becomes more nearly linear with time instead of proportional to the square root of time.[102] Examples of these changes are shown in Fig. 1. This increase in rates results at least in

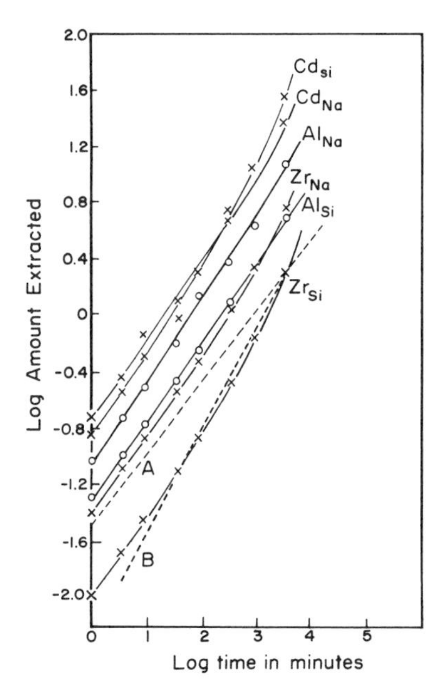

Fig. 1 Extraction of alkali and silica from ternary sodium silicate glasses at 100°C, the third cation being given on the lines. The dashed lines *A* and *B* have slopes of $\frac{1}{2}$ and 1, respectively. From Ref. 102.

part from the dissolution of the glass and the resultant shorter diffusion distance caused by the moving interface. The effective diffusion coefficients may also be lowered as the layer becomes more hydrated. It is not clear if the dissolution rate becomes precisely linear with time, or continues to increase with time.

The pH of the attacking solution affects the rate of dissolution because of reaction 11. The effect of pH on the rate of dissolution is not changed very much in the range from pH 1 to 8, as long as the pH of the solution is maintained constant by renewing the solution or by buffering it. In very alkaline solution, for example, 5% sodium hydroxide at 100°C, the differences in durability of silicate glasses largely disappear,[106] so that fused silica and soda-lime glass dissolve at about the same rate. In these strongly alkaline solutions ion exchange in the glass no longer occurs or is not important, since the surface is rapidly dissolving by reaction 11. Some glasses containing fairly large amounts of lead or boron oxides dissolve somewhat more rapidly, apparently because of the greater susceptibility of these oxides to alkaline attack. Addition of alumina to the glass lowers the rate of dissolution in sodium hydroxide. The reason for this effect is not clear, since in basic solution aluminum oxide acts as a weak acid and should behave similarly to silica. It may be that the aluminum-oxygen-silicon bond is more stable to reaction with water than the silicon-oxygen-silicon, but this would be surprising in view of the higher reactivity of the former bond to hydrogen, as mentioned in a previous section.

In highly acidic solution below pH 0, or solutions more concentrated than 1 normal, the initial rate of dissolution of soda-lime glass is more rapid than in less acidic solutions, but the total amount of glass dissolved in a fixed volume of solution is lower.[107] These results were the same for sulfuric and hydrochloric acid, indicating that specific reactions with the anion of the acids were not important. The authors interpret these changes as resulting from formation of a surface layer of some sort on the glass by the action of the acids. Reflection spectra in the infrared showed a decrease of the silica bands at around 1000 cm^{-1} by the acid attack. It is not clear what kind of surface layer would give these results, and more work needs to be done on these phenomena.

The relative rate of attack of silicate glasses by acids generally parallels rates in neutral solution, except that glasses containing substantial quantities of lead and boron oxides are attacked more rapidly, just as in basic solution. Glasses containing alumina are attacked more rapidly in acidic solution, in contrast to their behavior in basic solution, apparently because hydrated aluminum oxide acts as a weak base in acidic solution, making it more soluble there.

Silicate glasses are rapidly dissolved by hydrofluoric acid, apparently because of the formation of silicofluoride complexes in acidic solution. The reaction of $Si(OH)_4$ molecules in solution with hydrofluoric acid can be written as

$$Si(OH)_4 + 6HF = H_2SiF_6 + 4H_2O \tag{13}$$

The fluorosilic acid H_2SiF_6 is a strong acid, being ionized in water to about the same extent as sulfuric acid.[108] It is also possible that fluoride ions react directly with silicon-oxygen bonds in the silicate lattice. As mentioned in the last chapter some surface hydroxyl ions on silicate glasses can be replaced by fluoride ions in acid solution.

It is curious that the salts of the transition metals with the silicofluoride ion are quite soluble in water, but those of the alkali and alkaline earths are not very soluble in water at room temperature.[108] Thus when a silicate glass containing alkali or alkaline-earth oxides is dissolved in hydrofluoric acid, the reaction products may not be soluble, and sometimes form a film on the glass surface, retarding further reaction. Vigorous agitation and renewal of the attacking acidic solution remedies this retardation.

The rate of dissolution of amorphous silica films in acidic fluoride solutions was studied by Judge.[109] He pointed out that in dilute solutions the significant species in these solutions are described by the following equilibria:

$$HF = H^+ + F^- \tag{14}$$

$$HF + F^- = HF_2^- \tag{15}$$

Judge found that his conclusions could be interpreted as resulting from reactions with the species HF and HF_2^-, so he expressed his rate of dissolution R of silica as

$$R = A[HF] + B[HF_2^-] + C \tag{16}$$

where A, B, and C are constants, and the brackets denote concentrations. However, based on the equilibria above his results would be equally well represented as

$$R = A'[H^+][F^-] + B'[H^+][F^-]^2 + C \tag{17}$$

Judge claimed that the dissolution rate was not related to fluoride ion concentration because at pH 7 it was insensitive to the total fluoride concentration. However, this may result simply because of the low hydrogen ion concentration. Reaction 17 seems more reasonable because from reaction 13 one would expect the reacting species to be hydrogen and fluoride ions. More work is needed to confirm the exact dependence of dissolution rate on the concentration of these ions.

Phosphoric acid attacks silicate glasses at temperatures above 200°C. After long treatment in the acid the glass surface becomes coated with silicon phosphate crystals, which protect the glass from HF attack.[110]

WEATHERING

The reaction of glass with the atmosphere is called weathering. For silicate glasses weathering is almost always caused by reaction of the glass with water in the atmosphere. At ambient temperatures the ion exchange of hydrogen ions from the water with alkali ions in the glass (reaction 12) is the predominant reaction. The sodium oxide formed from this reaction reacts rapidly with more water from the atmosphere, forming sodium hydroxide.

Fig. 2 Electron micrograph of crystals on the surface of a $Na_2O \cdot 2SiO_2$ sodium disilicate glass after fracture and short exposure to the atmosphere. Magnification 42,000 ×.

Since sodium hydroxide is deliquescent, it picks up more water and forms a thin film of sodium hydroxide solution, which attacks the glass. It is also possible for the sodium oxide to react with carbon dioxide from the atmosphere, forming sodium carbonate.

In Fig. 2 crystals are shown that have grown on the surface of a $Na_2O{\cdot}2SiO_2$ sodium disilicate glass after short exposure to the atmosphere. The crystals were not identified, but were probably sodium carbonate, possibly hydrated.

REFERENCES

1. R. V. Adams and R. W. Douglas, *J. Soc. Glass Tech.*, **43**, 147 (1959).
2. V. Garino-Canina, *C.R.*, **239**, 705 (1954).
3. V. Garino-Canina and M. Priqueler, *Phys. Chem. Glasses*, **3**, 43 (1962).
4. R. H. Doremus, in *Reactivity of Solids*, J. W. Mitchell, R. C. DeVries, R. W. Roberts, and P. Cannon, Eds., Wiley-Interscience, New York, 1969, p. 667.
5. A. J. Moulson and J. P. Roberts, *Trans. Far. Soc.*, **57**, 1208 (1961).
6. T. Drury and J. P. Roberts, *Phys. Chem. Glasses*, **4**, 79 (1963).
7. G. J. Roberts and J. P. Roberts, *Phys. Chem. Glasses*, **5**, 26 (1964).
8. G. Hetherington and K. H. Jack, *Phys. Chem. Glasses*, **3**, 129 (1962).
9. N. Scholze, *Glastech. Ber.*, **32**, 81, 142, 278 (1959).
10. H. Franz and H. Scholze, *Glastech. Ber.*, **36**, 347 (1963).

10a. J. Götz, *Glastech. Ber.*, **45**, 14 (1972).

11. H. Franz, *Glastech. Ber.*, **38**, 54 (1965).
12. R. Bruckner, *Glastech. Ber.*, **37**, 536 (1964).
13. T. Bell, G. Hetherington, and K. H. Jack, *Phys. Chem. Glasses*, **3**, 141 (1962).
14. R. W. Lee, *J. Chem. Phys.*, **38**, 448 (1963).
15. R. H. Doremus, unpublished work.
16. S. P. Faile and D. M. Roy, *Mat. Res. Bull.*, **5**, 385 (1970); *J. Am. Ceram. Soc.*, **54**, 533 (1971).
17. K. H. Beckman and N. J. Harrick, *J. Electrochem. Soc.*, **118**, 614 (1971).
18. R. H. Doremus, in *Symp. on Nucleation and Crystallization in Glasses and Melts*, American Ceramic Society, Columbus, Ohio, 1962, p. 119.
19. R. Zsigmandy, *Dingler's Polytech. J.*, **266**, 364 (1887); **306**, 68, 91 (1897); *Sprechsaal*, **27**, 123, 147, 175, 201, 227 (1894).
20. A. W. Bastress, *J. Am. Ceram. Soc.*, **30**, 52 (1947).
21. J. T. Randall and R. E. Leeds, *J. Soc. Glass Tech.*, **13**, 16T (1929).
22. R. L. Green and K. B. Blodgett, *J. Am. Ceram. Soc.*, **31**, 89 (1948).
23. J. L. Barton and M. Morain, *J. Noncryst. Solids*, **3**, 115 (1970).
24. E. A. Weaver, K. W. Heckman, and E. L. Williams, *J. Chem. Phys.*, **47**, 4891 (1967).
25. R. H. Doremus, *J. Electrochem. Soc.*, **115**, 181 (1968).
26. G. E. Toop and C. S. Samis, *Trans. AJME*, **224**, 878 (1962).
27. R. W. Douglas, P. North, and A. Paul, *Phys. Chem. Glasses*, **6**, 216, (1965).
28. F. Irmann, *J. Am. Chem. Soc.*, **74**, 4767 (1952).
29. E. J. Michal and R. Schuhmann, *J. Metals*, **4**, 723 (1952).
30. W. D. Johnston, *J. Am. Ceram. Soc.*, **48**, 184 (1965).
31. P. North and R. W. Douglas, *Phys. Chem. Glasses*, **6**, 197 (1965).
32. A. Paul and R. W. Douglas, *Phys. Chem. Glasses*, **6**, 207, 212 (1965).
33. A. Paul and R. W. Douglas, *Phys. Chem. Glasses*, **7**, 196 (1960).

34. H. J. Tress, *Phys. Chem. Glasses*, **1**, 196 (1960).
35. P. Winchell, *High Temp. Sci.*, **1**, 200 (1969).
36. R. H. Doremus, in *Modern Aspects of the Vitreous State*, Vol. 2, J. D. Mackenzie, Ed., Butterworths, London, 1962, p. 1.
37. H. O. Mulfinger and H. Meyer, *Glastech. Ber.*, **36**, 481 (1963).
38. H. O. Mulfinger and H. Franz, *Glastech. Ber.*, **38**, 235 (1965).
39. J. Kelen and H. O. Mulfinger, *Glastech. Ber.*, **41**, 230 (1968).
40. E. L. Swarts, *J. Can. Ceram. Soc.*, **38**, 155 (1969).
41. H. Franz, *J. Can. Ceram. Soc.*, **38**, 89 (1969).
42. M. L. Pearce, *J. Am. Ceram. Soc.*, **47**, 342 (1964).
43. C. Kroger and D. Lummerzheim, *Glastech. Ber.*, **38**, 229 (1965).
44. Z. Strnad, *Phys. Chem. Glasses*, **12**, 152 (1971).
45. E. Preston and W. E. S. Turner, *J. Soc. Glass Tech.*, **18**, 143 (1934).
46. M. L. Pearce, *J. Am. Ceram. Soc.*, **48**, 175 (1965).
47. C. J. B. Fincham and F. D. Richardson, *Proc. Roy. Soc.*, **A223**, 40 (1954).
48. E. T. Turkdogan and M. L. Pearce, *Trans. AIME*, **227**, 940 (1963).
49. H. Meier Zu Kocker and D. Chandra, *Glastech. Ber.*, **45**, 139 (1972).
50. R. W. Douglas and J. O. Isard, *J. Soc. Glass Tech.*, **33**, 289T (1949).
51. C. H. Greene and R. F. Gaffney, *J. Am. Ceram. Soc.*, **42**, 271 (1959).
52. C. H. Greene and J. Kitano, *Glastech. Ber.*, **32K**, 44 (1959).
53. C. H. Greene and H. A. Lee, *J. Am. Ceram. Soc.*, **48**, 528 (1965).
54. R. H. Doremus, *J. Am. Ceram. Soc.*, **43**, 655 (1960).
55. E. M. Firth, F. W. Hodkin, and W. E. S. Turner, *J. Soc. Glass Tech.*, **10**, 3T (1926); **11**, 190T (1927).
56. C. Kuhl, H. Rudow, and W. Weyl, *Glastech. Ber.*, **16**, 37 (1938).
57. V. T. Slavyanskii, *Gases in Glass*, Moscow, 1957.
58. M. Cable and M. A. Haroon, *Glass Tech.*, **11**, 48 (1970).
59. M. Cable, A. R. Clarke, and M. A. Haroon, *Glass Tech.*, **9**, 101 (1968); **10**, 15 (1969).
60. *Gmelin Handbuch der Anorganischen Chemie: Antimon, System No. 18*, Part B, Gmelin-Verlag GmbH., Clausthal-Zellerfeld, 1949, pp. 360–378.
61. *Gmelin Handbuch der Anorganischen Chemie: Arsen, System No. 17*, Verlag Chemie GmbH., Weinheim/Bergstrasse, 1952, pp. 258–276.
62. A. K. Lyle, in *Travaux du IV Congres Int. du Verre*, Imprimerie Chaix, 20, rue Bergere, Paris, 1957, p. 93.
63. A. K. Lyle, *J. Am. Ceram. Soc.*, **33**, 300 (1950).
64. F. L. Harding, *Glass Tech.*, **13**, 43 (1972); *J. Am. Ceram. Soc.*, **55**, 368 (1972).
65. D. Brown and R. W. Douglas, *Glass Tech.*, **6**, 190 (1965).
66. R. W. Douglas and M. S. Zaman, *Phys. Chem. Glasses*, **10**, 125 (1969).
66a. C. H. Greene and D. R. Platts, *J. Am. Ceram. Soc.*, **52**, 106 (1969).
67. F. L. Harding and R. J. Ryder, *Glass Tech.*, **11**, 54 (1970).
68. *Symp. sur L'Affinage du Verre*. Union Scientifique Continentale du Verre, Charleroi, Belgium, 1956.
69. J. C. Potts, *J. Soc. Glass Tech.*, **23**, 129 (1939); *J. Am. Ceram. Soc.*, **24**, 43 (1941).
70. K. Schwerdtfeger, *J. Phys. Chem.*, **70**, 2131 (1966).
71. M. Truchlarova and O. Verprek, *Glastech. Ber.*, **40**, 257 (1967); **42**, 9 (1969).
72. J. Hlaváć and H. Nademlýnská, *Glass Tech.*, **10**, 54 (1969).
73. M. Cable and D. Martlew, *Glass Tech.*, **12**, 142 (1971).
74. B. S. Samaddar, W. D. Kingery, and A. R. Cooper, *J. Am. Ceram. Soc.*, **47**, 249 (1964).
75. A. R. Cooper, *Glastech. Ber.*, **37**, 137, 189 (1964).
76. K. G. Kreider and A. R. Cooper, *Glass Tech.*, **8**, 71 (1967).

77. J. H. Partridge, *Glass-to-Metal Seals*, Society of Glass Technology, Sheffield, England, 1949.
78. A. J. Andrews, *Porcelain Enamels*, 2nd ed., Gerrard Press, Urbana, Illinois, 1961.
79. J. A. Pask, in *Modern Aspects of the Vitreous State*, Vol. 3, J. D. Mackenzie, Ed., Butterworths, London, 1964.
80. M. P. Borom and J. A. Pask, *Phys. Chem. Glasses*, **8**, 194 (1967).
81. F. D. Gaidos and J. A. Pask, in *Advances in Glass Technology*, Plenum Press, New York, 1967, p. 548.
82. M. P. Borom and J. A. Pask, *J. Am. Ceram. Soc.*, **49**, 1 (1966).
83. D. G. More and H. K. Thornton, *J. Res. Nat. Bur. Stand.*, **62**, 127 (1959).
84. D. M. Mattox, *J. Appl. Phys.*, **37**, 3613 (1966).
85. S. D. Stookey, *J. Am. Ceram. Soc.*, **32**, 346 (1949).
86. J. A. Pask and R. M. Fulrath, *J. Am. Ceram. Soc.*, **45**, 592 (1962).
87. A. W. Hull and E. E. Burger, *Physics*, **5**, 384 (1934).
88. D. J. Pomeranz, *Anodic Bonding*, U.S. Patent 3,397,278, 1968.
89. G. Wallis and D. J. Pomeranz, *J. Appl. Phys.*, **40**, 3946 (1969).
90. G. Wallis, *J. Am. Ceram. Soc.*, **53**, 563 (1970).
91. G. Wallis, J. Dorsey, and J. Beckett, *Bull. Am. Ceram. Soc.*, **50**, 958 (1971).
91a. P. B. De Nee, *J. Appl. Phys.*, **40**, 5396 (1969).
92. G. W. Morey, *The Properties of Glass*, Reinhold, New York, 1954, pp. 101 ff.
93. L. Holland, *The Properties of Glass Surfaces*, Chapman and Hall, London, 1964, pp. 122 ff.
94. H. Walters and P. B. Adams, *Appl. Optics*, **7**, 845 (1968).
94a. V. Gottardi, Ed., *The Chemical Durability of Glass*, International Commission on Glass, Institut du Verre, Paris, France, 1972.
95. G. B. Alexander, W. M. Heston, and R. K. Iler, *J. Phys. Chem.*, **58**, 453 (1954); **61**, 1539 (1957).
96. S. S. Jorgensen, *Acta Chem. Scand.*, **22**, 335 (1968).
97. T. L. O'Conner and S. A. Greenberg, *J. Phys. Chem.*, **62**, 1195 (1958).
98. W. Stober, in *Advances in Chemistry No. 67*, American Chemical Society, Washington, D.C., 1967, p. 161.
99. G. C. Kennedy, *Econ. Geol.*, **45**, 629 (1950).
100. R. J. Fournier and J. J. Rowe, *Am. Minerologist*, **47**, 897 (1962).
101. M. A. Rana and R. W. Douglas, *Phys. Chem. Glasses*, **2**, 179 (1961).
102. C. R. Das and R. W. Douglas, *Phys. Chem. Glasses*, **8**, 178 (1967).
103. G. Eisenman, in *Glass Electrodes for Hydrogen and Other Cations*, G. Eisenman, Ed., M. Dekker, New York, 1967, Chapter 5, p. 133.
104. Z. Boksay, G. Bouquet, and S. Dobos, *Phys. Chem. Glasses*, **8**, 140 (1967).
105. P. Ehrmann, M. deBilly, and J. Zarzycki, *Verres Refrac.*, **18**, 169 (1964).
106. M. E. Nordberg, Ref. 93, p. 131.
107. B. Wollast and P. Brennet, *Verres Refrac.*, **24**, 251 (1970).
108. N. V. Sidgwick, *The Chemical Elements and Their Compounds*, Vol. 1, Clarendon Press, Oxford, England, 1950, p. 614–615.
109. J. S. Judge, *J. Electrochem. Soc.*, **118**, 1772 (1971).
110. N. H. Ray, *J. Noncryst. Solids*, **5**, 71 (1970).

14

Ion Exchange and Potentials of Glass Electrodes

Although glass is not usually thought of as an ion exchanger, many of the chemical properties of glasses are determined by ion exchange, as shown by the discussions in the preceding two chapters of hydrogen-alkali ion exchange in the reactions of water with glass. Ion exchange also gives rise to the potentials of glass electrodes, and the introduction of larger ions into glass by ion exchange results in a compressive surface stress, thus strengthening the glass (see Chapter 17). This chapter contains sections on measurements of the distribution of ions between glass and solutions, experimental results on distribution coefficients, theories of the selectivity of anionic groups in glass for different cations, potentials of glass electrodes, and tests of the ion exchange theory of glass electrode potentials. My review on ion exchange in glasses[1] should be consulted for more details on earlier work. Helfferich's treatise on ion exchange[2] emphasizes work on organic exchangers, but gives some discussions applicable to glasses.

MEASUREMENT OF IONIC DISTRIBUTION

An ion exchanger can be treated as a separate phase. The exchange of an ion B in a solution with an ion A in the exchanger (glass) is represented by the following equation:

$$A(g) + B(s) = A(s) + B(g) \tag{1}$$

where (s) represents the solution and (g) is the glass. At equilibrium the ions are distributed between the two phases in a fixed ratio. A number of coefficients have been used to represent this equilibrium, as discussed in

Ref. 2, pp. 152 ff. Here the thermodynamic equilibrium constant K is used whenever possible:

$$\frac{a_s b_g}{a_g b_s} \tag{2}$$

where the lower-case letters represent the thermodynamic activities of the A and B ions. The absolute value of K depends on the reference functions and states chosen to define the activities. For aqueous solutions the most usual reference state is the concentration at infinite dilution. However, for fused salts it is more convenient to use the mole fraction N of a component in the pure state; that is,

$$\lim_{N \to 1} \frac{a}{N} = 1$$

The latter reference state is also used for the glassy exchanger in this work. If the exchanger is homogeneous and has only one type of site, K is constant with changing composition at constant temperature and pressure. Inhomogeneous exchangers are considered below.

The value of K reflects the selectivity of the exchanger. It is a quantitative measure of the preference of the exchanger for one ion over another in solution with it. The surface potential of a glass electrode also depends on K, as described below.

Two methods have been used to estimate K in glass. One technique is a direct measurement of the distribution of ions between the solution and glass; the second method is an indirect inference of the selectivity of a glass electrode from measurement of its electrical potential. The first method is preferable and can be used in a molten salt at higher temperatures, where considerable exchange takes place. When a glass electrode is used in aqueous solution, the region of exchange at the glass surface is limited and direct measurement of ionic concentrations becomes difficult. Therefore, almost all results on equilibria between glass and ions in aqueous solution come from potential measurements.

In making a direct measurement of K for exchange between a molten salt and a glass, powdered glass can be equilibrated with a salt bath of desired composition and then analyzed for the exchanging constituents. If the rate of diffusion in the glass is too slow for a bulk sample, such as a tube or rod, to come to equilibrium with the melt, it is still possible to find K by measuring the concentrations of the exchanging ions near the glass surface.[1,3] To find the surface concentration, layers of the glass are etched off with hydrofluoric acid and analyzed for the exchanging ions. The concentrations are then plotted as a function of distance from the glass surface and extrapolated back to the surface. To ensure that the exchange process at the surface has reached equilibrium, measurements can be made at different times of immersion in the molten salt.

In some glasses such as fused silica the concentration of exchanging ions is so low that conventional analysis is difficult. In this case it is possible to calculate the exchange coefficient from the conductivity of the glass in equilibrium with melts of different concentration.[4] The total conductivity σ_T of the glass is then made up of contributions from the two ions:

$$\sigma_T = N_A \sigma_A + N_B \sigma_B \tag{3}$$

where σ_i is the conductivity of the glass containing only one kind of ion, and N_i is the mole fraction of this ion ($N_A + N_B = 1$). In using this equation it is assumed that the total ionic concentration in the glass is constant and uniform, and also that the mobility ratios of the ions do not depend on ionic concentration.

If the thermodynamic activities of the ions in the salt are known, tentative selectivity coefficients K' can be calculated from them and the measurements of the concentration ratio C_B/C_A in the glass:

$$K' = \frac{a_s C_B}{b_s C_A}$$

If the K' values are constant with changing solution concentration, the ionic activities in the glass should be equal to the concentrations, as required for an ideal solution. If the Ks vary with solution concentration, the pairs of ions in the glass behave nonideally, and the calculated Ks provide a measure of this nonideality.

To find exchange coefficients between aqueous solutions and the hydrated layer on glass, Eisenman measured the uptake of radioactive ions, as described in Ref. 1, p. 8, and Ref. 5, Chapter 5.

The potential of a glass electrode can also be used to examine the selectivity of a glass for different ions. The electrical potential V of a glass electrode immersed in a solution containing A and B ions with thermodynamic activities a and b is given by (see section on glass electrode below)

$$V = V_0 + \frac{RT}{F} \ln\left\{a + K\left(\frac{u_B}{u_A}\right)b\right\} \tag{4}$$

when the ions in the glass behave ideally (activities can be equated to concentrations), and their mobility ratio, u_B/u_A, in the glass is constant with composition. In Eq. 4 R is the gas constant, T is the absolute temperature, F is the Faraday, and V_0 is invariant with the composition of the solution. The product $K(u_B/u_A)$ can be found from measurements of the glass electrode potential (membrane potential) as a function of ionic activities in solution. Since it is relatively easy to measure the potential of a glass electrode in aqueous solution, many such measurements have been made, and this dependence of potential on K has given information on K values in many systems for which no direct measurements are available. However, there are unknown factors in the relationship between potential and K, so these results must be analyzed carefully. For example, only if values of the mobility ratios u_i/u_j are similar for different pairs of ions for a particular glass, or for the same ions for different glasses, can relative values of K be deduced from potential measurements. Such similarities exist for some of the glasses for which measurement of mobility ratio have been made; however, in other cases large differences in mobility ratios are found for different ions and different glasses.[6]

Many empirical equations have been proposed to describe the variation of selectivities of ion exchangers with solution concentration.[7–9] In the present notation the equations are essentially the means for describing the activity ratio, b_g/a_g, of the ions in the glass. The most successful of these equations for a variety of ion exchangers was first proposed by Rothmund and Kornfeld[10]:

$$\frac{b_g}{a_g} = \left(\frac{N_B}{N_A}\right)^n \tag{5}$$

where n is a constant, usually equal to or greater than 1, and N_i is the mole fraction of ion i in the exchanger. Eisenman has shown that this equation is particularly useful for deduction of K from potential measurements.[11] When $(N_B/N_A)^n$ is substituted for b_g/a_g in Eq. 2, a plot of $\log(b_s/a_s)$ against $\log(N_B/N_A)$ is characterized by a slope of n if K is constant with concentration. This formulation is used here to describe experimental results and is related to theories of selectivity described below.

EXPERIMENTAL RESULTS ON DISTRIBUTION COEFFICIENTS

Results of measurements of K values for various glasses are summarized in Tables 2 to 4. The compositions of the glasses are given in Table 1. Unless otherwise stated these results were found by analysis of equilibrated glass powders.

Activity coefficients γ of the ions in the nitrate melts were calculated from formulas of the form

$$\ln \gamma_A = \frac{EX_B{}^2}{RT}$$

Table 1 Compositions of Glasses

	A	*B*	*C* (Pyrex)	*D*	*E*	*F*	*G*	*H*	*J*
SiO_2	69	73	81		71.5	79.7	66.7	64.1	69
B_2O_3		0.8	13	75				18.7	
Na_2O	15.2	15.2	4	25			16.3	11.4	27
Li_2O					11.4	5.9			
K_2O	3.6	0.7							
CaO	7.4	4.6							
MgO	0.4	3.6				4.1			
Al_2O_3	4.4	1.7	2		16.5	11.5	13.2	5.8	4
TiO_2						3.8	3.8		
Fe_2O_3	0.4								

where E is independent of concentration and temperature, and X_B is the mole fraction of B in solution. For sodium-silver nitrate solutions, for example, E was 840 cal/mole.[4]

The sequence of preference of soda-lime glass for ions from nitrate melts (Table 2) is $Ag^+ \gg Na^+ > K^+$. In lithium aluminosilicates the order is $Ag^+ > Li^+ > Na^+$, with not nearly so strong a preference for silver. In borosilicates and fused silica the preference for silver over sodium is further reduced.

In chloride melts the attraction of the glass for silver is dramatically lower, undoubtedly because of the strong bonding between chloride and silver ions in the melt. From Stern's work[16] the sequence in fused silica would appear to be $Na^+ > Li^+ \gg Ag^+$, but van Reenan et al.[17] found sodium and lithium about equally favored by fused silica. This difference may result from the use by the latter authors of a zinc chloride melt, although the size of the difference is surprising. These authors also found the sequence $Li^+ > K^+ > Na^+$ for a soda-lime glass and Pyrex borosilicate glass. One questionable aspect of both studies of ionic distributions with chloride melts is that the authors apparently used solution concentrations in calculating their exchange coefficients; that is, they assumed the solutions were ideal and the activities were equal to the concentrations. It is not at all clear that this assumption is valid and it may be responsible for some of the peculiarities of the results.

Results on the exchange of ions in Pyrex borosilicate glass were interpreted with a two-phase model because the calculated K values were not constant, nor did they fit Eq. 5, as shown in Table 5 for silver-sodium

Table 2 Equilibrium Coefficients for Ionic Distribution between Nitrate Melts and Glasses

Glass	Refs. to Table 1	Ion in solution (*B*)	Ion in glass (*A*)	Temperature (°C)	*K*	*n*	Remarks	Refs.
Soda-lime	*A*, *B*	Ag^+	Na^+	320	120	1.08	From	3, 12
Soda-lime	*B*	K^+	Na^+	350	0.05		surface conc.	12
Sodium borate	*D*	Cs^+	Na^+	550	$1.5(10)^{-3}$		Ions	13
Sodium borate	*D*	Sr^{2+}	Na^+	550	$2(10)^2$		dilute	
Sodium borate	*D*	Eu^{3+}	Na^+	550	$4(10)^3$		in $NaNO_3$	
Lithium aluminosilicate	*E*	Na^+	Li^+	400	0.28	1.9		14
Lithium aluminosilicate	*E*	Ag^+	Li^+	300	3.2	2.2		14
Sodium aluminosilicate	*G*	K^+	Na^+	500	0.94	1.2		14
Lithium aluminosilicate	*F*	Na^+	Li^+	400	0.27	3.2		14
Sodium borosilicate	*H*	Ag^+	Na^+	300	2.6	1.4		14
Fused silica (General Electric 204)		Ag^+	Na^+	337	2.1	1.0	From Eq. 3	15
Fused silica (General Electric 204)		K^+	Na^+	823	2.1		From Eq. 3	16a

Table 3 Equilibrium Coefficients for Ionic Distribution between Chloride Melts and Glasses

Glass	Ion in solution (B)	Ion in glass (A)	Temperature (°C)	K	n	Remarks	Refs.
Fused silica	Ag^+	Na^+		$2(10)^{-4}$		From	16
(General Electric 204)	Ag^+	Li^+		$1.2(10)^{-3}$		electrode	
	Ag^+	K^+		$1.7(10)^{-4}$		potentials	
	K^+	Na^+	815	1.1		From Eq. 3	16a
			560	1.9			
Fused silica	Li^+	Na^+	600	1.7	1.8	In molten	17
Soda-lime	Li^+	Na^+	450	81	2.1	zinc	
Soda-lime	K^+	Na^+	450	5.5	1.7	chloride	
Pyrex	Li^+	Na^+	550	14	1.2		
Pyrex	K^+	Li^+	550	0.6	1.2		

Table 4 Equilibrium Coefficients for Ionic Distributions between Nitrate Melts and Pyrex Borosilicate Glass Assuming Two Glassy Phases

Ion in solution (B)	Ion in glass (A)	Temperature (°C)	K_1	N_1	K_2	Remarks	Refs.
Ag^+	Na^+	335	230	0.05	1.6	Surface conc.	18
K^+	Na^+	380	$7(10)^{-7}$	0.13	1.1		
K^+	Na^+	435	$5(10)^{-4}$	0.08	1.4		19
K^+	Na^+	465	$5(10)^{-3}$	0.03	1.4		

Table 5 Exchange of Silver Ions in a Sodium Nitrate Melt with Pyrex Glass[20]

Mole fraction of $AgNO_3$ in melt	Exchange coefficient K
0.398	1.97
0.310	1.87
0.153	1.98
0.0475	2.05
0.0022	7.06
3×10^{-6}	12.2

exchange. This glass separates into two phases as described in Chapter 4, in agreement with the two-phase model. In the following paragraphs this model is briefly discussed and applied to results on Pyrex glass.

An ion exchanger is considered to have two ideal phases, I and II, that exchange ions A and B with the melt according to Eq. 1. A K for each phase can be defined as

$$K_{\mathrm{I}} = \frac{a_s c_{\mathrm{I}B}}{b_s c_{\mathrm{I}A}}, \qquad K_{\mathrm{II}} = \frac{a_s c_{\mathrm{II}B}}{b_s c_{\mathrm{II}B}}$$

where a_s and b_s are the activities of ions A and B in the melt, the cs refer to the concentrations of ions in the two phases, and the volume of the two-phase glass is taken as the unit basis for computation. In each phase the total concentration, c_{I} or c_{II}, of exchanging ions is assumed to be constant:

$$c_{\mathrm{I}} = c_{\mathrm{I}A} + c_{\mathrm{I}B} \qquad c_{\mathrm{II}} = c_{\mathrm{II}A} + c_{\mathrm{II}B}$$

The mole fraction N_i of exchanger sites in each phase is then

$$N_{\mathrm{I}} = \frac{c_{\mathrm{I}}}{(c_{\mathrm{I}} + c_{\mathrm{II}})} \qquad N_{\mathrm{II}} = \frac{c_{\mathrm{II}}}{(c_{\mathrm{I}} + c_{\mathrm{II}})}$$

The actual measured distribution coefficient K for the two phases is thus

$$K = \frac{a_s \sum c_{iB}}{b_s \sum c_{iA}} = \frac{a_s(c_{\mathrm{I}B} + c_{\mathrm{II}B})}{b_s(c_{\mathrm{I}A} + c_{\mathrm{II}A})}$$

When the equations above are combined, an expression for K as a function of the melt activities a_s and b_s, the two constant Ks, and the constant mole fraction of sites (N_{I} and N_{II}) in the two phases results:

$$K = \frac{N_{\mathrm{I}} K_{\mathrm{I}} + N_{\mathrm{II}} K_{\mathrm{II}} + K_{\mathrm{I}} K_{\mathrm{II}}(b_s/a_s)}{1 + (b_s/a_s)(N_{\mathrm{I}} K_{\mathrm{II}} + N_{\mathrm{II}} K_{\mathrm{I}})} \tag{6}$$

This equation gives a satisfactory description of results for Pyrex borosilicate glass.

Nicolsky et al. have also presented equations for ion exchange with two different sites in glass.[21,22] Both their treatment and the present one contain several adjustable parameters, so it is not possible to distinguish between them with the data of Table 5. Nicolsky et al. consider that the ions and sites in the hydrated glass can be associated or disassociated, as can ions in aqueous solution. In my treatment the Ks represent an average for all counterions and sites in the exchanger.

The results for this model as applied to Pyrex borosilicate glass are given in Table 4. Reasonable agreement between two different investigations on N_i, the mole fraction of ions in phase I, gives some confidence in the model.

Phase I is identified with the high-silica phase of Pyrex, which contains only about 5% of the total sodium ions, although it occupies 86% of the volume of the glass. Phase II is the sodium borosilicate phase, dispersed as fine particles about 25 Å in diameter in the silica matrix (see Chapter 4). In the borate phase the exchange coefficients are similar to those found in a sodium borosilicate glass and fused silica (Table 2), whereas the silica-rich phase differentiates strongly between the ions in the sequence $Ag^+ \gg Na^+ \gg K^+$. Possible reasons for these sequences are given in the next section.

Deviations from single-phase behavior of K in Pyrex become marked only when one exchanging ion has a low concentration in the melt, as shown in Table 5. Thus in several studies of Pyrex glass these deviations were not noticed and the results were interpreted in terms of a single-phase model. For some purposes this model may be useful, but for studies at low-melt concentration, for example, with tracer ions, it is necessary to use the two-phase model.

Eisenman (Ref. 5, Chapter 5) investigated the equilibrium between samples of a hydrated-soda aluminasilicate glass (glass G, Table 1) and solutions of potassium ions by the method described in the preceding section. He found K values of 90 and 102 (glass prefers potassium) for two different samples of about the same composition.

Since few direct measurements of K with aqueous solutions are available, it is important to try to supplement these data from estimates of K values from potential studies. As described above, the product $K(u_B/u_A)$, where the us are ionic mobilities, can be calculated from measurements of membrane potentials across glasses. The utility of this approach is dependent on knowledge of the value of u_B/u_A. From measurements of mobility ratios in dry glass it seems likely that certain generalities about the properties of this ratio can be made. For example, the mobility ratios of sodium to potassium ions seems to be about the same in a variety of glasses whose only univalent cation is sodium (soda glasses). In lithia or potash glasses this ratio is lower than in soda glasses. Thus the selectivity of different glass electrodes made of sodium silicates may be compared, but comparisons of these with glasses containing lithium or potassium ions should be less reliable. These conclusions are based on mobility measurements in dry glasses; mobilities in hydrated glasses are much higher, and it is quite possible that some of the differences in mobility ratios found in dry glass are reduced in hydrated glasses. A more definitive analysis awaits further mobility measurements on hydrated glasses.

Results of numerous measurements of selectivities of glass electrodes for ions in aqueous solution, mainly in connection with the "alkaline error" of pH electrodes, are reviewed in Ref. 5, particularly Chapters 3, 6, 7a, and 9. Since most of these measurements were not specifically designed to examine the role of glass composition, discussion of this work is limited to the extensive experiments of Eisenman.[11]

In Eisenman's experiments the conditions on one side of the glass were held constant. The potential on this side of the glass, the electrode potentials, and junction potentials of salt bridges are constant with changing ionic concentrations and contribute only to V_0, so the experiments should provide a measure of the product $K(u_B/u_A)$. For an ideal solution in the glass, Eq. 4 applies. If the glass is not ideal and the activities are given by Eq. 5, the potential is (see section on electrode potentials below)

$$V = V_0 + \frac{nRT}{F} \ln \left\{ a^{1/n} + \left[K_{AB} \left(\frac{u_B}{u_A} \right) b \right]^{1/n} \right\} \tag{7}$$

Since hydrogen ions are always present in aqueous solution, Eisenman measured the selectivities of the glass for alkali ions by determining the potential as a function of pH for a constant concentration (0.1 M) of alkali cation B. At higher pH the potential becomes constant because the glass is sensitive only to the alkali ion; the difference between this constant potential and one at a pH where the glass is sensitive only to hydrogen ions is equal to $(RT/F)[\ln K_{HB}(u_B/u_A)]$. If ion B is different from the alkali ion in the glass

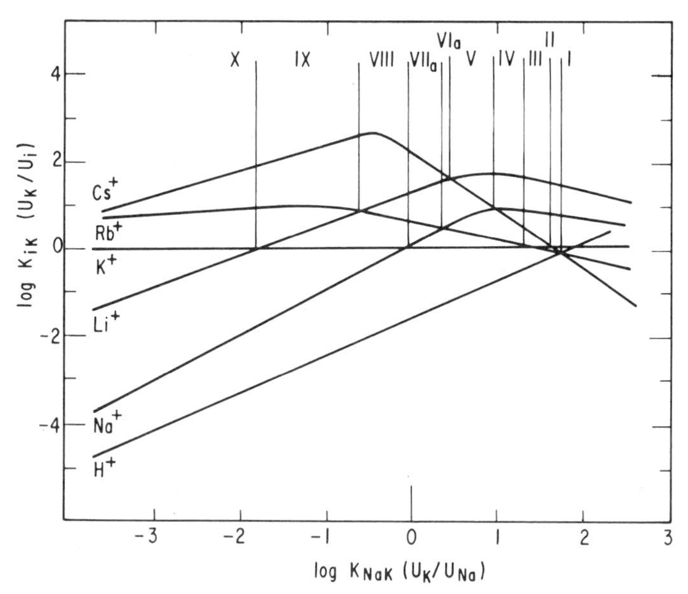

Fig. 1 Selectivity of sodium aluminosilicate glasses for monovalent cations. The roman numerals refer to the selectivity sequences in Table 6. A particular glass composition would be represented by a vertical line on the figure. From Eisenman.[11]

(usually sodium), the system contains three cations. However, as long as this ion is not in the solution, it contributes only a constant additional diffusion potential to V_0 and does not affect this method of calculating $K_{HB}(u_B/u_A)$. The selectivity of the glass for any pair of ions can be found from the results of these measurements, since

$$K_{AB}\left(\frac{u_B}{u_A}\right) = \frac{K_{HB}(u_B/u_H)}{K_{HA}(u_A/u_H)}$$

Results of this kind are presented schematically for a variety of glasses in Fig.1, taken from Ref. 11. The results are plotted in terms of the sodium-potassium selectivity of the glasses, therefore the abscissa is ln $K_{NaK}(u_K/u_{Na})$, and the ordinate is the same term for any pair of ions, ij, ln $K_{ij}(u_j/u_i)$. At the right of the figure the lines are plotted in terms of the cesium-potassium selectivity. Although the data scatter considerably, a definite qualitative pattern in the selectivity of the ions emerges. Eisenman analyzed these patterns into the 11 sequences shown in Table 6 and Fig. 1. The data in the figures fit almost entirely into these few sequences, out of the 120(5!) that are possible for the five ions. Thus the sodium-potassium selectivity provides a basis for estimate of the selectivity sequences of the glass.

The selectivity sequences are related in a definite way to the composition of the glass. For example, data for silicate glasses without boron and aluminum fall off of Fig. 1 to the left, whereas the borosilicates fall to the right of the figure. In the sodium aluminosilicates the important factor determining selectivity is the sodium/aluminum atom ratio. The theoretical significance of these results is discussed below.

Table 6 Eisenman's Selectivity Orders among the Alkali Ions

Increasing field strength (top to bottom)

I	Cs > Rb > K > Na > Li		
I	Cs > Rb > K > Na > Li		
IIa	Cs > K > Rb > Na > Li	or II	Rb > Cs > K > Na > Li[a]
IIIa	K > Cs > Rb > Na > Li	or III	Rb > K > Cs > Na > Li[a]
IV	K > Rb > Cs > Na > Li		
V	K > Rb > Na > Cs > Li		
VI	K > Na > Rb > Cs > Li		
VII	Na > K > Rb > Li > Cs		
IX	Na > K > Li > Rb > Cs		
X	Na > Li > K > Rb > Cs		
XI	Li > Na > K > Rb > Cs		

[a] Calculated for closely spaced sites.

Other monovalent ions, such as silver, thallium, and ammonium, seem to fit into the selectivity sequences in a regular way.[23]

Values for n from Eq. 5 and the potential data for hydrogen-alkali ion pairs were found to depend on the glass composition, and strongly on the cation, being higher (up to 6 for cesium) the larger the cation.

SELECTIVITY THEORIES

The simple empirical relation of Eq. 5 between activity and concentration of ions in the glass fits data on many different types of ion exchangers.[8,9] Garrels has shown that this description is related to the solution theory of binary mixtures.[1,24]

In the regular solution model the activity coefficient γ_A of species A of a binary solution is given by[25]

$$\ln\left(\frac{\gamma_A}{\gamma'_A}\right) = -\left(\frac{E}{RT}\right)(N_B^2 - N_B'^2)$$

where N_B is the mole fraction of component B, the primes refer to the reference state, and E is a net or excess interaction energy of neighboring A and B ions (for a rigorous definition of E, see Ref. 25) and is independent of composition.

One of the simplest models of a binary solution results from quasichemical or nearest-neighbor assumptions about bonding in the solution. Somewhat more generally one can define a simple mixture as one that has a molar Gibbs-free energy of mixing ΔG_m with the composition dependence[25]:

$$\Delta G_m = EN_A N_B + RT(N_A \ln N_A + N_B \ln N_B)$$

where E can depend on temperature, but not composition, and the Ns are mole fractions. The second term results from the ideal entropy of mixing expression. From this equation the activity coefficients γ of the two components are given by

$$\ln \gamma_A = -EN_B^2/RT$$
$$\ln \gamma_B = -EN_A^2/RT \tag{8}$$

Here the reference state for activity is pure A for component A and pure B for B. Then the ratio of the activity coefficients is

$$\ln\frac{\gamma_A}{\gamma_B} = \frac{E(N_A^2 - N_B^2)}{RT} = \frac{E(N_A - N_B)}{RT}$$

To compare this with Eq. 5, it is assumed that the solution theory can be applied to the glass being considered as a rigid matrix with a mixture of A and B ions on the anionic sites. From Eq. 5

$$\ln \frac{\gamma_A}{\gamma_B} = (n - 1) \ln \left(\frac{N_A}{N_B} \right)$$

If the logarithm is expanded,

$$\ln \left(\frac{N_A}{N_B} \right) = 2 \left[(N_A - N_B) + \left(\frac{N_A - N_B}{3} \right)^3 + \left(\frac{N_A - N_B}{5} \right)^5 + \cdots \right]$$

For N_A between 0.1 and 0.9, where most ion-exchange data are obtained, only the first term in the brackets is important. A comparison of the two earlier equations gives

$$n = 1 - \frac{E}{2RT} \tag{9}$$

Thus the empirical Eq. 5 of Rothmund and Kornfeld has the same form as is derived from the simple solution model. When one or the other component is very dilute, the n representation of Eq. 5 is unrealistic for an exchanger with identical sites, since it requires the activity coefficient of the dilute component to be a function of its concentration even at great dilution. Under these conditions the simple solution equations should be used, but at intermediate compositions the simple n equation is often more convenient.

In the nearest-neighbor solution model a positive value of E denotes a net attraction between unlike atoms, whereas a negative value indicates repulsion. Since Eisenman found n values greater than or equal to 1, deviations from ideality in his glasses were caused by repulsion of the alkali and hydrogen ions. One would expect that as the size difference between the exchanging ions becomes greater, the nonideality (repulsion) or n value should become greater; this functionality was found, except for sodium-hydrogen exchange, which is nearly ideal ($n \approx 1$) for all glasses studied. Glasses with small selectivity between ions were closer to ideal than those with greater selectivity, possibly because the former are more hydrated. From this simple model one would expect more ideal behavior as the site density decreased; however, Fig. 5 of Ref. 11 does not seem to show any such dependence. Apparently the deviations from ideality depend on several factors in a complicated way.

The data of Eisenman, summarized in Fig. 1, show that the selectivity of glasses for ions has a definite dependence on the composition of the glass. The most important factor in this dependence is the type of anionic site in the glass. Thus the alkali silicate glasses containing no other glass former than silica show sequences X or XI (Table 6), with a strong preference for hydrogen and the smaller alkali ions. With aluminum or boron in the glass, however, the sequence is different and depends in a regular way on the glass composition. Eisenman proposed the following theory, based on the structure of the anionic sites in the glass, to explain these dependences.[11]

The SiO^- anionic groups are more compact—the negative charge is less "spread out"—than for the $AlOSi^-$ site. Therefore the field strength, or attractive force for cations, is greater for the SiO^- site, giving it a smaller equivalent anionic radius. The interaction energies of these sites with alkali cations were calculated by Eisenman with a simple electrostatic model and compared to the free energies of formation of the hydrated ions in aqueous solution.[11] The SiO^- site showed the selectivity pattern XI, whereas the $AlOSi^-$ site showed pattern I. It is therefore clear qualitatively why the glasses with only SiO^- sites prefer small ions, whereas the $AlOSi^-$ sites (and similarly the $BOSi^-$ sites) tend to show a preference for larger ions. Eisenman defined the equivalent anionic radius of a site to be the radius which yields the same electrostatic energy with each cation as does the actual site. On this basis the silicate site has an effective radius of about 0.9 Å, whereas the aluminosilicate site has one of about 2.0 Å.

It is also possible to estimate the effective anionic radius of a group from the dissociation behavior of an acid containing it. For example, Eisenman observed that silicic acid is very weak, whereas the aluminosilicate group gives a strong acid. He calculated anionic radii comparable to those of the preceding paragraph by comparing the dissociation constants of these groups with those of halide acids.[11] Boric acid is quite weak, so the selectivity of sodium borate glass should be similar to that of sodium silicate glass. The attraction of borate glass for higher valent ions, as opposed to the behavior of silicate glasses, may possibly result from a shorter distance between neighboring anionic groups in the former. This behavior is also found in amorphous and polymeric phosphates. In borosilicates, on the other hand, the $BOSi^-$ group appears to be similar to the $AlOSi^-$ group.

Eisenman extended his calculations on isolated dry sites by considering the effects of varying site density and extent of hydration. As the sites become more dense, the calculations show that the effective field strength of each site increases, shifting the selectivity patterns and their magnitude, but not changing their sequence. This predicted behavior does not seem to be borne out by the experimental data, however, perhaps because other effects intervene. As the glass becomes more hydrated, the only calculated change is to decrease the magnitude of selectivity [value of $K(u_B/u_A)$], and the sequence of selectivity remains the same. Some of the glasses to the right of Fig. 1, which imbibe more water than the others, seem to show this predicted effect.

Nicolsky et al.[22] showed that both the SiO^- and $AlOSi^-$ sites contribute to the potentials measured in glass electrodes made with sodium aluminosilicate glasses containing less than about 3 wt. % Al_2O_3; however, above this composition only the $AlOSi^-$ sites seem to be effective. In hydrated glass the SiO^- site has H^+, rather than Na^+, associated with it. Apparently the presence of the $AlOSi^-$ sites supresses exchange of other ions with the SiO^-H^+ site or, in terms of Eisenman's model, increases the field strength of the silica site.

In comparing experimental data derived from electrode potentials with the calculations, the effect of changes in mobility ratio have been completely ignored. As mentioned previously, this assumption may be reasonable for certain groups of glasses, for example, for sodium silicates, but is uncertain for others. The qualitative success of Eisenman's considerations indicates that in hydraded glasses the equilibrium selectivity effects are possibly larger than changes in mobility ratios. In a more refined treatment, however, these changes in mobility ratio must be taken into account.

Eisenman's selectivity theory can also be applied to the results on distributions between glass and fused salts, although in this case data are not available for hydrogen, rubidium, or cesium ions. The strong selectivity of soda-lime glass for sodium over potassium results from the small anionic radius of the SiO^- groups, just as for hydrated glasses. Silver is anomalous for its size because of its strong polarizability, which causes it to be attracted strongly by the SiO^- anion.

In the aluminosilicates the preference for silver is reduced because of the larger anionic radius of the aluminosilicate group. The selectivity of sodium over potassium found in the soda-lime glasses is lost, although the smaller lithium ion is still preferred over sodium. In the borosilicate phases there is little selectivity between silver, sodium, and potassium ions, again reflecting the larger anionic size of the borosilicate groups. In fused silica the monovalent ions are associated with the alumina groups in the glass, so a greater relative attraction of these groups for lithium than for sodium, and little selectivity between sodium and silver, is reasonable. More definitive interpretation of selectivity in different dry glasses awaits results on selectivity between ions of widely different sizes.

POTENTIALS OF GLASS ELECTRODES

Until the last decade the origin of potentials in glass electrodes was not well understood. However, recognition that the glass electrode is analogous to an ionic membrane, and studies of ionic behavior in glass, in both aqueous solution and fused salts have led to clarification of this problem. The historical development of the understanding of glass electrode potential is summarized in Ref. 5, Chapters 3 and 5, and Ref. 23. The treatment of the ion exchange theory given here is taken from Chapter 4 of Ref. 5, to which the reader is referred for further details.

The experimental arrangement that is used to measure electrode potentials is shown schematically in Fig. 2. A thin piece of glass separates two

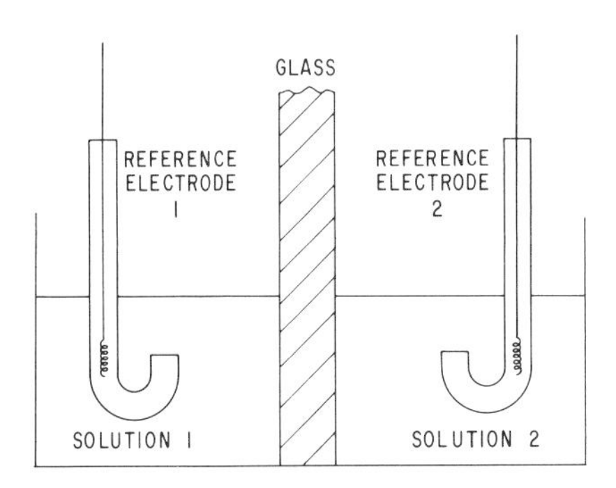

Fig. 2 Schematic diagram of a cell used to measure potentials in glass.

sources of ions. These sources could be gases, liquids, or solids; however, only two types of sources have been studied: aqueous solutions and mixed fused salts. These sources are referred to as solutions. A reference electrode is placed in each solution, and the total potential between these electrodes is read by an external instrument, such as a potentiometer or electrometer.

Two types of measurements are of interest here. In the first type the conditions in solution 2 are held fixed, and the potential is varied by changing the conditions in solution 1. The potential between reference electrode 1 and solution 1 must also be constant under these changing conditions. Then the only changes in the system occur at the solution 1-glass interface. No attempt is made to interpret the absolute value of the potential measured; only its changes with respect to an arbitrary value are considered. This is the type of measurement made in conventional pH meters and in most measurements of the concentrations of other ions with a glass electrode. In the pH application the glass is formed into a bulb, with solution and reference electrode 2 on the inside and solution and electrode 1 outside. Reference electrode 2 is usually a calomel or silver-silver chloride electrode, and solution 2 a buffered chloride solution.[26] Reference electrode 1 can be a calomel electrode connected to solution 1 by a saturated potassium chloride salt bridge. Under these conditions, variations in the pH of solution 1 change the total cell potential by changing the conditions at the solution-glass interface.

In a second type of measurement the absolute value of the cell potential is studied. Usually solvents 1 and 2 are the same, and the reference electrodes are also the same, therefore their potentials cancel or are calculable. Then the total membrane potential (measured directly or calculated) is that caused by different ionic concentrations in solutions 1 and 2, giving rise to different conditions at the two solution-glass interfaces.

The total membrane potential V of a glass electrode can be separated into two additive parts: the boundary or surface potential V_s and the diffusion potential V_D. To interpret measurements of the membrane potential, equations for both V_s and V_D are necessary. First the equations for the surface potential are described.

Consider an ionic solute in two different phases at constant temperature and pressure. If both positive and negative ions can exchange between the two phases, considerations similar to those for a molecular solute apply. However, if only ions of one sign can exchange between the two phases as with silicate glasses, electrical fields can arise. At equilibrium between the two phases the total (or electro-) chemical potentials μ of the exchanging cations in them must be equal. However, this equality cannot be achieved by gross changes of the cation concentration in the glass because the concentration is fixed by the number of anionic groups in the glass network. As the cations leave or enter the glass, giving an excess or deficient total charge, an electrical-potential difference between the two phases builds up. At equilibrium this electrical-potential difference $(\chi' - \chi'')$ just balances the difference in thermodynamic activity of the cations between the two phases. In equation form,

$$\mu' - \mu'' = \phi' - \phi'' + RT \ln \left(\frac{a'}{a''}\right) + Z(\chi' - \chi'')F = 0$$

where F is the Faraday constant, Z is the valence (assumed to be $+1$) of the exchanging ions, ϕ is a function of temperature only and is often called the standard chemical potential, and one prime represents the solutions and two primes the glass.

In the experimental arrangement of Fig. 2, an equation of this type applies to both solution-glass boundaries, so the overall boundary potential V_s for the presence of one cation is

$$V_s = (\chi'_2 - \chi''_2) - (\chi'_1 - \chi''_1) = \frac{[(\phi'_1 - \phi''_1) - (\phi'_2 - \phi''_2)]}{F} + \left(\frac{RT}{F}\right) \ln \left(\frac{a'_1 a''_2}{a'_2 a''_1}\right) \tag{10}$$

where solution 2 is considered positive.

If the glass is uniform and homogeneous, $\phi_1'' = \phi_2''$ and $a_1'' = a_2''$. If both solvents are the same and the same reference state for activity of the ion is used for them, then $\phi_1' = \phi_2'$, and

$$V_s = \left(\frac{RT}{F}\right) \ln \left(\frac{a_1'}{a_2'}\right) \tag{11}$$

so that the boundary potential is a direct measure of the activity ratio of the exchanging ion in the two solutions. In the normal use of glass electrodes, the conditions in solution 2 are held constant, and the relative variation of V_s then measures the activity of the ion in solution 1. If only one ion is present, there is no diffusion potential in the glass.

Purists may object to the individual ionic activities used in the above derivation, since they cannot be measured separately. However, if other components are carried along throughout the derivation, the results are the same. Only ratios of ionic activities, which can be measured, appear in the final results. A further discussion of this point is given by Hefferich (Ref. 2, pp. 140 369 ff.).

From Eq. 2, for both surfaces 1 and 2,

$$\frac{a'}{a''} = \frac{(a' + Kb')}{(a'' + b'')}$$

so from Eq. 10 the boundary potential is

$$V_s = \frac{RT}{F}\left(\ln \frac{a_1' + Kb_1'}{a_2' + Kb_2'} - \ln \frac{a_1'' + b_1''}{a_2'' + b_2''}\right)$$

If the sum of the activities in the glass is the same at both surfaces, the total boundary potential is related to the solution activities by

$$V_s = \left(\frac{RT}{F}\right) \ln \left[\frac{(a_1' + Kb_1')}{(a_2' + Kb_2')}\right] \tag{12}$$

However, if Eq. 5 is followed and n is not unity, the sum of the activities at the two faces is generally not the same, and the second term must be retained. An alternative form for this case is given by Eisenman,[11] and also below (Eq. 20), including the diffusion potential.

In the limit of small concentrations of either A or B in both solutions, Eq. 12 goes to Eq. 11. Similar, but more complicated expressions, are valid when more than two ions exchange.

Now equations for the diffusion potential are derived from the Nernst-Planck equations for interdiffusion of two ions, as described in Chapter 9.

From Eq. 19, Chapter 9, the electrical potential set up when two monovalent ions interdiffuse is

$$E = \frac{RT}{F} \frac{u_B - u_A}{c_A u_A + c_B u_B} \frac{\partial c_A}{\partial x} \frac{\partial \ln a}{\partial \ln c_A} \tag{13}$$

and the diffusion potential V_D is, again considering solution 2 positive,

$$V_D = \int_1^2 E\,dx = \frac{RT}{F} \int_{c_A(1)}^{c_A(2)} \frac{(u_B/u_A) - 1}{c_B(u_B/u_A) - c_A} \frac{\partial \ln a_A}{\partial \ln c_A} dc_A \tag{14}$$

Therefore the diffusion potential depends only on the ionic concentrations at the two surfaces $c_A(2)$ and $c_A(1)$, the mobility ratio u_B/u_A, and the thermodynamic factor $\partial \ln a_A/\partial \ln c_A$. The latter can be written

$$\frac{\partial \ln a_A}{\partial \ln c_A} = 1 + \left(\frac{c_A}{\gamma_A}\right)\left(\frac{\partial \gamma_A}{\partial c_A}\right)$$

since the activity coefficient γ is defined by $a = \gamma c$. Therefore this term measures the change in activity coefficient with concentration; if this coefficient is constant, the term is unity. If the ion exchange follows Eq. 5, then $(\partial \ln a)/(\partial \ln c_A) = n$. With this condition and a constant mobility ratio, Eq. 14 can be integrated to

$$V_D = \frac{nRT}{F} \ln \frac{c_A(2) + c_B(2)(u_B/u_A)}{c_A(1) + c_B(1)(u_B/u_A)} \tag{15}$$

where use is made of the constancy of the total ionic concentration $c_A + c_B$. The diffusion potential can be related to the ionic concentrations in the external solution through the coefficient K of Eq. 2. For the same solvent in the two solutions and $n = 1 (a''/b'' = c_A/c_B)$, the result is[27]

$$V_D = \frac{RT}{F} \left[-\ln \frac{a_2' + K(u_B/u_A)b_2'}{a_1' + K(u_B/u_A)b_1'} + \ln \frac{a_2' + Kb_2'}{a_1' + Kb_2'} \right] \tag{16}$$

The total membrane potential V for $n = 1$, including both boundary and diffusion contributions, is the sum of Eqs. 12 and 16 (see Ref. 27):

$$V = \frac{RT}{F} \ln \frac{a_1' + K(u_B/u_A)b_1'}{a_2' + K(u_B/u_A)b_2'} \tag{17}$$

If the mobility ratio and thermodynamic term are functions of concentration, Eq. 14 cannot be integrated without knowledge of this functionality. However, even in this case the potential V_D is not a function of time, but only of the surface concentrations, for a given dependence of the mobility ratio and thermodynamic term on ionic concentration. Therefore as soon as equilibrium between the solutions and glass surfaces is established the diffusion potential is fixed and does not change with time, independent of the concentration profile of diffusion ions in the glass. The diffusion potential should change with time only if there is some change in structure of the glass with time or position, giving changes of mobility ratio or thermodynamic factor with time and position. Examples of this kind of change are progressive hydration of the glass in contact with water, internal variations in the glass structure, and stress changes in the glass.

Various complications in potentials of glass electrodes are treated in the following references: multicationic systems (Refs. 28 and 29); pressure or stress gradients (Ref. 28); inhomogeneous glass (Refs. 28 and 29); and phase separation (Ref. 30).

EXPERIMENTAL MEASUREMENTS OF POTENTIALS OF GLASS ELECTRODES

A vast number of measurements of potentials of glass electrodes have been made, with objectives ranging from the development of practical analytical tools to the understanding of theories for the electrode potential. Earlier measurements are reviewed in Refs. 5, 26, 31, and 32. In this section a few more recent measurements of special kinds are singled out for comment.

Changes of glass electrode potentials and electrical resistance in aqueous solutions as a function of time have been measured under a variety of conditions.[33–35] These experiments are important in determining the time required before the electrodes are acceptably stable. Response times for glass electrodes have also been studied.[36,37]

The selectivity of Pyrex electrodes to various ions in fused nitrate has been deduced from their potentials,[38,39] and a selectivity sequence of $Na^+ > Ag^+ > Li^+ > K^+ > NH_4^+ \approx Tl^+$ was found. However, these results represent changes in the product $K(u_B/u_A)$ and not in K itself, since the mobility ratios were not measured. These ratios can be quite different from unity in dry glass, so the above sequence must be considered tentative. Furthermore, ionic activities in molten ammonium nitrate, which was used as a solvent in some experiments, were not measured, leading to further uncertainties in the meaning of the selectivities found.

EXPERIMENTAL CONFIRMATION OF POTENTIAL EQUATIONS

Many theories, in addition to the ion exchange theory given above, have been suggested to explain the origin of potentials of glass electrodes.[23,31,32,40] Dole classified the theories into phase-boundary, ion-adsorption, and diffusion-potential categories. In virtually all of these theories it is found that the potential V of a glass electrode when it is sensitive to only one ion A is given by

$$V = V_0 + \frac{RT}{F} \ln a_s \tag{18}$$

where V_0 is independent of concentration, and a_s is the activity of the ion in solution in contact with the electrode. This equation is the same as Eq. 11 with $V_0 = -RT/F \ln a_2'$. V_0 can be different than this value if electrode compartment 2 has different conditions (for example, if different ions are in it), but as long as conditions in this compartment are held constant V_0 will be constant. In this single-ion case there is no diffusion potential. A large number of results agree with Eq. 18; for example, pH measurements with a glass electrode.

When another ion B also has influence on the potential, the following equation is derived in most theories when the glass phase is ideal (ions have constant activity coefficients in it):

$$V = V_0 + \frac{RT}{F} \ln (a_s + K'b_s) \tag{19}$$

where b_s is the activity of B in the solution, and K' is a coefficient independent of ionic concentration. This equation has the same form as Eq. 17 with conditions constant in one compartment and $K' = K(u_B/u_A)$. If the glass is not ideal, but activities in it are given by Eq. 5, then the potential is given by Eq. 7.

Several investigators have found agreement between Eq. 19 and data on electrode potentials and claimed this agreement was proof of the theory they espoused. However, since different theories can lead to Eq. 19, it is essential to test values of K' with independent data to decide which theory is preferable. Only a few such tests have been made. In the ion exchange theory derived above $K' = K(u_B/u_A)$, the product of the equilibrium distribution coefficient and the mobility ratio of the ions.

Eisenman measured the potentials of hydrated glass electrodes selective to sodium and potassium ions in aqueous solutions of these ions (Ref. 5, Chapter 5) and calculated values of $K(u_B/u_A)$ from these measurements. He also measured the mobility ratio and distribution coefficient independently from the penetration of radioactive tracer ions into the hydrated layer. The product $K(u_B/u_A)$ calculated in these two different ways was nearly the same, confirming the ion exchange theory for hydrated glass electrodes in aqueous solution.

I tested the ion exchange theory for dry fused silica electrodes immersed in molten nitrate mixtures at 333°C.[4] Again Eq. 19 was obeyed for the membrane potentials, from which $K' = 0.18$ was calculated for sodium-silver nitrate ion solutions. The ratio of silver-to-sodium mobility was found to be 0.073 from the conductivities of the glass containing only one ion or the other, and K was measured from electrolysis experiments, as described in the first section, and was found to be 2.1, giving $K' = 0.15$ in good agreement with the value measured from membrane potentials. Thus the ion exchange theory is judged to be valid in this case also.

Measurements of K' for Pyrex glass tubes immersed in molten nitrates have also been made[19,30]; however, the two-phase behavior of this glass complicates interpretation of the results. For concentrations of silver nitrate above about 10%, K' was calculated to be about 1.1 from measurements of the membrane potential, although these were deviations of up to 10 mV from the expected potential.[30] In this range of melt concentrations $K \approx 2.0$, and the mobilities of silver and sodium were about equal, giving $K' = 2$. This agreement within a factor of two is further evidence for the validity of the ion exchange model.

Garfinkel measured membrane potentials, mobility ratios, and distribution coefficients for sodium-potassium nitrate melts exchanging with Pyrex glass.[14,19,41] He found evidence for change in individual ionic mobilities as a function of concentration, as well as other concentration dependencies resulting from the two phases, and thus was unable to make an exact comparison between his result and the ion exchange theory.

van Keenan et al. measured membrane potentials of glass electrodes in zinc chloride melts containing sodium and lithium ions. Their results for ionic distribution coefficients and mobility ratios were derived from electrolysis measurements, and were somewhat uncertain because they did not take account of activity coefficient changes in the melt, as mentioned above. Furthermore, they neglected the two-phase nature of the glass. Nevertheless they achieved reasonable agreement between measured potentials and those calculated from the mobility ratio, which was a function of ionic concentration in the glass, and distribution coefficients. It seems likely that compensating errors in measurements of mobility ratios and distribution coefficients were at least partly responsible for this agreement, since the authors reported that these two factors seemed to be related and not independent. In spite of these uncertainties these results provide further evidence that the ion exchange theory is valid for glass electrodes.

Tadros and Lyklema found that the surface charge of hydrated sodium-sensitive glass electrodes in solutions of various alkali chlorides was the same for several different cations (lithium, sodium, potassium, cesium, and tetraethyl ammonium).[42] The surface charge at a certain pH was calculated from the amount of acid or base that had to be added to a solution plus powdered glass to reach this pH, less the amount that had to be added to the same solution without glass to reach the desired pH. Thus the surface charge was actually a measure of the adsorption of monovalent cations on the surface, associated with the negatively charged SiO groups (see the section on adsorption from solution, Chapter 12). These authors concluded that the surface potential V_s for all these ions was the same, since the affinity of the hydrated layer for the ions was the same. They therefore concluded that the value of K (with respect to hydrogen ion) for all these cations was the same, and differences in the selectivity of the glass resulted entirely from different mobility ratios and diffusion potentials, and not from differences in surface potentials. Such a result is not inconsistent with the ion exchange theory, but is contrary to Eisenman's finding that mobilities of sodium and potassium ions in the hydrated layer differed only by factors of five to twelve, and that most of the selectivity of these electrodes resulted from different K values and surface potentials.

The resolution of this discrepancy may reside in the distinction between adsorption just at the glass surface and the selectivity of groups inside the hydrated layer. The experiments of Tadros and Lyklema measured adsorption at the outermost portions of the glass surface, since in their experiments there was insufficient time for interdiffusion of ions very far into the hydrated layer. On the other hand, the equilibrium membrane potential of a glass electrode in a solution of two ions, both of which are present in appreciable amounts in the hydrated layer, results from the selectivity of the bulk of the hydrated layer, not from its surface groups. A longer time is required for these ions to interdiffuse and equilibrate some appreciable distance (several molecular layers) into the hydrated layer than for adsorption at its outer surface. This requirement for longer equilibration time is not contradicted by the rapid response time of a glass electrode in solutions of ions that the electrode selects strongly, for example, in pH electrodes. In this case the surface is saturated with the ion to which the electrode is responding, and when the concentration of this ion in solution changes, the electrode potential changes rapidly because no interdiffusion in the hydrated layer is necessary for establishment of the characteristic membrane potential. Thus Tadros and Lyklema were measuring selectivities of the outermost surface SiO groups for hydrated ions in solution, which can be very different from the selectivities for the unhydrated ions in the bulk of the hydrated layer on the glass.

Different amounts of surface adsorption do not affect overall membrane potentials as long as the adsorbed layer is in equilibrium with both the glass exchanger (hydrated layer) and the surrounding solution. If the adsorbed layer is at equilibrium it can have no concentration gradients and therefore no diffusion potential, and this layer will contribute no additional phase-boundary potential to the measured membrane potential, since the electrochemical potential of ions at the membrane surface will still equal their potential in solution.

The insensitivity of membrane potential to surface adsorption was shown by experiments on fused silica membranes. If the silica surface was treated with nitric acid its surface was highly hydrated, and it strongly adsorbed silver ions from nitrate melts containing silver ions, as shown by radioactive tracer measurements. When the silica surface was treated with dilute hydrofluoric acid the hydrated layer was removed, and the adsorption of silver was negligible. However, in both these samples the measured membrane potential was the same for the same contacting solutions, as was the coefficient K.

REFERENCES

1. R. H. Doremus, in *Ion Exchange*, J. Marinsky, Ed., M. Dekker, New York, 1969, p. 1.
2. F. Helfferich, *Ion Exchange*, McGraw-Hill, New York, 1962.
3. G. Schulze, *Ann. Physik*, **40**, 335 (1913).
4. R. H. Doremus, *J. Phys. Chem.*, **72**, 2877 (1968).
5. G. Eisenman, Ed., *Glass Electrodes for Hydrogen and Other Cations*, M. Dekker, New York, 1967.
6. R. H. Doremus, Ref. 5, Chapter 4.
7. Ref. 2, pp. 193 ff., where there are other references.
8. H. F. Walton, in *Ion Exchange, Theory and Application*, F. C. Nachod, Ed., Academic, New York, 1949, p. 1.
9. E. Hogfeldt, *Acta Chem. Scand.*, **9**, 151 (1955).
10. V. Rothmund and G. Kornfeld, *Z. Anorg, Allgem. Chem.*, **103**, 129 (1918): **108**, 215 (1919).
11. G. Eisenman, *Biophys. J.*, **2**, 259 (1962).
12. R. H. Doremus, unpublished work.
13. M. H. Rowell, *J. Inorg. Chem.*, **4**, 1802 (1965); **5**, 1828 (1966).
14. H. M. Garfinkel, *J. Phys. Chem.*, **72**, 4175 (1968).
15. R. H. Doremus, *Phys. Chem. Glasses*, **10**, 28 (1969).
16. K. Stern, *J. Phys. Chem.*, **74**, 1323 (1970).

16a. O. R. Flinn and K. H. Stern, *J. Phys. Chem.*, **76**, 1072 (1972).

17. T. J. van Reenan, M. Niekerk, and W. J. deWet, *J. Phys. Chem.*, **75**, 2815 (1971).
18. R. H. Doremus, *Phys. Chem. Glasses*, **9**, 128 (1968).
19. H. M. Garfinkel, *J. Phys. Chem.*, **73**, 1766 (1969).
20. R. H. Doremus, *J. Phys. Chem.*, **72**, 2665 (1968).
21. B. P. Nicolsky and T. A. Tolmacheva, *Zh. Fiz. Khim.*, **10**, 504 (1937).
22. B. P. Nicolsky, M. M. Schultz, E. A. Materova, and A. A. Balijustin, *Akad. Nauk. SSSR*, **140**, 641 (1961).
23. G. Eisenman, *Adv. Anal. Chem. Instr.*, **4**, 305 (1965).
24. R. M. Garrels and C. L. Christ, *Solutions, Minerals, and Equilibria*, Harper and Row, New York, 1965, pp. 272 ff.
25. E. A. Guggenheim, *Thermodynamics*, North-Holland, Amsterdam, 1959, pp. 250 ff.
26. R. G. Bates, in *Reference Electrodes*, D. J. G. Ives and G. J. Janz, Eds., Academic, New York, 1961, p. 231.
27. G. Karreman and G. Eisenman, *Bull. Math. Biophys.*, **24**, 413 (1962).
28. R. H. Doremus, Ref. 5, Chapter 4.
29. F. Conti and G. Eisenman, *Biophys. J.*, **5**, 247 (1965).
30. R. H. Doremus, *J. Elect. Soc.*, **115**, 924 (1968).
31. M. Dole, *Glass Electrode*, Wiley, New York, 1941.

32. K. Schwabe and H. D. Suschke, *Angew. Chem.*, **76**, 39 (1964), and earlier papers by Schwabe and co-workers.
33. M. J. D. Brund and G. A. Rechnitz, *Anal. Chem.*, **41**, 1788 (1968).
34. R. P. Buch, *J. Electroanal. Chem.*, **18**, 363 (1968).
35. A. Wikby and G. Johansson, *J. Electroanal. Chem.*, **23**, 23 (1969); **33**, 145 (1971).
36. G. A. Rechnitz and G. C. Kugler, *Anal. Chem.*, **39**, 1682 (1967).
37. A. E. Bottom and A. K. Covington, *Anal. Chem.*, **24**, 251 (1970).
38. N. Notz and A. G. Kennan, *J. Phys. Chem.*, **70**, 662 (1966).
39. A. G. Keenan, N. Notz and F. L. Wilcox, *J. Phys. Chem.*, **72**, 1085 (1968).
40. M. Dole, *J. Am. Chem. Soc.*, **53**, 4620 (1931).
41. H. M. Garfinkel, *Phys. Chem. Glasses*, **11**, 151 (1970).
42. J. F. Tadros and J. Lyklema, *J. Electroanal. Chem.*, **22**, 9 (1969).

Part Four

STRENGTH

The design strength of ordinary glasses is 500 to 1000 times lower than their theoretical strength. The possibility of improving the design strength of glass is therefore one of the most enticing challenges in glass science. Some progress in improving strength has been made, but much can still be done. Advances in increasing strength by various treatments of glass are discussed in Chapter 17. The decrease in strength caused by reaction with water in the atmosphere, called static fatigue, is an important cause of lowered strength, especially in high-strength glasses, and is considered in Chapter 16. The theoretical cohesive strength of glass and the understanding of glass fracture in terms of surface flaws or cracks are treated in Chapter 15. Anelasticity of glass is not discussed in this part, but is included in Chapter 11 because at low temperatures it is often related to ionic transport.

15

Fracture of Glass

Mechanical failure of glass is caused by brittle fracture. Whether or not plastic deformation plays a role in low-temperature fracture of glass is still being argued. The ability to make hardness measurements, and "furrows" or pile-up on crack sides are cited as evidence for plastic deformation.[1–3] However, hardness measurements can be understood in terms of "densification" or anelastic behavior, and when pile-up of material occurs it can be attributed to frictional heating.[4] Other arguments in support of plastic deformation in glass are refuted by Hillig[4] and Ernsberger.[5] There is argument about the possibility of dislocations in glassy structures,[6,49] and although it is possible to consider the leading edge of a crack as a sort of line defect or "dislocation" of the glass structure, this defect will not behave in the same way that dislocations in crystals do. The viscosity of glasses at temperatures well below the glass transition temperature is too high to allow any appreciable viscous flow in them. Thus there is no mechanism for plastic deformation in glass at low temperatures and only uncertain experimental evidence for it.

The following subjects are discussed in this chapter: calculations and measurements of the theoretical cohesive strength of glass; flaw theories to explain actual measured strengths, including the dependence of strength on flaw size, sample size, and temperature; the surface energy approach to fracture; and velocity of propagating cracks. There are several valuable reviews of glass strength.[7–14]

THEORETICAL COHESIVE STRENGTH

The bonding forces between atoms in a material hold it together. Thus to calculate the force that must be overcome to tear apart atomic planes, some model of the forces between atoms is needed. For glasses several different models have been proposed; the calculated strengths do not depend critically on the model chosen.

A reasonable interatomic potential for covalently bonded atoms is the Morse potential[15,13]:

$$U = U_0[e^{-2B(x-a)} - 2e^{-B(x-a)}] \tag{1}$$

where U is the potential energy per unit surface area of two fracture surfaces separated by a distance x, U_0 and B are constants related to the properties of the material being considered, and a is the equilibrium separation of the planes or the interatomic distance of the atoms being separated. This potential function is illustrated in Fig. 1. The binding force per unit area, or

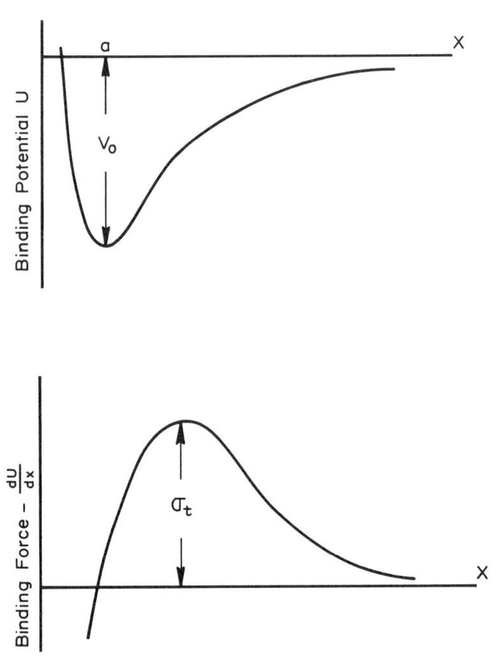

Fig. 1 Schematic diagram of the potential function U for atomic bonding and the binding force or stress dU/dx as a function of atomic separation.

stress, is given by dU/dx, as also shown in the figure:

$$\frac{dU}{dx} = U_0[-2Be^{-2B(x-a)} + 2Be^{-B(x-a)}] \tag{2}$$

When $x = a$, the equilibrium atomic separation, the binding potential is $-U_0$, and the stress is zero. The stress necessary to separate the two planes is the maximum stress between the two planes, and is called the theoretical fracture stress σ_t. This maximum occurs when $dU^2/dx^2 = 0$, or when $x - a = \ln 2/B$; at this distance

$$\frac{dU}{dx} = \sigma_t = \frac{U_0 B}{2} \tag{3}$$

To relate the parameters U_0 and B to properties of the material, it is assumed that Hook's law holds at the equilibrium separation a, so that at $x = a$ the change in stress is proportional to the change in strain:

$$d\left(\frac{dU}{dx}\right)_{x=a} = \frac{E\,dx}{a} \tag{4}$$

where E is Young's modulus, or

$$\frac{E}{a} = 2U_0 B^2 \tag{5}$$

From Eqs. 3 and 5

$$\sigma_t^2 = \frac{EU_0}{8a} \tag{6}$$

which is the equation from which the theoretical cohesive strength σ_t can be estimated from Young's modulus and U_0. The potential energy per unit surface area at equilibrium separation, U_0, can be calculated from bond energies and the atomic density of the material. U_0 can also be considered to be equal to 2γ, where γ is the surface energy of the material.

In silicate glasses the most important cohesive force is the silicon-oxygen bond in the silicon-oxygen network. Therefore U_0 for these glasses can be calculated from the silicon-oxygen bond energy of 106 kcal/mole and the density of these bonds per unit area in the glass. The parameters and results of the calculation for fused silica are shown in Table 1. The calculated value of theoretical strength is similar to values found by others and is about 1.7 times the highest measured value for the fracture strength of fused silica.[16,17] It is interesting that the separation distance for maximum stress is about 1.6 Å, the distance to which two planes of atoms must be separated before they will stay apart.

From the present point of view the difference between the theoretical strength of fused silica and of other silicate glasses should come in three factors: Young's modulus, the density of silicon-oxygen bonds, and the strength of these bonds. Young's modulus is not a strong function of glass composition for most silicate glasses, decreasing no more than about 40% for high-alkali glasses.[18] The bond density per unit surface area is proportional to the two-thirds power of silica concentration, and is therefore

Table 1 Theoretical Strength of Silica Glass

Si—O Bond energy	106 kcal/mole = $8.9(10)^5$ J/mole
Bond density	$7.9(10)^{14}$ molecules/cm^2 = $1.31(10)^{-5}$ moles/m^2
Surface energy = $\frac{U_0}{2}$	$2.9(10)^3$ ergs/cm^2 = 2.9 J/m^2
Young's modulus	$1.05(10)^7$ psi = $7.2(10)^{10}$ N/m^2
Bond distance	1.62 Å = $1.62(10)^{-10}$ m
Theoretical strength σ_t	$2.6(10)^6$ psi = $1.8(10)^{10}$ N/m^2
Measured strengths	
at −196°C (Ref. 16)	$2.0(10)^6$ psi = $1.35(10)^{10}$ N/m^2
at −269°C (Ref. 17)	$2.1(10)^6$ psi = $1.47(10)^{10}$ N/m^2

also not a strong function of glass composition. The effect of composition on strength of the silicon-oxygen bond is not known quantitatively; there are indications of reductions in bond strengths in some silicate glasses from the silica values (see Chapters 6 and 13). All these factors tend to decrease the strength, but since σ_t depends on the square root of all of them, one would expect a decrease of no more than a factor of two in σ_t for high-alkali silicate glasses, as compared to fused silica. No measurements of strengths of high-alkali glasses have exceeded 10^{10} N/m^2 ($1.4(10)^6$ psi). Hosegawa et al.[19] reported strengths of fire-polished sodium borosilicate glass (20 wt. % Na_2O, 10% B_2O_3, 70% SiO_2) up to 10^{10} N/m^2 (N/m^2 = Newtons/meter^2).

The effect of temperature on the factors governing theoretical strength is not large. The calculations above were made for room temperature, but none of the factors changes much up to the transition temperature, perhaps a reduction of up to 25% in σ_t being expected between room temperature and 500°C. Only a small uniform change with lowered temperature would be expected. These expectations are in conflict with experimental results on strengths to be described in the next section.

FLAW THEORIES OF FRACTURE

The theoretical strengths calculated in the preceding section are much greater than strengths actually measured for ordinary glasses, which range from about $3(10)^7$ to 10^8 N/m^2 (5,000 to 15,000 psi). This reduction is caused by flaws in the surface of the glass, which concentrate applied stresses and cause failure at much lower than theoretical applied loads, although the actual effective stress separating atoms must be that calculated in the preceding section. These flaws are introduced into the glass surface by normal handling and abrasion. If a rod of glass, such as fused silica, is drawn out in a flame and protected from contact with other materials, it can have very high strength up to the values recorded in Table 1. A number of experiments, as described by Shand[7] and many others,[8–10] have shown conclusively that the lowered strength of glass is caused by surface flaws introduced by chance abrasion. Hillig showed that stress-induced thermal fluctuations or flaws in a glass surface are unlikely.[10]

For glass rods or bars there is an effect of sample diameter on strength, since the smaller the surface-to-volume ratio the smaller is the chance of a flaw that would cause fracture, given the same sample length. Griffith[20] measured the dependence of strength on the size of glass fibers and found that the smaller the fiber the stronger it was, and by extrapolating to zero-fiber diameter he found a maximum strength of about $1.1(10)^{10}$ N/m^2 ($1.6(10)^6$ psi), close to values actually measured for the maximum strength of glass (see Table 1). Several other authors have measured various dependencies of glass strength on sample size.[10]

Some authors have presented explanations of the size effect in terms of intrinsic flaws, but Hillig has shown that these explanations are not tenable.[10] Hillig has also shown how surface flaws can be revealed by etching a glass with hydrofluoric acid. Thus there is no intrinsic increase of strength as the fiber becomes smaller, only fewer flaws. Hillig showed that a large fiber ($\approx$ 1 mm in diameter) can have very high strength if flaws are removed from its surface.

Different functions for the distribution of the effectiveness of flaws in causing fracture of glass have been assumed. These functions may be useful in describing the flaw distribution on samples abraded in a systematic way, but in presumably pristine samples various bizarre distributions have been found, for example, with two different maxima of strengths.[21] Actually what is desired is not the distribution of flaw effectiveness but the spacial distribution of the most effective flaws; that is, those that actually cause fracture.

In Fig. 2 fracture data at 76°K on centerless-ground FN glass (66% SiO_2, 24% B_2O_3, 3% Al_2O_3, 4% Na_2O, 3% K_2O) rods are given as the fraction of rods that break below a given strength.[22] The probability P of finding a sample of strength s is given fairly well by a gaussian distribution:

$$P = \frac{h}{\sqrt{\pi}} e^{-h^2(s-s_m)^2} \qquad (7)$$

where s_m is the strength of greatest probability ($P_m = h/\sqrt{\pi}$), and h is a measure of the spread of the distribution. The integral of Eq. 7 gives the fraction F of rods that breaks below a certain load s, as plotted in Fig. 2:

$$F = \tfrac{1}{2}[1 + \operatorname{erf} h(s - s_m)] \qquad (8)$$

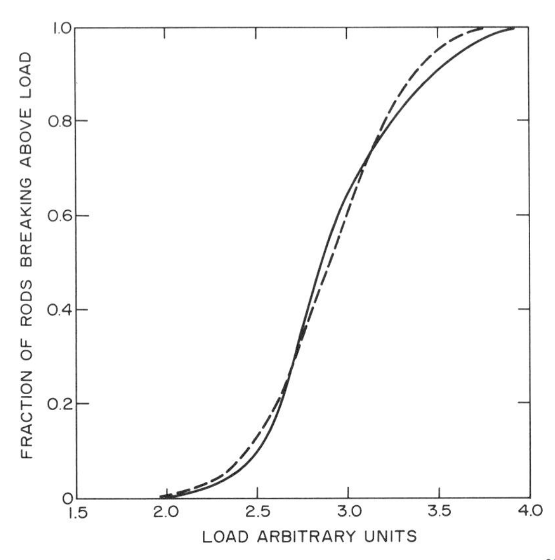

Fig. 2 Distribution of fracture stress at 76°K for abraded rods of FN glass.[22] Solid line, data; dashed line, from Eq. 8.

This function is compared to the data in Fig. 2 and agrees fairly well. However, with other surface treatments other distributions of strength are possible.

What is the nature of the surface flaws responsible for the fracture of glass? It is generally assumed that they are narrow cracks with small radii of curvature at their tips, where applied stresses are concentrated. Such a flaw morphology is consistent with the etching experiments of Hillig, but very few direct observations of these flaws have been made. Most deductions about them come from fracture experiments and distributions of strengths on bulk samples.

If the flaws are assumed to be fine cracks, the stress at their tips can be calculated from elasticity theory. Inglis[23] carried out this calculation for a straight crack with an elliptical cross-section with the applied stress perpendicular to the crack, as shown in Fig. 3. He found that the stress σ at the crack tip in the direction of the applied stress S was

$$\sigma = S\left(1 + \frac{2L}{a}\right) \tag{9}$$

where L is the crack length (semimajor axis of the ellipse) and a half the crack width (semiminor axis of the ellipse). In terms of the radius of curvature $\rho = a^2/L$ at the crack tip, and assuming $L \gg a$,

$$\sigma = 2S\sqrt{\frac{L}{\rho}} \tag{10}$$

Even if the crack is not elliptical in shape Eq. 10 should give a good approximation to the concentration of stress in terms of the length of the crack and the radius of curvature at its tip. A number of other methods of calculating the stress field around the crack tip have been devised,[24–26] but all give Eq. 10 with the assumptions made here.

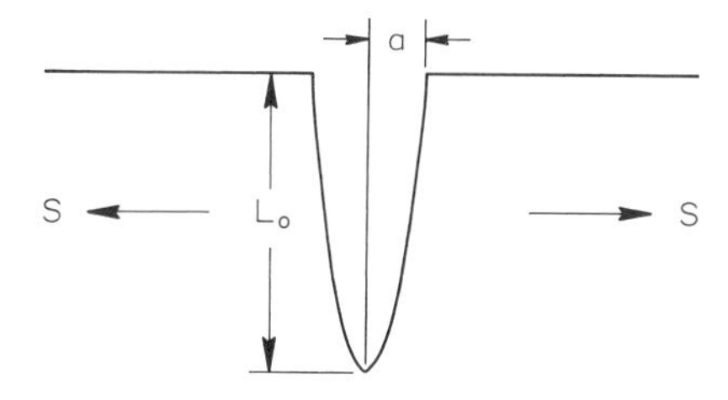

Fig. 3 Elliptical cross-section of a crack with applied tensile stress S.

It is difficult to apply Eq. 10 to actual fracture problems because the radius ρ of the crack tip is probably of atomic dimensions, and has therefore not been measured directly. Many workers have assumed that ρ is a constant for cracks of different lengths, so from Eq. 10 the fracture stress S_f should be inversely proportional to the square root of the length of the crack. This same dependence was found by Griffith in a somewhat different way, as described in the next section.

There are few experimental data available for testing the dependence of the fracture stress on crack length. Griffith provided some of the most convincing evidence for the dependence of breaking stress on crack length.[20] He put cracks right through thin glass bulbs and took their length as twice the effective crack length. Then he pressurized the insides of the bulbs and calculated the stress at fracture. This experiment should be a reasonably good approximation to the crack of Fig. 3, although Griffith's cracks were complete ellipses, not half an ellipse as usually assumed. His results are given in Table 2 and show that the product of the fracture stress and the square root of crack length was indeed constant. From Griffith's data and Eq. 10 the radius of curvature of the crack tip is about 7 Å. The data of Shand[7] on the fracture of thin glass strips with cracks put all the way through the strips are consistent with those of Griffith, but the scatter of the results is so great that the fracture stress could equally well depend on the reciprocal of the length as on the square root of the reciprocal. Anderson[8] plots other results on the dependence of fracture stress on crack length, but these results are

Table 2 Results of Griffith on the Dependence of Fracture Stress on Crack Length in Glass

Crack length $2L$ (in.)	Fracture stress S_f (lb/in.^{2})	$S_f\sqrt{L}$
0.15	864	237
0.27	623	228
0.54	482	251
0.89	366	244

based on uncertain estimates and extrapolations. Mould and Southwick[27] found a reciprocal dependence of fracture stress on square root of crack length in abraded soda-lime glass rods, and their value of $S_f\sqrt{L}$ was about 290 lb/in.$^{3/2}$, as compared to Griffith's average value of 240 lb/in.$^{3/2}$ for a different glass. Thus there is some evidence for a constant $S_f\sqrt{L}$ product, but more results would be desirable to confirm this relationship and the hypothesis that the radii of curvature of the crack tip are constant for different crack lengths.

Some investigators have used the reaction of glass with sodium vapor to reveal surface flaws.[28,29] However, Argon[30] and Ernsberger[31] concluded that most of the cracks visible after the sodium vapor treatment resulted from stresses in the glass surface caused by the reaction of the sodium with glass and subsequent cooling, so this technique does not uniquely reveal preexisting flaws.

Ernsberger devised an ingenious technique for measuring the fracture stress of glass not subjected to external influences.[32] He formed oblate bubbles in glass by sealing off capillary tubing. The tubing was first etched with hydrofluoric acid, cleaned, sealed off with air in the bubble, heated to round the bubble, then deformed to give an oblate bubble, and finally annealed and then cooled slowly. Compression and tension tests were made on bubbles in a soda-lime (Kimble R6) and a borosilicate (Corning 7740) glass (see Chapter 6 for analyses of these glasses). An illustration of one end of a bubble in the soda-lime glass that cracked in compression is shown in Fig. 4.

In compression the soda-lime glass failed at about $4.3(10)^9$ N/m^2 ($6.3(10)^5$ psi), as calculated from elasticity theory for shear stresses around an oblate bubble. This stress is somewhat below the theoretical cohesive strength, as calculated above, although the strength of the silicon-oxygen bonds in a soda-lime glass is uncertain, as mentioned before. No cracks were formed around bubbles in the borosilicate glass; instead a densified layer formed at the bubble surface above a stress of about $7(10)^9$ N/m^2 (10^6 psi). After heating for 90 min at 150°C this layer was not visible.

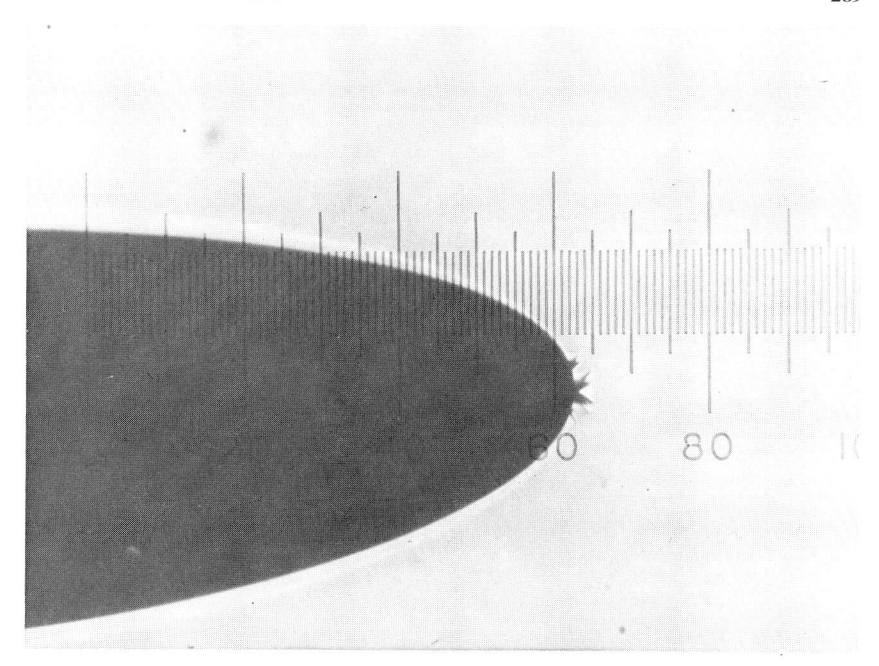

Fig. 4 Cracks formed on an oblate bubble in soda-lime glass in compression.[32]

Ernsberger found a small decrease in strength with time, although the decrease was much slower than found for exposure to ambient air of usual humidities, as described in the next chapter. Thus it is possible that there was a small amount of water in the bubbles, which may have come from out-gassing of the glass at the forming temperature. Ernsberger studied the fracture at internal bubble surfaces of the soda-lime glass in tension as a function of temperature; the results are shown in Fig. 5.

Other investigators have found a similar increase in strength of high-strength glasses at low temperature (Proctor et al.[33] on fused silica, and Cameron[34] on E glass, composition in wt. %: 54.5, SiO_2 ; 14.5, Al_2O_3 ; 16.5, CaO; 4.5, MgO; 9.5, B_2O_3 ; 0.5, Na_2O). On the other hand, the strength of abraded glasses of lower strength shows very little dependence on temperature, at least for short times of loading, and theoretically from Eq. 6 one would not expect much change in the ultimate strength with temperature. Ernsberger considered several explanations for this temperature dependence. The presence of a particular region of temperature in which the strength drops, together with the known sensitivity of strength to reaction with water, as described in the next chapter, strongly suggests to me that this temperature dependence involves a reaction of the glass with water, possibly controlled by the diffusion of water from the glass to its surface, or the

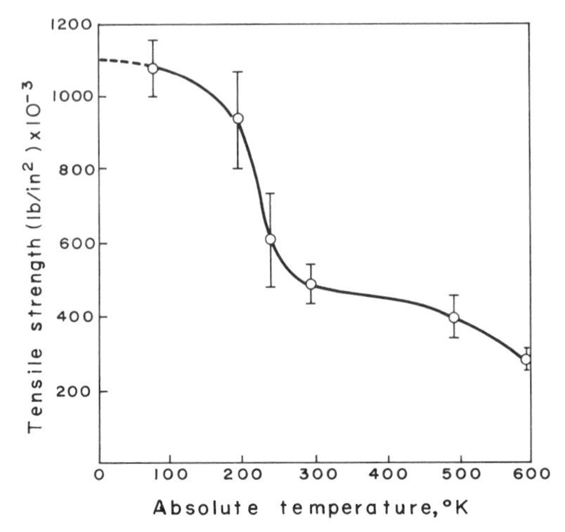

Fig. 5 Temperature dependence of the fracture strength of soda-lime glass as determined by the oblate bubble technique.[32]

surface of the bubble. This possibility needs further testing, as do the explanations offered by Ernsberger.

No effect of hydrostatic pressure on fracture stress up to 20 kbar could be detected by Wiederhorn and Johnson.[35]

SURFACE ENERGY APPROACH

In his original treatment of brittle fracture, Griffith assumed that the total strain energy around a crack should equal the surface energy resulting from the formation of two surfaces by propagation of the crack all the way through a material. From this assumption he calculated the relationship between fracture stress, surface energy, and crack length, as shown below. Hillig argued that this equation provides a necessary but not a sufficient criterion for fracture[10]; the surface energy cannot be larger than the strain energy, but can be less. The more fundamental requirement for crack propagation is that the stress be equal to the ultimate cohesive value σ_t at the crack tip, since if the stress is less than this value, the crack will not propagate, even if the strain energy exceeds the surface energy. This view is supported by the first law of thermodynamics or a simple energy balance: some of the strain energy released by propagation of the crack can appear as heat; it need not all be converted to surface energy.

From the stresses around the crack of Fig. 3 the strain energy W is found to be

$$W = \frac{2\pi L^2 S^2}{E} \tag{11}$$

per unit length of a crack with depth L. E is Young's modulus, and S is the applied stress. The energy to form two surfaces per unit crack length is $4\gamma L$, where γ is the surface energy, and if this energy is equated to the strain energy of Eq. 11, and the result solved for S,

$$S = \sqrt{\frac{2E\gamma}{\pi L}} \tag{12}$$

which is the Griffith criterion for failure. The fracture stress given by this equation can be equated to that of Eq. 10, and the result solved for the radius of curvature ρ of the crack tip:

$$\rho = \frac{8E\gamma}{\sigma_t^2} \tag{13}$$

Thus the Griffith criterion in which surface and strain energy are equated is equivalent to assuming that ρ is given by Eq. 13. Furthermore, any calculation of surface energy from crack propagation data requires an implicit assumption about the value of ρ. In the same way relations between stress intensity factors[36,26,13] and surface energy also involve an implicit assumption about ρ.

Bikerman has discussed the validity of surface energies calculated from fracture experiments.[37]

PROPAGATION OF CRACKS

The discussion so far has concentrated on the requirements for initiating fracture. The speed and direction of propagating cracks have also been studied intensively and give some clues about stress distribution in the solid, as well as being intrinsically interesting.

The surface of glass after fracture shows certain regular characteristics. The radial region closest to the initiating flaw is smooth and is called the mirror region. Beyond it there is a misty region where the surface roughens, and further away a region of gross roughening is called "hackle." The mist results from the beginnings of branching of the crack, and the hackle results from more extensive branching.

A variety of methods have been used to measure the velocity of cracks and are reviewed by Field.[38] The first method used depends on line markings on the mirror surface known as "Wallner" lines, after the man who first explained them,[39] and was developed by Smekal.[40] These lines are formed by reaction of the propagating crack with stress waves resulting from the collision of the crack with extraneous defects. These lines are shown schematically in Fig. 6, from Field.[38] The circular lines in the figure are the

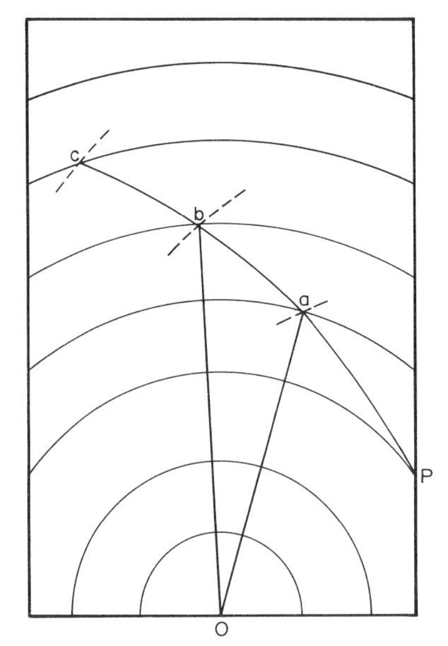

Fig. 6 Schematic representation of Wallner line formation on the fracture face.[38] Crack originates at 0 and reacts with defect at *P*.

crack front at successive times after its origination at 0. If the crack reacts with a defect at *P*, the stress wave interacts with the crack front successively at *a*, *b*, and *c*, forming a Wallner line *Pabc*. The dotted lines show the positions of the stress front.

If the velocity of the stress wave is known, as well as points 0 and *P*, then the crack velocity can be calculated. At higher velocity the line *Pabc* is steeper.

It is also possible to modulate the growing fracture with ultrasonic waves introduced perpendicular to the face.[41,38] The crack velocity is then calculated from the spacing of the resulting ripple marks and the known frequency of the waves.

Various techniques of high-speed photography, such as rotating mirror and multiple-spark cameras, have also been used to measure crack velocity.

Another method involves measurement of the electrical conductivity of metal strips painted on the glass that are successively broken by the crack front.

A number of theories for the propagation of cracks have been reviewed by Erdogan.[42] A simple equation for crack velocity v as a function of crack length L, based on the work of Mott,[43] Berry,[44] and Dulaney and Brace,[45] is

$$v = v_m\left(1 - \frac{L_0}{L}\right) \tag{14}$$

where L_0 is the initial length of the crack, and v_m is the maximum velocity. This maximum velocity is approached asymptotically, as is found experimentally. From the theories v_m is equal to $mv_e = m\sqrt{E/\rho}$, where v_e is the

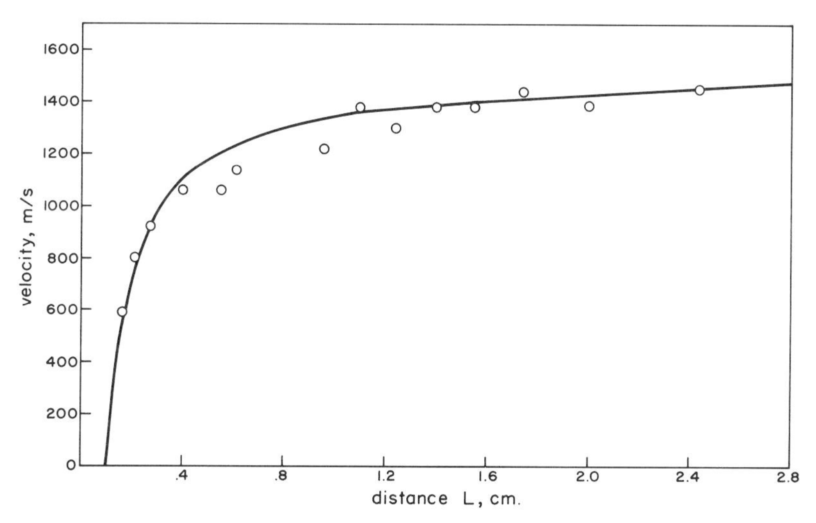

Fig. 7 Velocity of a propagating crack in a soda-lime glass as a function of crack length. Fracture stress $6.7(10)^6$ N/m^2 (10^3 psi). Line, from Eq. 14; points, from Fig. 11, Ref. 38.

longitudinal wave velocity, E is Young's modulus, ρ is the mass density of the material, and m is about 0.4 to 0.5.[42] Equation 14 is compared with some experimental results on the velocity of a crack in a soda-lime glass[38] in Fig. 7. The values assumed for Eq. 14 were $v_m = 1500$ m/sec and $L_0 = 1$ mm. The fit is fairly good, showing that the functional form of Eq. 14 is approximately correct. Congleton and Petch also compared Eq. 14 to crack propagation data and found poorer agreement,[46] as did Kerkhof and Richter.[47]

The variation of the maximum velocity v_m with the glass composition was summarized by Schardin.[41] He found that v_m was only approximately proportional to $\sqrt{E/\rho}$, or put in another way, that the parameter m varied for different glass compositions. He found a regular decrease in m from 0.55 to 0.37 as the ratio of lead to silica or alkali to silica in alkali lead silicate glasses increased. Schardin also observed a good correlation between fracture velocity and $\sqrt{H/\rho}$, where H is the microhardness, and ρ is the density.

The terminal velocity is apparently not affected by the fracture stress.[41]

The distance at which crack branching, resulting in hackle on the fracture surface, appears is inversely proportional to the square of the initial fracture stress in soda-lime glass, and apparently does not depend on the velocity of the crack.[46,48] Thus branching is more likely the higher the stress in the material.

Much lower velocities of cracks have been measured and are probably associated with corrosion of the glass with water. These phenomena are discussed in the next chapter.

REFERENCES

1. D. M. Marsh, *Proc. Roy. Soc.*, **282A**, 33 (1964).
2. K. W. Peter, *J. Noncryst. Solids*, **5**, 103 (1970).
3. E. Dick, *Glastech. Ber.*, **43**, 16 (1970).
4. W. B. Hillig, General Electric Research and Development Center Report 67-C-194, 1967.
5. F. M. Ernsberger, *J. Am. Ceram. Soc.*, **51**, 545 (1968).
6. W. B. Hillig and R. J. Charles, *J. Appl. Phys.*, **32**, 123 (1961).
7. E. B. Shand, *J. Am. Ceram. Soc.*, **37**, 52, 559 (1954).
8. O. L. Anderson, in *Fracture*, B. L. Averbach, D. K. Felbeck, G. T. Hahn, and D. A. Thomas, Eds., Tech. Press, Massachusetts Institute of Technology, and Wiley, New York, 1959, p. 331.
9. R. J. Charles, in *Progress in Ceramic Science*, Vol. I, J. E. Burke, Ed., Pergamon, Oxford, 1961, p. 1.
10. W. B. Hillig, in *Modern Aspects of the Vitreous State*, J. D. Mackenzie, Ed., Butterworths, London, 1962, p. 152.
11. C. J. Phillips, *Am. Scientist*, **53**, 20 (1965).
12. R. E. Mould, in *Fundamental Phenomena in the Materials Sciences*, Vol. 4, L. J. Bonis, J. J. Duga, and J. J. Gilman, Eds., Plenum, New York, 1967, p. 119.
13. S. M. Wiederhorn, in *Mechanical and Thermal Properties of Ceramics*, J. B. Wachtman, Ed., Bureau of Standards, Special Publications 303, 1969, p. 217.
14. N. H. MacMillan, *J. Mat. Sci.*, **7**, 239 (1972).
15. J. J. Gilman, in *The Physics and Chemistry of Ceramics*, C. Klingsberg, Ed., Gordon and Breach, New York, 1963, p. 240.
16. W. B. Hillig, *J. Appl. Phys.*, **32**, 741 (1961).
17. B. A. Proctor, J. Whitney, and J. W. Johnston, *Proc. Roy. Soc.*, **297A**, 534 (1967).
18. G. W. Morey, *The Properties of Glass*, Reinhold, New York, 1954, pp. 295 ff.

19. H. Hosegawa, K. Nishihama, and M. Imaoka, *J. Noncryst. Solids*, **7**, 93 (1972).
20. A. A. Griffith, *Phil. Trans. Roy. Soc.*, **221A**, 163 (1921).
21. R. E. Mould, *J. Appl. Phys.*, **29**, 1263 (1958).
22. J. E. Burke, R. H. Doremus, W. B. Hillig, and A. M. Turkalo, in *Ceramics in Severe Environments*, W. W. Kriegel and H. Palmour, Eds., Plenum, New York, 1971, p. 435.
23. C. E. Inglis, *Trans. Inst. Naval Arch.*, **55**, 219 (1913).
24. N. J. Muschelischivili, *Akad. Wiss. USSR* (1935); see also G. N. Savin, *Stress Concentration around Holes*, Pergamon, N.Y., 1961.
25. H. M. Westergaard, *Trans. ASME*, **61A**, 49, 53 (1939).
26. J. N. Sneddon and M. Lowengrub, *Crack Problems in the Classical Theory of Elasticity*, Wiley, New York, 1969.
27. R. E. Mould and R. D. Southwick, *J. Am. Ceram. Soc.*, **42**, 542 (1959).
28. E. N. Andrade and L. C. Tsien, *Proc. Roy. Soc.*, **159A**, 346 (1937).
29. J. E. Gordon, D. M. Marsh and M. E. Parratt, *Proc. Roy. Soc.*, **249A**, 65 (1959).
30. A. S. Argon, *Proc. Roy. Soc.*, **250A**, 472 (1959).
31. F. M. Ernsberger, *Proc. Roy. Soc.*, **257A**, 213 (1960).
32. F. M. Ernsberger, *Phys. Chem. Glasses*, **10**, 240 (1969).
33. B. A. Proctor, J. Whitney, and J. W. Johnson, *Proc. Roy. Soc.*, **297A**, 534 (1967).
34. N. M. Cameron, *Glass Tech.*, **9**, 14 (1968).
35. S. M. Wiederhorn and H. Johnson, *J. Appl. Phys.*, **42**, 681 (1971).
36. G. R. Irwin, *J. Appl. Mech.*, **24**, 361 (1957); also in *Structural Mechanics*, J. N. Goodier and N. J. Hoff, Eds., Pergamon, New York, 1960, p. 557.
37. J. J. Bickerman, *Phys. Stat. Sol.*, **10**, 3 (1965).
38. J. E. Field, *Contemp. Phys.*, **12**, 1 (1971).
39. H. Wallner, *Z. Phys.*, **114**, 368 (1939).
40. A. Smekal, *Glastech. Ber.*, **23**, 57 (1950).
41. H. Schardin, in *Fracture*, B. L. Averbach, D. K. Felbeck, G. T. Hahn, and D. A. Thomas, Eds., Tech. Press, Massachusetts Institute of Technology, and Wiley, New York, 1959, p. 297.
42. F. Erdogan, in *Fracture*, Vol. II, H. Liebowitz, Ed., Academic, New York, 1968, p. 498.
43. N. F. Mott, *Engineering*, **165**, 16 (1948).
44. J. P. Berry, *J. Mech. Phys. Solids*, **8**, 195 (1960).
45. E. N. Dulaney and W. F. Brace, *J. Appl. Phys.*, **31**, 2233 (1960).
46. J. Congleton and N. J. Petch, *Phil. Mag.*, **16**, 749 (1967).
47. F. Kerkhof and H. Richter, in *Fracture 1969*, P. L. Pratt, Ed., Chapman and Hall, London, 1969, p. 463.
48. S. R. Anthony, J. P. Chubb, and J. Congleton, *Phil. Mag.*, **22**, 1201 (1970).
49. J. J. Gilman, *J. Appl. Phys.*, **44**, 675 (1973).

16

Static Fatigue

The strength of glass deteriorates when it is held under stress in atmospheric air. This deterioration depends on the amount of water in the air,[1] so it apparently results from reaction of water with the glass. In this chapter experimental results on delayed failure of glass as a function of stress, surface treatment, temperature, and glass composition are summarized. Theories to explain this failure are then described. They involve stress-induced corrosion at the tips of cracks, caused by reaction of the glass at the crack tip with water. Experiments on slow propagation of cracks are then shown to be influenced by water, and therefore are presumably related to delayed failure. The value of proof tests for predicting life of stressed glass is finally treated.

EXPERIMENTAL RESULTS

Mould and Southwick studied the delayed failure of microscope slides of soda-lime glass.[2] They used different methods of abrading the glass surface to give different degrees of damage. They found standard deviations in strengths of from 5 to 9% of the average value for surfaces struck by fine particles or "grit-blasted," and deviations of from 8 to 12% with hand abrasions with emery cloth. To obtain these deviations it was necessary to age the unstressed glass in water for 24 hr. Mould and Southwick found that there was no change in strength with time at liquid nitrogen temperature, so the breaking strength S_N at this temperature provided a normalizing factor. When the fraction S/S_N was plotted as a function of log failure time at stress S for different surface treatments, a series of curves of the same shape were found, as shown in Fig. 1. The average failure time at $S/S_N = \frac{1}{2}$, called $t_{1/2}$, was a strong function of surface treatment, as recorded in Table 1; for a particular kind of abrasion this time was shorter when the glass was stronger (had a higher S_N value). However, $t_{1/2}$ was also a strong function of

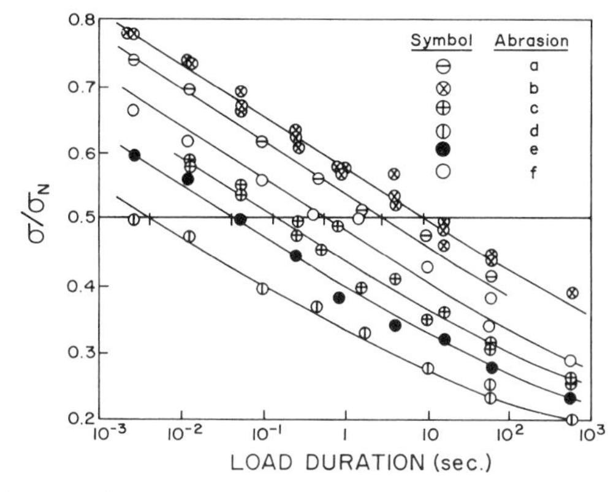

Fig. 1 Fracture time as a function of reduced strength S/S_N for different surface treatments of soda-lime glass.[2] See Table 1 for key to abrasions.

the type of abrasion. The most striking result found by Mould and Southwick was that all the curves had the same shape, so if the data were plotted as a function of log $t/t_{1/2}$ they all fell on one line, the "universal fatigue curve."

Ritter and Sherbourne[3] studied delayed failure of soda-lime glass rods and found different slopes on the S/S_N versus log failure time curves for etched rods, compared to the results of Mould and Southwick. Values of $\beta\sigma_t$, the reciprocal of the slope of the linear portion of the S/S_N versus log time plot, are summarized in Table 2. These authors also found a strong effect of humidity on $t_{1/2}$, but not on the slope.

Table 1 S_N **and** $t_{1/2}$ **for Soda-Lime Glass with Different Abrasions**[a]

	Abrasion	S_N [N/m² (lb/in.²)]	$t_{1/2}$ (sec)
a	Severe grit blast	13.6×10^3	2.9
b	Mild grit blast	12.4	8.8
	Emery cloth, perpendicular to stress		
d	600 Grit	19.5	0.0043
e	320 Grit	13.8	0.039
f	150 Grit	10.1	0.56
	Parallel to stress		
c	150 Grit	23.9	0.14

[a] From Ref. 2.

Table 2 $\beta\sigma_t$ for Various Glasses with Different Surface Treatments

Glass	Surface treatment	Breaking stress at 77°K, S_N, kpsi	$\beta\sigma_t$	Refs.
Soda-lime	Grit blast and Emery cloth	10.6–21.8	31	2
Soda-lime	Etched with 15% HF + 15% H_2SO_4	470	77	3
FN Borosilicate	Abraded	11.0	31	4
FN Borosilicate	Centerless ground	14.8	43	4
Fused silica	Centerless ground	15.1	31	4
Fused silica	Flame-polished	2040	72	8

Delayed failure experiments on General Electric FN borosilicate glass (Corning 7052: 66% SiO_2, 24% B_2O_3, 3% Al_2O_3, 4% Na_2O, 3% K_2O) were reported by Burke et al.[4] and gave the distribution in log failure times shown in Fig. 2. In the figure this distribution is compared with the integrated gaussian curve:

$$F = \frac{1}{2}\left\{1 + \text{erf}\left[h \log\left(\frac{t_m}{t}\right)\right]\right\} \tag{1}$$

where F is the fraction of sample that breaks before time t, t_m is the time at $F = \frac{1}{2}$, and h is a measure of the spread of the distribution. The fracture stress at liquid nitrogen temperature was described by a similar distribution in the preceding chapter (Fig. 2 and Eq. 8). The comparison is good except at short times, where the measurements are probably less reliable. Plots of log failure time as a function of reduced stress gave the slopes recorded in Table 2 for different abrasions. The data for centerless-ground samples of FN glass gave a different slope than those for the soda-lime glass, whereas other samples of FN glass abraded in various ways showed about the same slope as for soda-lime glass and a smaller $t_{1/2}$ value than for the centerless-ground samples.

Experimental results on delayed failure of fused silica rods are collected in Fig.3. The time to failure for a given reduced stress is much longer for this glass than for the others. Furthermore, the slope of the S/S_N versus log failure time and $t_{1/2}$ are strongly affected by the surface treatment, as shown

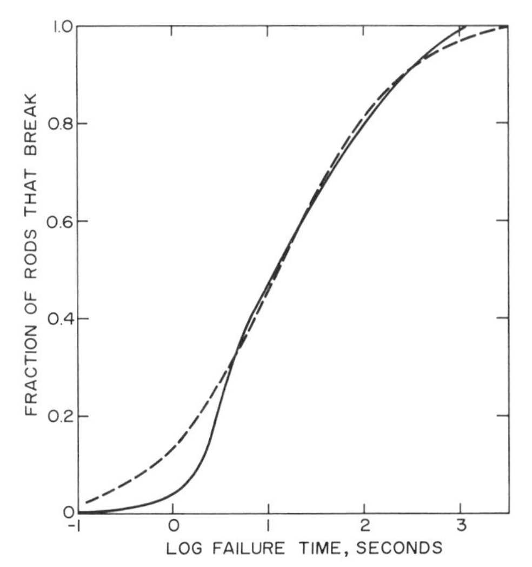

Fig. 2 Distribution in log failure times for FN borosilicate glass (line) as compared with the distribution calculated from Eq. 1, with $h = 0.7$.

in Fig. 3 and Table 2. Of course the absolute stress for the flame-polished samples is much higher than for the abraded or etched samples.

Cameron[9] studied the effect of environment on the strength of E-glass (wt. %: SiO_2, 54.5; Al_2O_3, 14.5; CaO, 16.5; MgO, 4.5; B_2O_3, 9.5; Na_2O, 0.5). He found that at room temperature the strength in vacuum was constant for about 100 min, and at longer times and a vacuum of less than about 0.1 mm Hg the strength slowly increased up to about 15% higher

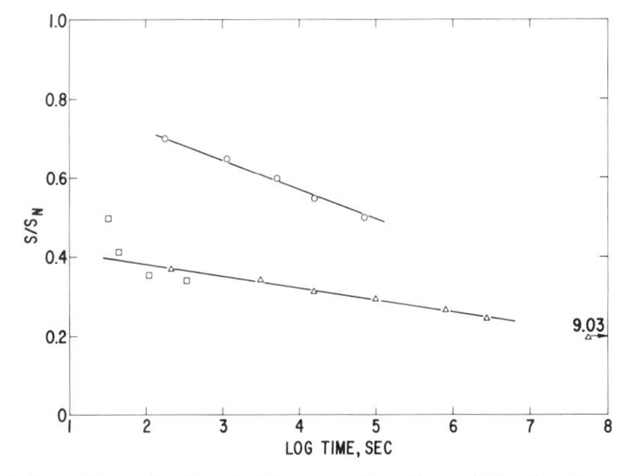

Fig. 3 Delayed failure times for fused silica as a function of the fraction of liquid-nitrogen breaking stress. ○, centerless ground[4]; △, flame-polished[8]; □, etched with 8% HF.[7]

values. Cameron attributed this increase to slow removal of tightly bound water near the glass surface, and it is difficult to find another reasonable explanation. However, when the E-glass fibers were heated above about 200°C, where out-gassing should be more rapid, the strength decreased instead of increasing. Apparently this decrease resulted from dust adhering to the sample surface and its chemical reaction at higher temperatures to form flaws.

Glass fibers of the basic E composition with 5% soda or potash added crack spontaneously when held in acid.[10] This cracking is not a result of corrosion accelerated by stress, as is delayed failure, but is caused by ion exchange between the alkali ions in the glass and hydrogen ions from the water, which results in a build-up of tensile stress in the glass surface.

THEORY OF STATIC FATIGUE

Hillig and Charles have developed a theory to explain delayed failure in glass in terms of stress-induced corrosion by water at crack tips.[11,12] The basic assumptions behind this theory are given here, together with the results derived from it.

Hillig and Charles assumed that the rate of reaction of water with glass controls the rate of change in the shape of the crack tip. This reaction is (see Chapters 12 and 13)

$$-\overset{|}{\underset{|}{\mathrm{Si}}}-\mathrm{O}-\overset{|}{\underset{|}{\mathrm{Si}}}- + H_2O = 2-\mathrm{SiOH} \tag{2}$$

This reaction of water with the silicate network leads to a corrosion of the load-bearing elements of the glass and thus to a change in the effective shape of the crack tip. Under a stress, reaction 2 is accelerated according to the following equation[13]:

$$v = v_0 \exp \beta\sigma \tag{3}$$

where v is the velocity at which the glass surface corrodes under a tensile stress σ, v_0 is the rate of corrosion with no stress, and β is the measure of the effect of stress on the corrosion reaction. β is sometimes written as equal to V^*/RT, where V^* is called the "activation volume." An exponential dependence of the rates of many chemical reactions on pressure has been found, and Eq. 3 is consistent with this dependence. The experiments of Wiederhorn[14,15] on the velocity of crack propagation show the exponential form of Eq. 3, as described in the next section.

A term involving the surface energy is sometimes added to Eq. 3.[11,12] The surface energy of the glass surface in a humid environment is probably much lower than the surface energy involved in forming two fresh glass surfaces by fracture; therefore this surface term may be negligible. In this treatment it is neglected; the available experimental results can be interpreted satisfactorily without it, and there are no results at lower stresses that could show whether it is important.

Failure of the glass is assumed to occur when the stress at the tip of a crack reaches the theoretical fracture stress of the glass. A long crack with an elliptical cross-section, as shown in Fig. 3, Chapter 15, is assumed. The shape of the crack tip is not critical to the treatment, since other shapes should give similar results. The crack depth L is equal to one-half of the major axis of the ellipse, and a is one-half the crack width or one-half of the minor axis of the ellipse. A uniform tensile stress S is applied perpendicular to the crack, as shown in the figure. The stress σ at the crack surface as a function of the coordinates X and Y is (see Refs. 18 to 21 of the preceding chapter)

$$\frac{\sigma}{S} = \frac{1 + 2L/a - (L+a)^2 X^2/a^4}{1 + (L^2 - a^2)X^2/a^4} \tag{4}$$

For a For a crack with $L \gg a$, this function shows that the stress at the tip $(x = 0)$ is

$$\sigma = S\left(1 + \frac{2L}{a}\right) \tag{5}$$

and that the stress drops sharply away from the tip. Equations 4 and 5 are strictly valid for an elliptical hole, and they should also be valid for the crack (a semi-ellipse) when L is more than a few times a. For most cracks $L \gg a$, therefore

$$\sigma \approx \frac{2LS}{a} \tag{6}$$

In any practical situation the orientation of cracks with respect to the stress can be variable. Thus the actual stress at the crack is reduced by factor $\cos \alpha$, where α is the angle between the crack length and the direction of the tensile stress. Furthermore, the crack may penetrate into the glass at some angle with its surface, again reducing the effective stress at its tip. On the other hand, for randomly oriented cracks of the same depth the highest stress occurs at the one oriented perpendicular to the stress, and this crack should be the first to propagate to failure.

Hillig and Charles used the rate Eq. 3 and stress distribution Eq. 4 together with geometric considerations to find the rate of change of the crack shape with time. They found that the radius of curvature of the crack tip $\rho = a^2/L$ changed much more than the length of the crack as corrosion proceeded. In terms of ρ,

$$\sigma \approx 2S\sqrt{\frac{L}{\rho}} \tag{7}$$

When ρ becomes small enough, the stress σ at the crack tip exceeds the theoretical cohesive strength σ_t of the glass, discussed in the preceding chapter, and the crack propagates rapidly to failure. Hillig and Charles found that the failure time t as a function of the reduced stress S/S_N, where S_N is the failure stress at liquid nitrogen temperature, and where there should be no corrosion reaction, was given by

$$\ln\left(\frac{t}{t_{1/2}}\right) \approx -\beta\sigma_t\left(\frac{S}{S_N} - \frac{1}{2}\right) \tag{8}$$

where $t_{1/2}$ is the failure time at $S/S_N = \frac{1}{2}$. Other terms were small for S/S_N greater than about 0.3.

If there is no external stress on the crack, a corrosion process tends to blunt it, rather than sharpen it. There is some small stress at which the shape of the crack tip is preserved as the corrosion proceeds, and at this stress the strength of the glass is constant with time. This situation is called the "fatigue limit." In the Hillig and Charles model without the surface term the fatigue limit is at $S/S_N = 1/(2\beta\sigma_t)$, or about 0.02 for the values of $\beta\sigma_t$ found experimentally.

Equation 8 can be derived more simply by assuming that the length L changes much more rapidly than a, the semiminor axis of the ellipse.[14] This assumption is equivalent to assuming proportionality of the radius of curvature ρ to $1/L$. This assumption may not be realistic for the actual conditions at the crack tip, since a small amount of corrosion just at the crack tip can change ρ much more than L, as assumed by Hillig and Charles.

COMPARISON OF THEORY WITH EXPERIMENT

Delayed failure data for various glasses agree with the functional form of Eq. 8 except at the lowest stresses and longest times, as shown in Figs. 1 and 3. More data are needed to show whether these deviations signal a fatigue limit at lower stresses or simply a breakdown of assumptions used in deriving Eq. 8.

Some results for soda-lime glass, fused silica, and FN borosilicate glass give values of $\beta\sigma_t$ which are about the same, as shown in Table 2. However, with different surface treatments other values are found, as shown in the table. Since both β, the stress-sensitivity factor in Eq. 3, and σ_t, the ultimate fracture stress, should be intrinsic properties of a particular material, the product $\beta\sigma_t$ should not change with different surface treatment of the glass. Such changes demonstrate some inadequacy of the models for delayed failure.

In Hillig and Charles' theory the time $t_{1/2}$ for failure at $S/S_N = \frac{1}{2}$ is about inversely proportional to S_N for a particular material with constant $\beta\sigma_t$. Mould and Southwick found that $t_{1/2}$ was about 100 times longer for glass abraded parallel to the direction of stress than for that abraded perpendicular to the direction of stress, as shown in Table 1. They also found a much greater dependence of $t_{1/2}$ on S_N for a particular mode of abrasion than predicted by the theory, indicating a further deficiency in the model. On the other hand, $t_{1/2}$ for the etched soda-lime glass of Ritter and Sherburne[3] was in the same range as found by Mould and Southwick for abraded soda-lime glass, with S_N more than 20 times smaller. In this case the $\beta\sigma_t$ values were different, and the theory would predict a much smaller $t_{1/2}$ for the more strongly etched glass. The much smaller $t_{1/2}$ for flame-polished fused silica, as compared to abraded silica, can be accounted for from the lower $\beta\sigma_t$ of the latter. More measurements of the effect of surface treatment on $t_{1/2}$ are needed to clarify its relation to S_N and crack shape and orientation.

CRACK PROPAGATION

The rate of slow propagation of cracks in glass is influenced by the humidity of the surrounding air.[15–18] Thus this rate may be related to static fatigue in glass. In these experiments the length of a crack is measured as a function of time under stress. A typical arrangement of the sample is shown in Fig. 4.

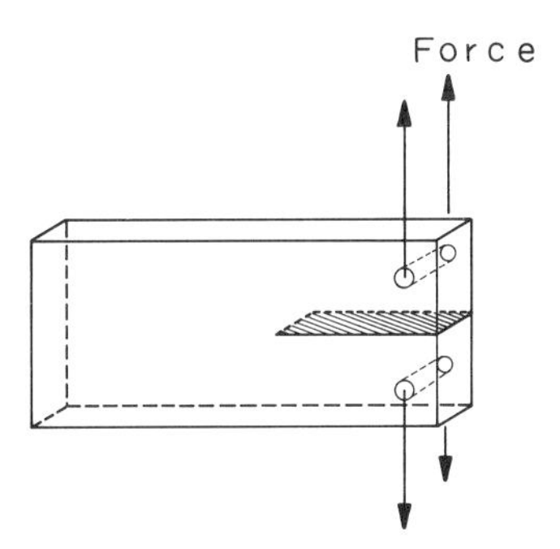

Fig. 4 Specimen for measurement of crack propagation in a microscope slide.[15]

The results of Wiederhorn[15] for the velocity of crack propagation as a function of load (stress) are shown in Fig. 5. The velocity is exponentially dependent on stress at lower stresses, as required by Eq. 3. At higher stress there appears to be a leveling off of the velocity dependence on stress, until the velocity no longer depends on stress. In this transition region (region II in Fig. 5) the velocity still depends on humidity. Wiederhorn suggested that

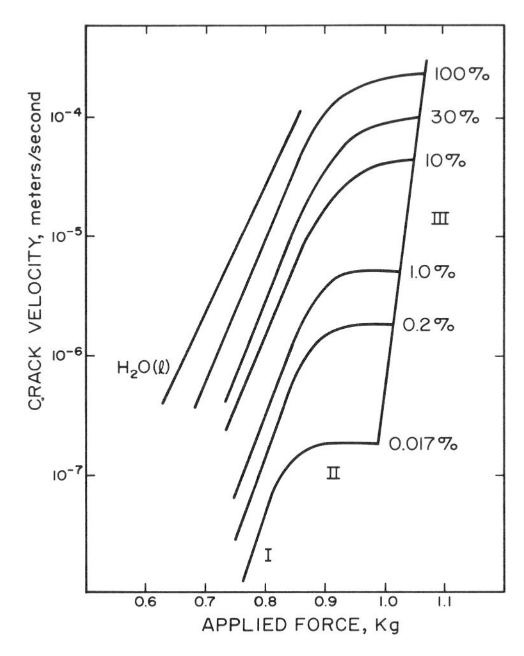

Fig. 5 Crack velocity as a function of applied force in a soda-lime glass at room temperature. Percent relative humidity is shown for each line.[15]

in this region the velocity of propagation of the crack is limited by the rate of transport of water vapor to the crack tip. At still higher stresses the velocity again increases exponentially with load and no longer depends on the humidity. These velocities are still much lower than the maximum velocity of propagation of 1560 m/sec, as described in the preceding chapter. In a later study Wiederhorn and Bolz[16] found an anomalous decrease in velocity at very low stresses, which perhaps was related to a fatigue limit, as discussed in the preceding section.

The dependence of crack velocity on humidity at the same stress in the lower stress region is shown in Fig. 6. At a humidity greater than about 10%

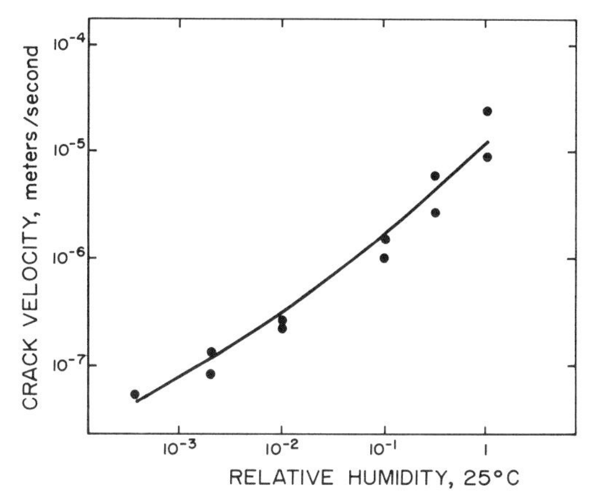

Fig. 6 Effect of humidity on crack velocity in a soda-lime glass at constant stress.[15]

the crack velocity increases linearly with stress; however, at lower humidity the velocity is about proportional to the square root of the humidity, although there are not enough data to determine this dependence with certainty.

Wiederhorn and Bolz measured crack velocity for several different glasses at different temperatures. The results are summarized in Table 3 in terms of v_0, as determined from Eq. 3. This parameter increased exponentially with reciprocal temperature, and activation energies for this dependence are given in Table 3. The composition of aluminosilicate I in wt. % was SiO_2, 57; B_2O_3, 4; Al_2O_3, 20; MgO, 12; CaO, 6; and Na_2O, 1; the composition of aluminosilicate II was SiO_2, 62; Al_2O_3, 17; MgO, 4; Na_2O, 12; and

Table 3 Parameter for Crack Velocities in Different Glasses[a]

Glass	$\ln v_0$	Activation energy (kcal/mole)	$\beta\sigma_t$
Silica	−1.3	33	70
Aluminosilicate I	5.5	29	52
Aluminosilicate II	7.9	30	62
Pyrex borosilicate	3.5	31	64
Lead alkali	6.7	25	39
Soda-lime	10.3	26	36

[a] From Ref. 16.

K_2O, 3; and of the lead alkali was SiO_2, 60; Al_2O_3, 4; PbO, 24; Na_2O, 10; and K_2O, 2.

From the values of v_0 in Table 3 it can be seen that the crack velocity is roughly related to the chemical durability of the glass; the more durable the glass, the slower is the velocity. The activation energy is about the same as for diffusion of alkali in silica, as described in Chapter 9, but is appreciably higher for crack velocity than for sodium diffusion for the other glasses. The activation energy for velocity is also higher than for diffusion of water molecules in the glass (see Chapter 8), and for dissolution of the glass (Chapter 13). A lower activation energy for slow crack propagation of about 14 kcal/mole was found by Schonert et al.[18]

Values of $\beta\sigma_t$ can be calculated from the crack propagation studies for comparison with delayed failure results. In making this comparison the surface energy approach to fracture, as described in the preceding chapter, is used. In terms of the parameters of Wiederhorn and Bolz[16]

$$\beta\sigma_t = \frac{bK_{IC}}{RT} \tag{9}$$

where K_{IC} is the stress intensity factor for fracture, and b is a coefficient of this factor in the velocity equation. K_{IC} is given by[19]

$$K_{IC}^2 = \frac{2\gamma E}{1 - \nu^2} \tag{10}$$

where γ is the fracture surface energy, E is Young's modulus, and ν is Poisson's ratio. The values of γ tabulated by Wiederhorn[19] and E and ν values from Morey[20] are used in Eq. 10 to find K_{IC}, which is then substituted into Eq. 9 together with the b values of Wiederhorn and Bolz. The values of $\beta\sigma_t$ calculated in this way are given in Table 3. The value for soda-lime glass agrees fairly well with that found from the data of Mould and Southwick on delayed failure of abraded samples, but for fused silica the value is closer to delayed failure data on flame-polished or etched samples than on abraded samples. The reasons for these differences are not entirely clear, but may result from inadequacies in the surface energy approach to fracture, as discussed in the preceding chapter.

Varner and Fréchette found marks on the fracture surface of soda-lime silicate glass, which they interpreted as resulting from crack propagation in region II where the crack velocity was not dependent on stress.[21] They ascribed these marks to the velocity-limiting effect of transport of water vapor to the crack tip. As the water vapor at the tip was depleted, the crack stopped because the stress was not high enough to cause fracture without stress corrosion. After build-up of the water vapor concentration again, the crack moved again. This stepwise motion resulted in the surface markings. The marks formed at a velocity of about $2(10)^{-4}$ m/sec in gas, with a dew point of 18°C, and at $5(10)^{-6}$ m/sec at a dew point of −85°C, showing the strong dependence on water vapor concentration.

ENGINEERING DESIGN

The conventional plot of S/S_N versus log t or log $t/t_{1/2}$ is satisfactory for discussions of mechanism of static fatigue, but a relationship between S and t is needed for engineering purposes. The most straightforward approach is to measure S_N and $t_{1/2}$ for the structural element in question, and then construct a graph of S versus t from the universal curve, or a similar curve obtained from specimens of simple geometry. Unfortunately such measurements are inconvenient, and in many cases impossible. Simpler approaches are needed.

The universal curve can be approximated up to intermediate times by a straight line:

$$\log t = -p\frac{S}{S_N} + \log t_{1/2} + k' = -p\frac{S}{S_N} + K \tag{11}$$

where $p = \beta\sigma_t$. If S_N is determined for the structural element, the constant K can be evaluated for the maximum time and S/S_N that the piece will be required to support. From the K a minimum value of S/S_N can be determined for any convenient short time, and a proof test can be performed to eliminate pieces that will not support that stress. In this approach a difficult measurement of S_N must be made, since from Eq. 11 a graph of log t versus S will have a slope that is dependent on S_N.

Charles[22] and Mould and Southwick[2] observed that many static fatigue data yield a straight line on a log-log plot at values of S/S_N less than about 0.6. In Fig. 7 the fit is good for data on soda-lime glass and fused silica. Neither set of results shows an indication of a static fatigue limit when plotted in this way. The straight line portion of this curve follows a power law of the type

$$\log t = -n \log S + C \tag{12}$$

where n is about 13. This relationship cannot hold at short times, since the maximum failure stress is S_N, but represents the data well at the longer times where prediction is desired.

This log-log plot is more convenient than that of Eq. 8 for design and prediction purposes, since curves of log t versus S are not influenced by

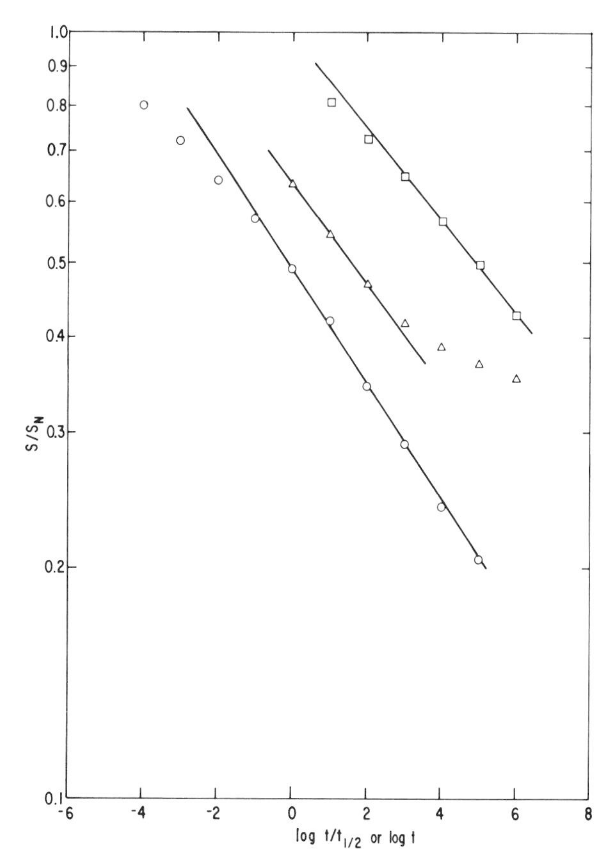

Fig. 7 Log failure time as a function of log S/S_N : □, fused silica; △, FN borosilicate glass[4]; ○, log $t/t_{1/2}$ for soda-lime glass.[2]

different values of S_N. If $n = 13$, one can compute, for example, that if a glass piece is to support a stress of 3000 psi for 10^6 sec (about 4 months), C is at least 51.2. With this value of C the piece should support a proof test of at least 7500 psi for 10 sec. All weaker specimens can be eliminated by such a test. If the value of n is uncertain, it can be found by measuring the time of fracture for specimens of simple geometry at two or three different stresses.

In the above treatment it is assumed that Eq. 12 fits the static fatigue behavior of the specimens at all stresses. If the failure time at low stresses becomes longer than predicted by Eq. 12, then there is a factor of safety over the safe time predicted by the proof test. At short times the proof stress derived from Eq. 12 may be above the expected breaking stress, as shown in Fig. 7. Again the proof test is conservative.

In the above method the stresses in the samples to be tested must be known. If a stress analysis is difficult, specimens can be held at two or three different loads until failure, defining the static fatigue curve of Eq. 12 in terms of load instead of actual stress. Then a proof test can be defined as before, using this curve as a guide.

REFERENCES

1. T. C. Baker and F. W. Preston, *J. Appl. Phys.*, **17**, 179 (1946).
2. R. W. Mould and R. D. Southwick, *J. Am. Ceram. Soc.*, **42**, 582 (1959).
3. J. E. Ritter and C. L. Sherburne, *J. Am. Ceram. Soc.*, **54**, 601 (1971).
4. J. E. Burke, R. H. Doremus, W. B. Hillig, and A. M. Turkalo, in *Ceramics in Severe Environments*, W. W. Kriegel and H. Palmour, Eds., Plenum, New York, 1971, p. 435.
5. W. B. Hillig and V. J. DeCarlo, unpublished results.
6. R. E. Mould, *J. Am. Ceram. Soc.*, **44**, 481 (1961).
7. R. H. Doremus and R. Allen, unpublished results.
8. B. A. Proctor, J. Whitney, and J. W. Johnston, *Proc. Roy. Soc.*, **A297**, 534 (1967).
9. N. M. Cameron, *Glass Tech.*, **9**, 14, 121 (1968).
10. A. G. Metcalfe, M. E. Gulden, and G. K. Schmitz, *Glass Tech.*, **12**, 15 (1971).
11. R. J. Charles and W. B. Hillig, in *Symp. on Mechanical Strength of Glass*, Union Sci. Continentale du Verre, Charleroi, Belgium, 1962, p. 511.
12. W. B. Hillig and R. J. Charles, in *High Strength Materials*, V. F. Jackey, Ed., Wiley, New York, 1965, p. 682.
13. D. A. Stuart and O. L. Anderson, *J. Am. Ceram. Soc.*, **36**, 416 (1959).
14. R. H. Doremus, in *Corrosion Fatigue*, O. Devereux, A. J. McEvily, and R. W. Staehle, Eds., Nat. Ass. Corrosion Engs., Houston, Texas, 1972, p. 743.
15. S. M. Wiederhorn, *J. Am. Ceram. Soc.*, **50**, 407 (1967).
16. S. M. Wiederhorn and L. H. Bolz, *J. Am. Ceram. Soc.*, **53**, 543 (1970).
17. F. Kerkhof and H. Richter, in *Fracture 1969*, P. L. Pratt, Ed., Chapman and Hall, London, 1969, p. 463.
18. K. Schonert, N. Umhauer, and W. Klemm, in *Fracture 1969*, P. L. Pratt, Ed., Chapman and Hall, London, 1969, p. 474.
19. S. M. Wiederhorn, in *Mechanical and Thermal Properties of Ceramics*, J. B. Wachtman, Ed., Bureau of Standards, Special Publication 303, 1969, p. 217.
20. G. W. Morey, *The Properties of Glass*, Reinhold, New York, 1954, pp. 302 ff.
21. J. R. Varner and V. D. Frechette, *J. Appl. Phys.*, **42**, 1983 (1971).
22. R. J. Charles, *J. Appl. Phys.*, **29**, 1549 (1958).

17

Strengthening of Glass

One of the greatest challenges in glass science is to increase the strength of glass. The factor of 100 or more between the theoretical cohesive strength and the engineering strengths of glass present tantalizing possibilities. Since the defects responsible for the loss of strength are on the glass surface, various surface treatments have been used to overcome their influence. Two types of treatments can be recognized. In the first method, the defects are removed, and the glass is protected from further damage, leaving it in the pristine state. Chemically etching or fire-polishing the glass surface are two ways to remove defects. The second type of treatment involves introducing a compressive stress in the glass surface, which adds to the normal fracture stress to increase total stress for fracture. Compressive stress can be introduced by rapid cooling, ion exchange, and surface crystallization. An entirely different method of strengthening is to introduce into the glass fine crystalline particles that form a second phase in the rigid glass. Each of these methods is discussed in turn.

Table 1 Factors by which Glasses Can Be Strengthened by Various Treatments

Treatment	Approximate maximum strengthening factor
Quench-hardening	6
Ion exchange	10
Surface crystallization	17
Ion exchange and surface crystallization	22
Etching	30
Fire-polishing	200
Second-phase particles	2

Uniformly crystallized glasses are usually stronger than the uncrystallized glass in which they form.[1–3] The mechanism of this strengthening is unclear; it may be related to the introduction of a compressive stress in the surface layer in much the same way that controlled surface crystallization develops stress. Alternatively there is some evidence that uniformly crystallized glasses have greater abrasion resistance, that is, greater resistance to the introduction of flaws, than do uncrystallized glasses. The strength increases as the average crystal size decreases[3]; the reason for this change is uncertain. More work is needed to understand the higher strengths of glass ceramics.

Schroder and Gliemeroth have reviewed strengthening of glass by surface treatment.[4]

FIRE-POLISHING

When a glass sample is heated above its softening point (the temperature where the viscosity is $10^{7.6}$ P), it flows fairly readily, and flaws on the surface are eliminated. If the glass is heated above the melting point of its crystalline forms, surface crystals, which can reduce strength, are removed. A convenient method of preparing pristine glass sample is to draw out a filament of glass in a flame. The resulting filament, if it is protected from damage, can have very high strength. Fire-polished rods of fused silica have strengths up to $1.4(10)^{10}$ N/m^2 (2 million psi) at -196°C, as shown in Chapter 15, Table 1. These pristine rods are very sensitive, however. If one is lightly touched with a finger, its strength is much lowered. Thus it takes little mechanical force, or perhaps only corrosion from the body fluids, to introduce an appreciable flaw. Drawn rods of other glasses are even more sensitive to damage or corrosion, as shown by the large scatter in the strengths of fire-polished soda-lime glass,[5] and the maximum strength of only $4(10)^9$ N/m^2 ($5.8(10)^5$ psi).

These results show that it is unlikely that bridges can be built from pristine glass. Multiple sources of damage preclude more than a temporary high strength. One thinks immediately of a protective coating. However, all coatings that have been tried are either too soft and easily penetrated, or too hard and lead to damage or stresses in the glass. Thus the elusive goal of high-strength glass has yet to be achieved by perfecting the glass surface.

CHEMICAL ETCHING

Surface flaws can be removed by chemical etching, the usual etching agent being hydrofluoric acid, because it dissolves silicates rapidly (see Chapters 12 and 13). Sodium and potassium fluoroborates dissolved in alkali nitrates have also been used to etch glass.[6] Etching also can round out the tips of cracks, leading to a lower stress concentration there and subsequent higher strength. In soda-lime, borosilicate, and silica glasses strengths up to 2.7 10^9 N/m^2 $(4(10)^5$ psi) are found after etching the glass with hydrofluoric acid.[7–9] However, for practical application the problem of easy damage is present for etched glass just as it is for fire-polished glass, and the maximum strength for etched samples is lower than for fire-polished.

QUENCH-HARDENING

When a rod or slab of glass is cooled rapidly from above its transformation temperature the surface cools first and reaches a temperature where it is rigid and will not flow. As the piece cools more the interior tries to shrink, but it is restrained from doing so by the rigid surface layer. These conditions lead to a compressive stress at the glass surface and a tensile stress in the interior. This effect can be enhanced by coating a glass with another one of lower thermal expansion; this method was introduced by Otto Schott in 1892.[4] The strength of alkali silicate glasses can be increased by a factor of four by quenching in air to room temperature.

ION EXCHANGE

When a large ion is exchanged for a smaller one at temperatures below the glass transition, a stress builds up in the exchanged region.[10] The profile of stress in a sodium aluminosilicate glass (composition mole %: SiO_2, 70; Al_2O_3, 7.5; Na_2O, 22.5), as measured by Burggraaf,[11] is shown in Fig. 1 after holding the glass in molten potassium nitrate for 24 hr at 350°C. The

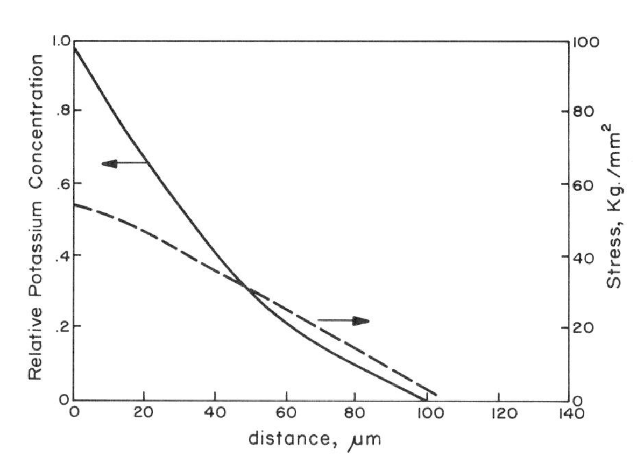

Fig. 1 Profile of stress and potassium ion concentration in a sodium aluminosilicate glass after exchange with potassium nitrate for 24 hr at 350°C.[11]

profile of potassium ion is also shown. With this method the profile and level of stress in the sample can be controlled more easily than for quenching. Abraded strengths of greater than ten times that of the glass before ion exchange have been reported.[12] When the exchanged sample is heated above about 250°C the strength starts to decrease because the ions interdiffuse and reduce the compressive stress at the surface.[13,13a]

It is possible to change the thermal expansion of the surface layer by ion exchange at temperatures above the glass transition; then as the glass cools a compressive stress develops on the surface. This effect can be enhanced by causing crystallization to occur in the surface layer as a result of the ion exchange.[14] This method has been used to strengthen various lithium-sodium aluminosilicate glasses by exchanging the sodium in them for lithium.[14]

Strengthening of glass by treatment in an atmosphere of sulfur dioxides, water, and oxygen results from the replacement of sodium ions by hydrogen ions in glass above its transition temperature.[15,16] (See also Chapter 13.)

The amount of strengthening by ion exchange depends on the glass composition, and time and temperature of exchange.[17,18] Usually the amount of compressive stress at the surface is proportional to the amount of exchange that has gone on, so the higher the temperature and the longer the time of exchange, the greater the stress and strengthening. As the diffusion coefficient of the ions increases, the strengthening at constant time and temperature increases; thus as aluminum is added to sodium silicate glasses the strengthening increases because the ionic diffusion coefficients increase (see Chapter 9).

If a small ion is exchanged for a larger one in the glass, tensile stresses are set up in the glass and it is weakened. Glasses in which lithium or hydrogen ions have replaced sodium ions often show fine cracks on the surface because of these tensile stresses.[19]

SURFACE CRYSTALLIZATION

As mentioned in the preceding section surface crystallization can result from ion exchange above the glass transition temperature. If these crystals cause the surface to have a lower expansion coefficient than the bulk, a compressive stress develops in the surface as the glass is cooled, strengthening it. In most glasses, surface crystallization or devitrification weakens the sample because it introduces tensile stresses or cracks; however, in certain lithium aluminosilicates surface crystallization can strengthen the glass.[20] For strengthening there was an optimum thickness of the surface layer of about 0.07 mm on $\frac{1}{2}$-cm diameter rods; if the crystallization was continued too long, the sample broke up.

COMBINATIONS OF METHODS

It is possible to combine various surface treatments to give both improved strength and abrasion resistance. Ray and Stacey introduced surface compression by ion exchange after rods of soda-lime glass were strengthened by etching.[6] The rods treated in this way and then abraded survived up to three times the stress that etched and abraded rods survived. Other combination treatments would undoubtedly lead to increased strength in certain glass compositions, and new ways to strengthen glass by surface treatment are probably possible. For example, Faile and Roy reported that the strength of fused silica increases slightly by impregnation with argon at 650°C and a pressure of above 1000 atm.[21] There is still, however, a large difference between practical strengths and the theoretical cohesive strength of glass.

ADDITION OF SECOND CRYSTALLINE PHASE

Composites of glass and alumina have been prepared by hot-pressing up to 800°C mixtures of glass powder (< 5 μm in size) with alumina powders from 35 to 44 μm in average size.[22,23] The strength of a borosilicate glass (70 wt. % SiO_2, 16% Na_2O, 14% B_2O_3) was increased up to a factor of two by addition of the particles. Lange explained the increase in strength as resulting from interaction of the crack front with the particles.[23] He hypothesized a pinning of the propagating crack by the second-phase particle, with consequent increase in the energy required for fracture. In terms of the Griffith model presented in Chapter 15, one would have to argue that the second-phase particles increased the theoretical strength of the composite material; the mechanism of this strengthening is not clear for this model. The glass was chosen to have about the same expansion coefficient as the alumina particles, and the composites were annealed, so only small residual stresses should have been present.

REFERENCES

1. P. W. McMillan, *Glass Ceramics*, Academic, London, 1964.
2. P. W. McMillan, B. P. Hodgson, and R. E. Booth, *J. Mat. Sci.*, **4**, 1029 (1969).
3. Y. Utsumi and S. Sakka, *J. Am. Ceram. Soc.*, **53**, 286 (1970).
4. N. Schroder and G. Gliemeroth, *Naturwiss.*, **57**, 533 (1970).
5. R. E. Mould, in *Fundamental Phenomena in the Materials Science*, Vol. 4, J. Bonis, J. J. Duga, and J. J. Gilman, Eds., Plenum, New York, 1967, p. 119.
6. N. H. Ray and M. H. Stacy, *J. Mat. Sci.*, **4**, 73 (1969).
7. B. A. Proctor, *Phys. Chem. Glasses*, **3**, 7 (1967).

8. C. Symmers, J. B. Ward, and B. Sugarman, *Phys. Chem. Glasses*, **3**, 76 (1962).
9. W. Brearley and D. G. Holloway, *Phys. Chem. Glasses*, **4**, 3 (1963).
10. S. S. Kistler, *J. Am. Ceram. Soc.*, **45**, 59 (1962).
11. A. J. Burggraaf, "The Mechanical Strength of Alkali Silicate Glasses after Ion Exchange," Thesis, Eindoven, 1965.
12. M. E. Nordberg, E. L. Mochel, H. M. Garfinkel, and J. S. Olcott, *J. Am. Ceram. Soc.*, **47**, 215 (1964).
13. H. M. Garfinkel, in *Symp. sur la Surface du Verre*, Union Sci. Continentale du Verre, Charleroi, Belgium, 1967, p. 165.
13a. O. A. Krohn, *Glass Tech.*, **12**, 36 (1971).
14. H. M. Garfinkel, D. L. Rothermel, and S. D. Stookey, in *Advances in Glass Technology*, Plenum, New York, 1962, p. 404.
15. R. W. Douglas and J. O. Isard, *J. Soc. Glass Tech.*, **33**, 289 (1949).
16. E. L. Mochel, M. E. Nordberg, and T. H. Elmer, *J. Am. Ceram. Soc.*, **49**, 585 (1966).
17. A. J. Burggraaf and J. Cornelissen, *Phys. Chem. Glasses*, **5**, 123 (1964).
18. H. M. Garfinkel, *Glass Ind.*, **50**, 28 (1969).
19. A. G. Metcalfe, M. E. Gulden, and G. K. Schmitz, *Glass Tech.*, **12**, 15 (1971).
20. J. S. Olcott and S. D. Stookey, in *Advances in Glass Technology*, Plenum, New York, 1962, p. 400.
21. S. P. Faile and R. Roy, *J. Am. Ceram. Soc.*, **54**, 532 (1971).
22. D. P. H. Hasselman and R. M. Fulrath, *J. Am. Ceram. Soc.*, **49**, 68 (1966).
23. F. F. Lange, *J. Am. Ceram. Soc.*, **54**, 614 (1971); *Phil. Mag.*, **22**, 983 (1970).

Part Five

OPTICAL PROPERTIES

This part consists of one chapter on optical absorption of glass. Other optical properties that are not considered, but are reviewed in the following references, are refractive index (Refs. 1 and 2); fluorescence and luminescence (Refs. 3 and 4); and scattering (Ref. 5).

REFERENCES

1. G. W. Morey, *The Properties of Glass*, Reinhold, New York, 1954, pp. 365 ff.
2. H. Scholze, *Glas*, Vieweg and Sohn, Braunschweig, Germany, 1965, pp. 111 ff.
3. W. A. Weyl, *Colored Glasses*, Society of Glass Technology, Sheffield, England, 1959, pp. 439 ff.
4. G. E. Rindone, in *Luminescence of Inorganic Solids*, P. Goldberg, Ed., Academic, New York, 1966, p. 419.
5. J. J. Hammel, in *Physics of Electronic Ceramics, Part B*, L. L. Hench and D. B. Dove, Eds., M. Dekker, New York, 1972, p. 963.

18

Optical Absorption in Glasses

Optical absorption in glass in the visible spectral region colors the glass, leading to applications in optics and to many decorative uses. Absorption and transmission in the infrared and ultraviolet are important in optical uses in instruments. Absorption in all three regions can be used to study short-range structure of glasses, that is, the immediate surrounding of the absorbing atom.

In this chapter the absorption of different glasses is discussed separately in the following order: fused silica, other network oxides, alkali and alkaline-earths oxides, transition- metal oxides, and chalcogenides. A discussion of absorption by small metallic particles in glass is also included.

General reviews of absorption spectroscopy in glasses have been written by Wong and Angell[1] and Kreidl.[2] Earlier, Simon reviewed infrared absorption in glasses,[3] and Weyl discussed colored glasses in a monograph.[4]

FUSED SILICA

Fused silica is useful as an optical material because of its high transmission of visible and ultraviolet light. The transmission edge in the ultraviolet depends on the purity and state of oxidation (stoichiometry) of the silica, and therefore on its method of manufacture. Figure 1 shows the optical absorption at intermediate wave lengths of different fused silicas that are listed in Table 1. The table also gives the methods of manufacture of the various types of fused silica and their approximate impurity contents. Reference to reviews on fused silica are given in Chapter 1.

Absorption by fused silica in the near ultraviolet can result from impurities such as iron, but the main source of absorption is probably a reduced center of some sort. The band at 0.24 μ has been associated with a

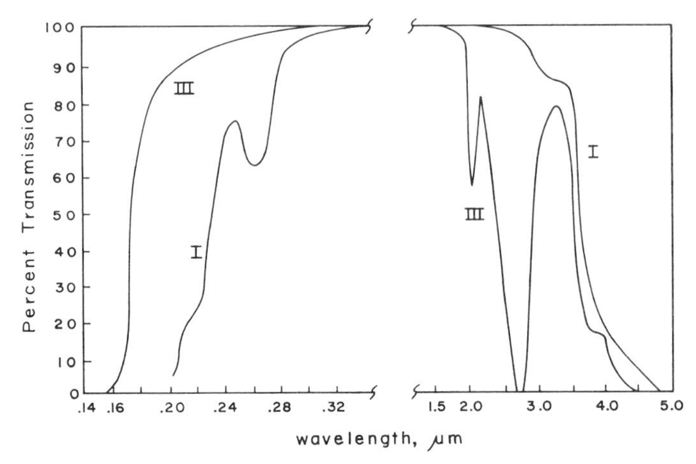

Fig. 1 Optical transmission of different types of fused silica (see Table 1) in the near ultraviolet, visible, and near infrared.

reduced silicogermanate site,[7] a reduced aluminosilicate site,[5] or with a trapped electron or hole near a vacancy.[8] It is not certain which if any of these interpretations is correct. It seems to me most likely that this reduced site is associated with aluminum impurity atoms, because the aluminum-oxygen bonds at an aluminosilicate site are known to be more reactive than normal silicon-oxygen bonds (see Chapter 13), and the intensity of 0.242 μ absorption is related to aluminum content. This band emits fluorescence radiation at 0.28 and 0.39 μ when it absorbs.[9]

The absorption band of fused silica in the near infrared at 2.7 μ is associated with hydroxyl groups in the silica.[10,11]

The absorption of vitreous silica in the infrared is shown in Fig. 2. The absorption bands are the same for different types of fused silica and can be considered to result from transitions of the silicon-oxygen network that are not disturbed by the impurities in fused silica. The band at about 1100 cm^{-1} is not found in the raman spectrum, and is attributed to the stretching vibration of the —Si—O—Si— bond.[3,1,12,13] The lack of a raman band indicates conditions of high symmetry, in agreement with the structure of fused silica suggested in Chapter 3 in which there is considerable short-range order in the arrangement of the silicon-oxygen tetrahedra. The band at 465 cm^{-1} has been assigned to "rocking" of the Si—O—Si bond.[12] The lower-intensity bands at intermediate frequencies involve various bending modes mixed with the stretching and rocking modes of the Si—O—Si bonds. Recent experimental work on raman spectra in silica was carried out by Hass[14] and Stolen.[15]

Table 1 Different Types of Fused Silica

Type[5]	Method of manufacture	Maximum impurity conc. (ppm)									Manufactures designations				
		Al	Fe	Ca	Mg	K	Na	Li	Cl	OH	General Electric	Thermal Syndicate	Heraeus	Corning	Quartz et Silice
I	Electrical fusion of quartz crystal	150	7	12	7	4	12	12	≈ 50	4	101 204 125	IR Vitreosil	Infrasil		Pursil
II	Flame fusion of quartz crystal	uncertain								400		OG Vitreosil	Herosil Homosil Vitrasil		
III	Flame hydrolysis of $SiCl_4$	10	6	4	3	2	2	1	60	1200[a]	151	Spectrosil	Suprasil I	7940	Tetrasil
IV	Vapor phase oxidation of $SiCl_4$	similar to III							500	low		Spectrosil WF	Suprasil W	7943	

[a] Lower values of impurity concentrations are frequently quoted, but are questionable. For alkalis see Ref. 6.

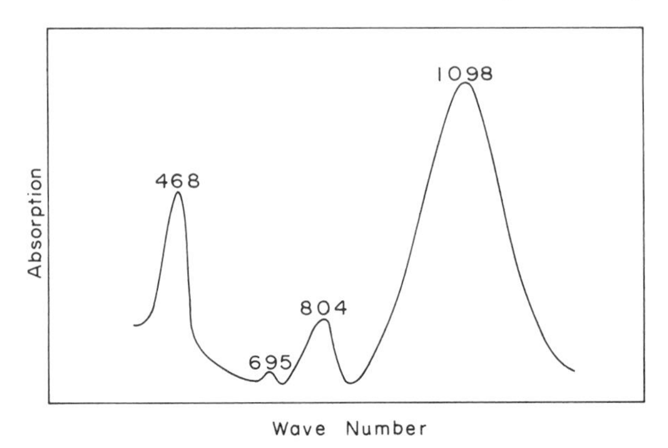

Fig. 2 Optical absorption of fused silica in the infrared.

Absorption and reflection of vitreous silica in the ultraviolet at frequencies higher than the absorption edge (wave lengths below about 0.16 μm) have been measured by Loh,[16] Philipp,[17] and Sigel.[18] The reflectance of fused silica and quartz at these higher energies is shown in Fig. 3. The similarity between these curves shows that optical absorption at these energies is determined by the short-range structure of the silicon-oxygen tetrahedra, which is the same in both amorphous and crystalline silica. This result is consistent with the random-network structure for fused silica discussed in Chapter 3, where it was shown that at short distances of the order of less than 10 Å the crystalline and amorphous structures are virtually identical.

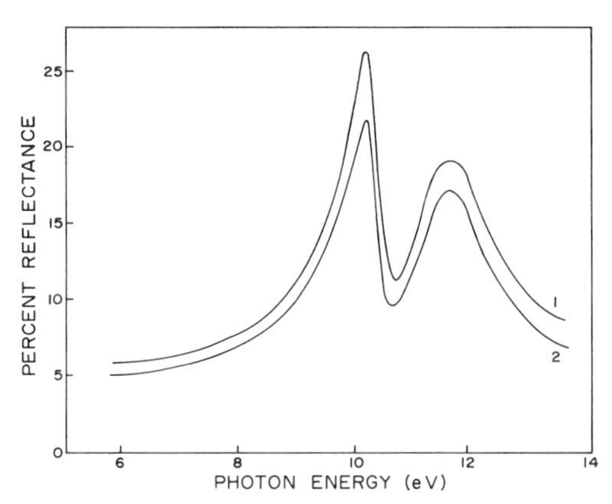

Fig. 3 The reflectance of fused silica and quartz in the ultraviolet.[18] The values for fused silica have been lowered by 5%.

By studying silicate glasses containing various amounts of alkali, Sigel showed that the 11.5-eV band is characteristic of the silicon-oxygen network and is not affected by these additionss.[18] However, the band at 10.2-eV is apparently not present in silicate glasses of the disilicate composition (67% SiO_2, 33% alkali oxide); in these glasses there is another band at 8.5 eV.

Theoretical calculations of the electronic structure of silica have employed valence band[19] and molecular orbital[20,21] approaches. The 11.5-eV band is interpreted to result from band-to-band electronic transitions[19] or atomic-like electronic transitions.[20] Loh and Ruffa suggest that the 10.2-eV band results from an "exciton" (a bound electron-hole pair). Sigel discussed this assignment as a possible way to explain the shift of the 10.2-eV band to 8.5-eV, which could result from the change in the effective dielectric constant of the glass as alkali is added.[18] However, this explanation is quite speculative, and the assignment of the 10.2-eV band to an exciton requires further work for its confirmation.

From a study of the ultraviolet reflectance of mixed silicon and oxygen films Philipp concluded that amorphous substances of all compositions between Si and SiO_2 can be formed, and that they are made up of random mixtures of Si—Si and Si—O bonds with the silicon atoms being four coordinated.[17] He found evidence that these materials are not mixtures of silicon and silica.

OTHER NETWORK-FORMING OXIDES

The approximate ultraviolet wave lengths at which the optical transmission drops sharply (the absorption edge) for simple oxide glasses in μ are SiO_2, 0.16; B_2O_3,[22] 0.17; and P_2O_5,[23] 0.27. The higher edge for P_2O_5 may occur because of the oxygen atoms doubly bonded to the phosphorous that are not part of the phosphorous-oxygen network.

The infrared spectrum of vitreous germania is similar to that of fused silica, but shifted to lower frequencies, with a strong absorption band at about 880 cm^{-1} resulting from —Ge—O—Ge— stretching vibration and bands at lower frequencies corresponding to those in silica.[24–26a] The infrared spectrum of amorphous germania is similar to that of the hexagonal crystalline form, but different from that of the tetragonal crystalline form. This result is consistent with a germanium coordination number of four in amorphous germania, since in the hexagonal form germanium is four coordinated, whereas in the tetrahedral it is six coordinated.

The infrared absorption spectrum of amorphous anhydrous boron oxide (B_2O_3) has been measured by several workers.[27–29] Strong absorption bands are found at 1260 cm^{-1}, and 718 cm^{-1}, with a broad band at lower frequencies, and a shoulder on the 1260-cm^{-1} band at about 1370 cm^{-1}, as

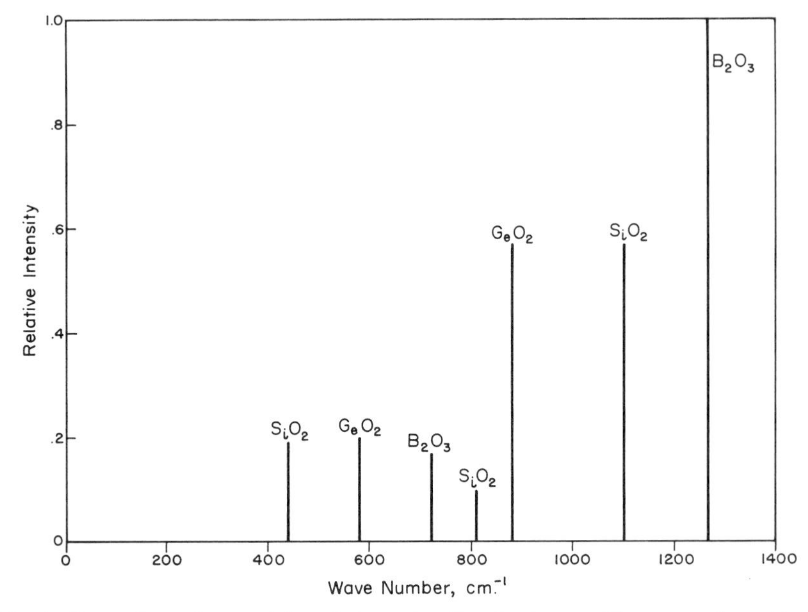

Fig. 4 Comparison of strong infrared absorption lines in vitreous silica, germania, and boric oxide.

shown schematically in Fig. 4. The raman spectrum[30] shows a very intense band at 808 cm^{-1} and a strong band at 1260 cm^{-1}. Krogh-Moe[31] has interpreted these results in terms of the boroxol structure of vitreous B_2O_3, which from x-ray measurements appears to be the correct one (see Chapter 3). In this structure planar boroxol rings with three boron and three oxygen atoms are linked together by other B—O bonds. The band at 1260 cm^{-1} is assigned to the stretching vibration of the B—O bonds linking the boroxol groups. In the ideal case the vibration giving rise to this band should be inactive, but because of its coupling with its surroundings the selection rules are relaxed. As water reacts with vitreous B_2O_3 the intensity of this band decreases.[27] The B—O bonds linking the boroxol groups should be the most reactive, consistent with this result.

The raman line at 808 cm^{-1} has a degree of depolarization which indicates that it results from a symmetrical vibration.[31] This vibration is considered to be a "breathing" motion of the oxygen atoms in the boroxol ring[32,33]; this vibration is inactive in the infrared. The band at 718 cm^{-1} is assigned to an out-of-plane vibration of the boroxol ring. These assignments are therefore consistent with the boroxol structure of vitreous B_2O_3 and strengthen the evidence for this structure.

The infrared spectrum of vitreous arsenic trioxide has been measured by Edmond et al.[33a] and discussed by Wong and Angell.[1] Assignments of the many absorption bands are difficult because of the many active vibrational modes in the crystalline modifications of As_2O_3.

GLASSES CONTAINING ALKALI OXIDE

As alkali oxide is added to silica the infrared absorption band at 1100 cm^{-1} decreases and new bands at 1075 and 940 cm^{-1} develop.[34–38] The band at 940 cm^{-1} has been assigned to the Si—O stretching vibration when the oxygen is not bonded to another silicon (nonbridging oxygen), and the 1075-cm^{-1} band is the 1100-cm^{-1} Si—O stretching band shifted by the presence of $SiONa^+$ groups. A raman band at 540 cm^{-1} can be assigned to bond-bending involving a nonbridging oxygen atom.[2,39]

The 940-cm^{-1} and 1075-cm^{-1} bands in alkali silicates have also been interpreted to result from aggregation of the alkali oxide, either as fine crystals or in separated phases. However, these bands occur in specimens for which there is no evidence for either of these second phases, so the previous interpretation seems more reasonable. Changes in the infrared and raman spectra of these glasses do occur when crystallization and phase separation can be demonstrated, so these spectra can be used to follow the phase changes.

Similar results for the infrared spectra of the alkaline-earth silicates have been found,[3,40] supporting the interpretation that the changes with respect to fused silica result from the introduction of nonbridging oxygen ions.

In the far infrared the absorption of a silicate glass (Corning 0211) was found[41] to increase with the square of the frequency in the range from 10 to 100 cm^{-1}. This absorption was interpreted to result from the excitation of lattice vibrations (phonons) by the alkali ions and other lattice-breaking ions in the glass.[41] In fused silica the absorption in this frequency range is much lower and apparently results from charged defects, either associated with impurities or intrinsic in the silica lattice.[41]

Sigel found that the optical density in the ultraviolet at wave lengths longer than the absorption edge (about 0.16 μ for fused silica) was directly proportional to the alkali concentration in the glass.[18] The intensity of absorption increased in the order of lithium, sodium, and potassium for a given alkali concentration. When aluminum was added to the alkali silicate glasses the absorption decreased. These results indicate that this absorption is associated with the nonbridging oxygen ions introduced with the alkali ions; when aluminum is added these ions are associated with it, rather than increasing the number of nonbridging oxygens. The size of the alkali ions influences the electric field at the nonbridging oxygen ions, changing the absorption. It is not clear if this absorption results from broadening of the 8.5-eV band or represents formation of a new band.

At shorter wave lengths reflection spectra of alkali silicate glasses indicated absorption bands at 11.5 and 8.5-eV, as mentioned in the preceding section.[18] The 11.5-eV band is also present in fused silica and is therefore interpreted to result from electronic transitions associated with the silicon-oxygen lattice. The 8.5-eV band is also found in alkaline–earth silicates, and its position is not much affected by the type lattice-breaking cation. This band could result from a shift of the 10.5-eV band in silica, as discussed in the preceding section.

References to measurements of infrared optical absorption of multicomponent alkali and alkaline-earth silicate glasses are given by Wong and Angell.[1] They show no particularly different features from absorption of binary silicates.

When alkali oxide is added to boron oxide, the infrared band at 1260 cm^{-1} decreases in itensity and a new peak forms[34,29] at about 1400 cm^{-1}. This change can be interpreted to result from the reaction of the boron-oxygen bonds linking the planar boroxol rings with alkali oxide to form boron atoms coordinated to four oxygen atoms, the excess charge of the boron being compensated by the alkali ions. This continuous change in the coordination number of boron with addition of alkali has been found with other techniques, as described in Chapter 3. Alkaline-earth borates show similar changes in the infrared spectrum as alkaline-earth oxides are added.[1]

When sodium oxide is added to boron oxide the ultraviolet absorption edge is not much affected up to about 15 mole % Na_2O.[22] This result is consistent with the interpretation above of the changing boron coordination number with addition of alkali, and contrasts with the increase in ultraviolet absorption in silicate glasses caused by the introduction of nonbridging oxygen ions. In the borates all oxygens are linked to boron atoms and there are no nonbridging oxygen ions. Above about 15 mole % Na_2O the ultraviolet absorption edge moves sharply to higher wave lengths. This increase in absorption is probably related to other changes in properties at this composition, known as the "borate anomaly." It seems likely that these changes are caused by separation of the glass into two amorphous phases,[42,43] as described in Chapter 4, but the exact relationship of the spectral results to the phase present has not been clarified in detail.

The infrared spectra of a variety of alkali and alkaline-earth phosphates were very similar,[1] indicating that the absorption bands resulted from the vibration of phosphorous-oxygen groups. These metaphosphates have a chainlike structure, and the spectra are consistent with calculations based on this structure.[1] In the ultraviolet the addition of alkali or alkaline–earth oxides to phosphorous oxide shifts the absorption edge to lower wave lengths, rather than higher as for other glass formers. This shift may occur because of tighter oxygen lattice.

The addition of alkali oxide to germanium induces a larger shift in the main infrared absorption band than for the same addition to silica.[24] The larger shift apparently results because the coordination number of germanium with oxygen is changing from four to six as alkali is added.[24]

TRANSITION METAL IONS

A variety of colors in glasses are caused by transition metal ions in them, often in small concentrations. Thus the green color observed for thick slabs of most commercial glasses results from absorption of impurity ferric ions in the glass, and some absorption of this ion in the ultraviolet persists even when it is present in concentrations of less than 1 part/million. Other familiar coloring ions for glass are chromium for green, cobalt for blue, and manganese for purple.

The absorption bands for several different transition metal ions in glass are summarized in Table 2. Also given in the table are the most common colors derived from each ion, its coordination number with oxygen, and the number of *d* electrons.

These visible absorption bands result from electron transitions of the 3*d* electrons of the ions. When the transition metal ions are coordinated with other ions the energy levels of these *d* electrons are split by the electric field of the coordinating ions, instead of being degenerate as in the free ion. The *d*-electron orbitals are strongly directional, so the splitting is sensitive to the arrangement of the surrounding ions, and the electronic transitions and resulting absorption spectra can be used to study the coordination numbers of the central ions. The theory of these effects, called "ligand field" theory, was first developed by H. Bethe, and its application to glasses has been discussed in detail by Bates.[44] Shorter treatments are also given in Refs. 1 and 2. Various experimental measurements on spectra of transition metal ions in glass are also reported in Refs. 45 to 54.

The assignment of coordination numbers given in Table 2 was mostly determined from the absorption spectra. In some cases (iron, cobalt, and nickel) both tetrahedral (4) and octahedral (6) coordination is possible, depending on the composition of the glass. Unequivocal determination of the coordination number from the optical spectrum alone is not always possible, for example, in the case of iron, so other techniques such as Mössbauer spectroscopy and electron spin resonance have been used to help in deciding on the correct coordination number.

Table 2 Visible Optical Absorption in Soda-Lime Silicate Glasses of Transition Metal Ions

Ion	No. of *d* electrons	Abs. max. (μ)	Coordination no. with oxygen	Color of glass	Remarks	Refs.
Cr^{3+}	1	0.66	6	Green		44
		0.45				
Ti^{3+}	1	0.57	6	Purple	Borosilicate glass	4
V^{4+}	1	1.12	6?	Red		45
V^{3+}	2	0.645	6	Green		45
		0.425				
Mn^{3+}	4	0.50	6	Purple		4
Mn^{2+}	5	0.435	4, 6?	Brown		4,51,52
Fe^{3+}	5	0.41	4?	Green	Potassium silicate	46,47
Fe^{2+}	6	1.10	4, 6?	Blue		46,53
Co^{2+}	7	0.56	6	Blue	Low alkali borate	48,50
		0.60	4		High alkali borate	
Ni^{2+}	8	1.33	6	Purple, grey	Low alkali borate	48
		0.76				50
		0.42				50a
		1.19				
		0.68–0.49	4		High alkali borate	
Cu^{2+}	9	0.79	6	Blue		45

CHALCOGENIDES

In the recent flurry of interest in semiconducting chalcogenide glasses optical absorption has been used to aid in determining the band structure of these materials (see Chapter 10). Tables listing measurements made before 1970 of the infrared spectra of a large number of different chalcogenide glasses are given in Ref. 1.

More recently there have been a series of studies of vitreous arsenious sulfide and selenide as two of the chemically simplest chalcogenide glasses.[55–58] In the ultraviolet these glasses show an absorption edge resulting from interband electronic transitions. The results indicate that a high density of electronic states spread into the forbidden band gap between the valence and conduction bands, as discussed in detail in Chapter 10. In the infrared, absorption bands in the region from 1500 to 3600 cm^{-1} are attributed to impurities and result from O-H and S-H stretching vibrations.[55] Bands and the increasing amount of absorption at lower frequencies are attributed to various lattice vibrations.[55,57]

METALLIC PARTICLES

Small metallic particles dispersed in glass absorb light and can develop striking colors. Best known of these glasses is gold ruby glass, which has been known since the seventeenth century. Faraday recognized that the color of gold ruby glass resulted from finely divided gold particles.[59] This glass is made by dissolving gold in the glass melt, as an ion, in which state the gold is retained when the glass is cooled rapidly.[60] To form the gold particles the glass is reheated to an intermediate temperature region. Certain agents, such as antimony oxide, in the glass aid nucleation of the particles, or they can be nucleated by ultraviolet, x-ray, or γ-radiation if a radiation-

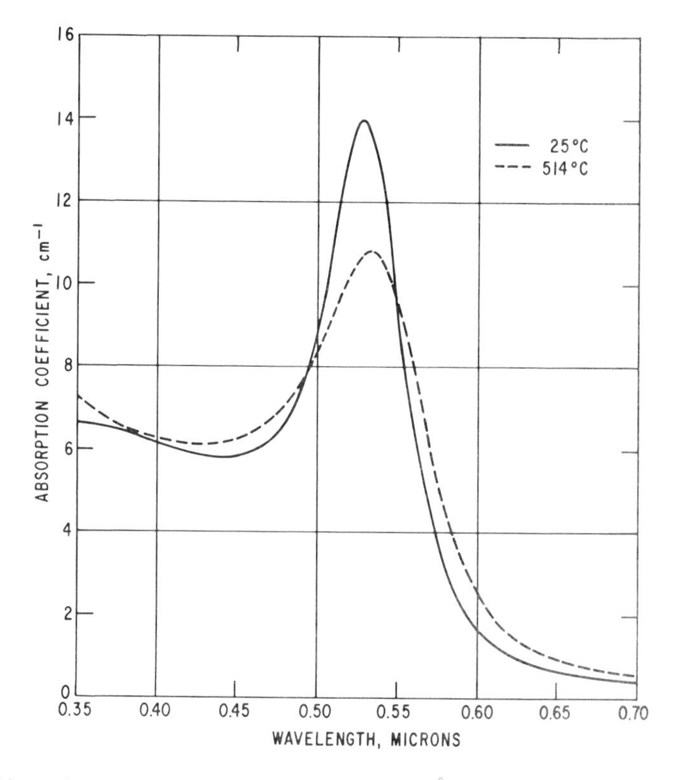

Fig. 5 Absorption spectrum of gold particles about 200 Å in diameter in glass.

sensitive ion such as cesium is present.[60] The growth of the particles takes place by diffusion of gold atoms or ions to the particles.[61]

The color of gold ruby glass results from an absorption band at about 0.53 μ, as shown in Fig. 5. This band comes from the spherical geometry of the particles and the particular optical properties of gold.[62] It can be considered as a "plasma-resonance" band, in which the free electrons in the metal are considered as a bounded plasma. These electrons oscillate collectively at a particular frequency in the bulk metal, known as the plasma-resonance frequency. The spherical boundary condition of the particles shifts this resonance oscillation to lower frequencies (longer wave lengths).

The size of the gold particles influences the absorption. For particles larger than about 200 Å in diameter the band shifts to longer wave length as the oscillation becomes more complex.[62] For smaller particles the bandwidth progressively increases because the mean free path of the free electrons in the particles is about 400 Å, and is effectively reduced.[63]

Silver particles in glass color it yellow, resulting from a similar absorption band at 0.41 μ.[64] It has been thought that the red color of copper ruby glass resulted from small copper particles in it, but other studies suggest that the color may arise from cuprous oxide particles in the glass.[65]

REFERENCES

1. J. Wong and C. A. Angell, *Appl. Spectrosc. Rev.*, **4**, 97 (1970).
2. N. J. Kreidl, in *Physics of Electronic Ceramics*, L. L. Hench and D. B. Dove, Eds., M. Dekker, New York, 1972, p. 915.
3. J. Simon, in *Modern Aspects of the Vitreous State*, Vol. I, J. D. Mackenzie, Ed., Butterworths, London, 1960, p. 138.
4. W. A. Weyl, *Colored Glasses*, Society of Glass Technology, Sheffield, England, 1959.
5. G. Hetherington, K. H. Jack, and M. W. Ramsay, *Phys. Chem. Glasses*, **6**, 6 (1965).
6. R. H. Doremus, *Phys. Chem. Glasses*, **10**, 28 (1969).
7. V. Garino-Canina, *Verres Refrac.*, **6**, 313 (1958).
8. W. H. Turner and H. A. Lee, *J. Chem. Phys.*, **43**, 1428 (1965).
9. A. Kats and J. M. Stevels, *Philips Res. Rep.*, **11**, 115 (1956).
10. V. Garino-Canina, *C.R.*, **239**, 705 (1954).
11. R. V. Adams and R. W. Douglas, *J. Soc. Glass Tech.*, **43**, 147 (1959).
12. R. J. Bell, P. Dean, and D. C. Hibbins-Butler, *J. Phys.*, **1**, 299 (1968); **3**, 2111 (1970); **4**, 1214 (1971).
13. J. B. Bates, *J. Chem. Phys.*, **56**, 1910 (1972).
14. M. Hass, *J. Phys. Chem. Solids*, **31**, 415 (1970).
15. R. H. Stolen, *Phys. Chem. Glasses*, **11**, 83 (1970).
16. E. Loh, *Solid State Commun.*, **2**, 269 (1964).
17. H. R. Philipp, *Solid State Commun.*, **4**, 73 (1966); *J. Phys. Chem. Solids*, **32**, 1935 (1971).
18. G. H. Sigel, *J. Phys. Chem. Solids*, **32**, 2373 (1971).
19. A. R. Ruffa, *Phys. Stat. Solids*, **29**, 605 (1968).

20. M. H. Reilly, *J. Phys. Chem. Solids*, **31**, 1041 (1970).
21. A. J. Bennett and L. M. Roth, *J. Phys. Chem. Solids*, **32**, 1251 (1971).
22. B. D. McSwain, N. F. Borrelli, and G. J. Su, *Phys. Chem. Glasses*, **4**, 1 (1963).
23. E. Kordes and E. Worster, *Glastech. Ber.*, **32**, 267 (1959).
24. E. R. Lippencott, A. V. Valkenburg, C. E. Weir, and E. R. Bunting, *J. Res. Nat. Bur. Stand.*, **61**, 61 (1958).
25. V. V. Obukhov-Denisov, N. N. Sobolev, and V. P. Chemisinov, *Opt. Spect.* (*USSR*), **8**, 267 (1960).
26. M. K. Murthy and E. M. Kirby, *Phys. Chem. Glasses*, **5**, 144 (1964).
26a. B. T. K. Chen and G. J. Su, *Phys. Chem. Glasses*, **12**, 33 (1971).
27. J. A. Siderov and N. N. Sobolev, *Opt. Spect.* (*USSR*), **3**, 560 (1957).
28. J. L. Parsons and M. E. Milberg, *J. Am. Ceram. Soc.*, **43**, 326 (1960).
29. N. F. Borrelli, B. D. McSwain, and G. J. Su, *Phys. Chem. Glasses*, **4**, 11 (1963).
30. E. N. Lotkova, V. V. Obukhov-Denisov, and N. N. Sobolev, *Opt. Spect.* (*USSR*), **1**, 772 (1956).
31. J. Krogh-Moe, *Phys. Chem. Glasses*, **6**, 46 (1965); *J. Noncryst. Solids*, **1**, 269 (1969).
32. J. Goubeau and H. Keller, *Z. Anorg. Chem.*, **272**, 303 (1953).
33. L. A. Kristionsen and J. Krogh-Moe, *Phys. Chem. Glasses*, **9**, 96 (1968).
33a. J. T. Edmond and M. W. Redfearn, *Proc. Phys. Soc.*, **81**, 378, 380 (1963)
34. P. E. Jellyman and J. P. Proctor, *J. Soc. Glass Tech.*, **39**, 173 (1955).
35. V. A. Florinskaya and R. S. Pechenkina, *The Structure of Glass*, Consultants Bureau, New York, 1958, p. 55; *J. Struct. Chem. USSR*, **4**, 850 (1963) (English translation).
36. G. J. Su, N. F. Borrelli, and A. R. Miller, *J. Phys. Chem. Glasses*, **3**, 167 (1962).
37. R. Hanna and G. J. Su, *J. Am. Ceram. Soc.*, **47**, 597 (1964).
38. D. Crozier and R. W. Douglas, *Phys. Chem. Glasses*, **6**, 240 (1965).
39. G. Wilmot, Ph.D. Thesis, Massachusetts Institute of Technology, 1954.
40. Y. G. Shteinberg and O. N. Setkina, *J. Appl. Chem.* (*USSR*), **36**, 712 (1963) (English translation); **38**, 1451 (1965).
41. W. Bagdage and R. Stolen, *J. Phys. Chem. Solids*, **29**, 2001 (1968).
42. W. Skatulla, W. Vogel, and H. Wessel, *Silikat Tech.*, **9**, 51 (1958).
43. R. R. Shaw and D. R. Uhlmann, *J. Am. Ceram. Soc.*, **51**, 377 (1968), and earlier articles referenced in this paper.
44. T. Bates, in *Modern Aspects of the Vitreous State*, Vol. 2, J. D. Mackenzie, Ed., Butterworths, London, 1962, p. 195.
45. S. Kumar, *Cent. Glass Ceram. Res. Inst. Bull.*, **6**, 99 (1959).
46. V. V. Vargeime and T. J. Weinberg, in *IVth Congres International du Verre*, Imprimerie Chaix, 20, Rue Bergere, Paris, 1957, p. 197.
47. C. R. Kurkjian and E. A. Sigety, *Phys. Chem. Glasses*, **9**, 73 (1968).
48. M. A. Aglan and H. Moore, *Trans. Soc. Glass Tech.*, **39**, 351 (1955).
49. H. Moore and H. Winklemann, *Trans. Soc. Glass Tech.*, **39**, 215 (1955).
50. A. Paul and R. W. Douglas, *Phys. Chem. Glasses*, **8**, 151, 233 (1967).
50a. W. H. Turner and J. A. Turner, *J. Am. Ceram. Soc.*, **55**, 201 (1972).
51. K. Bingham and S. Parke, *Phys. Chem. Glasses*, **6**, 224 (1965).
52. S. Kumar and P. Nath, *Trans. Ind. Ceram. Soc.*, **25**, 12 (1966).
53. A. Bishay, *J. Am. Ceram. Soc.*, **42**, 403 (1959).
54. C. Bamford, *Phys. Chem. Glasses*, **1**, 159, 165 (1960); **2**, 163 (1961); **3**, 54, 189 (1962).
55. P. A. Young, *Solid State Phys.*, **4**, 93 (1971).
56. L. B. Zlatkin and E. K. Ivanov, *J. Phys. Chem. Solids*, **32**, 1733 (1971).
57. L. B. Zlatkin and Y. F. Markov, *Phys. Stat. Solidi*, **4a**, 391 (1971).
58. R. Zallen et al., *Phys. Rev. Letters*, **26**, 1564 (1971); *Solid State Commun.*, **10**, 293 (1972).

59. M. Faraday, *Phil. Mag.*, **14**, 401, 512 (1857).
60. S. D. Stookey, *J. Am. Ceram. Soc.*, **32**, 246 (1949).
61. R. H. Doremus, in *Nucleation and Crystallization in Glasses and Melts*, American Ceramic Society, Columbus, Ohio, 1967, p. 117.
62. G. Mie, *Ann. Physik.*, **25**, 377 (1908).
63. R. H. Doremus, *J. Chem. Phys.*, **40**, 2389 (1964).
64. R. H. Doremus, *J. Chem. Phys.*, **41**, 414 (1965).
65. A. Ram and S. N. Prasad, in *Advances in Glass Technology*, Plenum, New York, 1962, p. 256.

Author Index

In most cases only the first citation of an article in a particular chapter is included, together with the page on which article is listed.

Abdel-Latif, A. I. A., 200, 202, 209
Acloque, P., 223, 228
Adam, G., 111, 112, 114
Adams, P. B., 223, 227, 243, 252
Adams, R. V., 229, 250, 320, 330
Adamson, A., 220, 227
Adler, D., 177, 188
Aggarwal, J. D., 115, 120
Aglan, M. A., 327, 331
Ainslie, N. G., 94, 97
Alexander, G. B., 243, 252
Allen, R., 299, 309
Allgaier, R. S., 13, 21
Alpert, D., 124, 144
Altemose, V. O., 138, 145
Anderson, O. L., 142, 145, 281, 294, 300, 309
Andrade, E. N., 288, 295
Andreev, N. S., 73
Andrews, A. J., 240, 252
Angell, C. A., 12, 21, 39, 43, 159, 175, 319, 330
Antal, J. J., 217, 227
Anthony, S. R., 294, 295
Antonini, J. R., 223, 227, 228
Appen, A. A., 195, 208
Argon, A. S., 288, 295
Armistead, C. G., 216, 227
Armstrong, R. A., 220, 227
Ashmore, P. G., 223, 227
Aver'yanov, V. I., 50, 73

Babcock, C. L., 154, 175
Bachman, G. S., 199, 209
Bagdage, W., 325, 331
Bagley, B. G., 40, 43, 187, 189
Baker, T. C., 296, 309
Balijustin, A. A., 260, 276
Balk, P., 169, 176
Bamford, C., 327, 331
Barrer, R. M., 121, 124, 144
Barry, T. J., 82, 87, 96
Bartenev, G. M., 5, 7, 104, 113, 120
Barton, J. L., 141, 145, 174, 176, 232, 250
Basilo, M. R., 219, 227
Bastress, A. W., 232, 250
Bates, J. B., 320, 330
Bates, R. G., 268, 276
Bates, T., 327, 331
Beall, G. H., 69, 73
Bearley, W., 312, 315
Beauchamp, E. K., 123, 144
Becker, R., 60, 73
Beckett, J., 242, 252
Beckman, K. H., 231, 250
Begeal, D. R., 121, 144
Bell, R. J., 26, 41, 320, 330
Bell, T., 231, 250
Bellardo, A., 222, 227
Benes, P., 224, 228
Bennett, A. J., 323, 331
Bergeron, C. G., 15, 22, 91, 97
Bermudez, V. M., 216, 227
Bernal, J. D., 40, 43
Berry, J. P., 293, 295
Bestul, A. B., 53, 72, 115, 120
Betts, F., 37, 42
Bickerman, J. J., 291, 295
Bienenstock, A., 37, 42

Bingham, K., 327, 331
Biscoe, J., 27, 34, 41, 42
Bishay, A., 327, 331
Bishop, S. G., 34, 42
Blackburn, D. H., 51, 72
Blair, H. E., 187, 189
Blau, H. H., 153, 175
Block, S., 56, 72
Blodgett, K. B., 232, 252
Bockris, J. O., 109, 113, 156, 175
Boksay, Z., 162, 167, 175, 244, 252
Bolchakov, O. J., 138, 145
Bolz, L. H., 303, 309
Bondi, A., 111, 113
Booth, R. E., 311, 314
Borelli, N. F., 323, 325, 331
Borisovski, E. S., 162, 175
Borom, M. P., 170, 176, 241, 252
Bottom, A. E., 272, 277
Bouquet, G., 167, 175, 244, 252
Boyer, R. F., 12, 21
Boynton, B. L., 179, 188
Brace, W. F., 293, 295
Brady, G. W., 35, 42
Bray, P. J., 31, 33, 34, 41, 42
Bre, M. M., 213, 226
Brehler, B., 39, 43
Brennet, P., 247, 252
Bresker, R. J., 195, 208
Breton, J. C., 200, 209
Bristow, R. H., 153, 175
Brodsky, M. H., 13, 21, 37, 42, 180, 188
Brosset, C., 29, 41
Brown, D. M., 170, 176, 239, 251
Bruckner, R., 5, 7, 28, 41, 231, 250
Brunauer, S., 219, 227
Brund, M. J. D., 272, 277
Buck, R. P., 272, 277
Bueche, F., 111, 114
Buff, H., 3, 6
Bunting, E. R., 323, 331
Bunton, G. V., 185, 188
Burger, E. E., 242, 252
Burggraaf, A. J., 168, 175, 312, 313, 315
Burke, J. E., 285, 295, 298, 309
Burlitz, R. S., 124, 144
Burn, I., 134, 145
Burnett, D. G., 50, 51, 72, 90, 96
Burt, R., 133, 145, 146, 174
Burton, W. K., 95, 97
Cable, M., 238, 240, 251
Cabrera, N., 95, 97
Cahn, J. W., 67, 73, 93, 95, 97
Callaerts, R., 182, 188
Cameron, N. M., 289, 295, 299, 309
Cameron, R. D., 196, 208
Cant, N. W., 222, 227
Cargill, G. S., 40, 43
Carter, A. C., 155, 175
Cerofolini, G. F., 221, 227
Cervinka, L., 37, 42
Chapman, J. D., 29, 41, 51, 72, 219, 222, 227
Chandra, D., 236, 251
Chang, S. S., 115, 120
Charles, R. J., 50, 51, 56, 57, 67, 72, 73, 159, 175, 192, 208, 281, 294, 300, 307, 309
Chaudhari, P., 28, 37, 41
Chemisinov, V. P., 323, 331
Chen, B. T. K., 323, 331
Chenykh, V. J., 173, 176
Chia-Chih, Li, 203, 209
Chie-Yüeh, Lo, 203, 209
Chipman, J., 170, 176
Christ, C. L., 264, 276
Christian, J. W., 53, 58, 72, 73
Chubb, J. P., 294, 295
Clarke, A. R., 238, 251
Clayton, G. T., 31, 42
Clinton, D., 82, 87, 96
Coenen, M., 199, 208
Cohen, M. H., 12, 17, 18, 21, 22, 111, 114, 183, 188
Collins, F. C., 64, 73
Collins, F. M., 186, 189
Congleton, J., 293, 294, 295
Conti, F., 272, 276
Cooper, A. R., 170, 176, 240, 251
Cooper, B. S., 27, 41
Copland, L. E., 219, 227
Copley, G. J., 200, 209
Cormia, R. L., 15, 22, 91, 96, 105, 106, 113
Cornelissen, J., 168, 175, 313, 315
Cottrell, A. H., 53, 72
Covington, A. K., 272, 277
Coward, L. A., 186, 189
Crank, J., 129, 144, 150, 174
Crowell, A. D., 220, 227
Crozier, D., 325, 331

Cusumano, J. A., 218, 227
Curtis, H. L., 225, 228

Dalton, R. N., 77, 96
Dalton, R. W., 224, 228
Das, C. R., 116, 175, 244, 252
Datta, R. K., 12, 21
Davies, L. B., 41, 43
Davis, E. A., 183, 188
Davydov, V. Y., 214, 226
Day, D. E., 162, 175, 200, 201, 203, 209
Dean, P., 26, 41, 320, 330
DeBilly, M., 147, 174, 252, 255
DeBoer, J. A., 216, 227
DeCarlo, V. J., 298, 309
DeConinck, R., 182, 188
Deeg, E., 201, 209
Deitz, V. R., 215, 227
DeLuca, J. P., 15, 22, 91, 97
Denager, M., 182, 188
DeNee, P. B., 242, 252
DeNeufville, J. P., 37, 42
Denney, D. J., 18, 22
DeNordwall, H. J., 18, 21
Denton, E. P., 178, 188
Derge, G., 170, 176
Deribere-Desgardes, M. L., 213, 226
Devi, A., 222, 227
DeWaal, H., 200, 201, 209
Dewald, J. F., 182, 185, 188, 189
DeWet, W. J., 257, 276
DeWitte, D. L., 115, 120
Dick, E., 281, 294
Dietzel, A., 26, 41
Douglas, R. W., 6, 7, 50, 51, 67, 72, 73, 90, 96, 105, 106, 108, 113, 120, 158, 166, 175, 192, 198, 199, 208, 209, 229, 233, 236, 239, 244, 250, 251, 252, 313, 315, 320, 325, 327, 330, 331
Dobos, S., 167, 175, 244, 252
Dole, M., 272, 276, 277
Domenici, M., 29, 41
Doolittle, A. K., 111, 114
Dorda, G., 196, 208
Dorsey, J., 242, 252
Drury, T., 134, 145, 230, 250
Duke, P. J., 198, 208
Dulaney, E. N., 293, 295
Dumbaugh, W. H., 5, 7
Dumbgen, G., 135, 245
Duwez, P., 12, 21, 40, 43

Eagan, R. J., 15, 22, 91, 97
Eaton, D. L., 187, 189
Edmond, J. T., 325, 331
Ehrmann, P., 147, 174, 245, 252
Eisenberg, A., 36, 42
Eisenman, G., 167, 175, 244, 252, 255, 256, 264, 271, 272, 276
Eitel, W., 5, 7
Eldridge, J., 169, 176
Ellis, B., 6, 7, 12, 21
Elmer, T. H., 219, 227, 313, 315
El-Shamy, T. M., 166, 175
Endow, N., 222, 227
Erdogan, F., 293, 295
Ernsberger, F. M., 168, 175, 213, 226, 281, 288, 294, 295
Eschbach, H. L., 125, 137, 140, 144
Estropiev, K. L., 35, 42
Evans, D. L., 26, 41
Evans, E. J., 187, 189
Ebstrop'ev, K. K., 147, 162, 174, 175, 207, 209
Ewell, R. E., 111, 113
Eyring, H., 109, 111, 113

Fagen, E. A., 182, 188
Faile, S. P., 12, 21, 231, 250, 314, 315
Faraday, M., 3, 6, 225, 228, 329, 332
Farnum, E. H., 150, 175
Feldman, C., 13, 21, 185, 189
Ferguson, J. B., 121, 144
Ferrier, R. P., 13, 21
Ferry, J. D., 111, 112, 114
Field, J. E., 292, 295
Fincham, C. J. B., 236, 251
Finney, J. L., 40, 43
Firth, E. M., 237, 251
Fitch, A. H., 198, 208
Fitzgerald, J. V., 154, 175, 199, 209
Fitzpatrick, J. R., 37, 42
Fleming, J. W., 201, 209
Flinn, O. R., 259, 276
Flood, E. A., 219, 227
Florinskaya, V. A., 325, 331
Flory, P. J., 38, 42
Folman, M., 222, 227
Fontana, E. H., 15, 22, 106, 113
Forland, 56

Forry, K. E., 199, 209
Fournier, R. J., 244, 252
Fowler, R. H., 125, 144
Frank, F. C., 64, 73
Frank, R. C., 121, 144
Franz, H., 230, 235, 250
Fraser, D. B., 198, 208
Frechette, V. D., 306, 309
Frenkel, J., 127, 142, 144
Frerichs, R., 12, 21
Frisch, H. L., 64, 73
Frischat, G. H., 140, 145, 150, 154, 166, 169, 174, 176
Fritzsche, H., 181, 182, 186, 188, 189
Frohnsdorff, G. J. C., 219, 227
Fry, D. L., 123, 144
Fulcher, G. S., 111, 114
Fulrath, R. M., 127, 144, 241, 252, 314, 315

Gaffney, R. F., 140, 145, 237, 251
Gaidos, F. D., 241, 252
Gammon, R. W., 115, 120
Garfinkel, H. M., 154, 166, 168, 175, 258, 259, 274, 276, 277, 313, 315
Garino-Canina, V., 229, 230, 250, 320, 330
Garrels, R. M., 264, 276
Geddes, S., 225, 228
Gee, G., 36, 42, 109, 113
Gerding, M., 12, 21
Gerth, K., 53, 72, 90, 96
Ghezzo, M., 170, 176
Ghoshtagore, R. N., 169, 176
Gibbs, J. W., 58, 73, 93, 111, 112, 114
Gilman, J. J., 281, 282, 291, 294
Glasser, F. P., 48, 72
Gliemeroth, G., 311, 314
Gobin, P., 200, 209
Goldschmidt, I. V. M., 24, 41
Goldstein, M., 72, 73, 112, 113, 114, 118, 120, 160, 175, 196, 205, 208, 209
Golub, H. R., 44, 72
Gordon, J. E., 288, 295
Gottardi, V., 243, 252
Gottwald, B. A., 221, 227
Götz, J., 230, 250
Goubeau, J., 324, 331
Gough, E., 147, 174, 195, 208
Gould, E. S., 36, 42
Graham, P. W. L., 200, 209
Graham, W. A. G., 38, 43
Graczyk, J. F., 28, 37, 41, 42
Grechanik, L. A., 179, 188
Green, R. L., 195, 208, 232, 250
Greenberg, S. A., 243, 252
Greene, C. H., 140, 145, 237, 239, 251
Greenwood, G. W., 66, 73
Greet, R. J., 18, 22, 91, 97
Gregory, A. G., 90, 96
Griffith, A. A., 3, 6, 205, 295
Grundy, P. J., 41, 43
Guggenheim, E. A., 125, 144, 264, 276
Gulden, M. E., 300, 309, 313, 315
Guntersdorfer, M., 186, 189
Gupta, Y. P., 49, 72

Hagel, W. C., 135, 145, 147, 174
Haggerty, J. S., 118, 120
Hair, M. L., 29, 41, 51, 72, 214, 216, 222, 223, 226, 227
Hakim, R. M., 161, 175, 193, 208
Haller, W., 51, 53, 67, 68, 72, 73
Hamann, S. D., 160
Hambleton, F. H., 214, 216, 222, 226, 227
Hammel, J. J., 44, 60, 72, 73, 317
Hanna, R., 325, 331
Harding, F. L., 239, 251
Harrington, R. V., 1, 6
Hansen, K. W., 178, 188, 203, 209
Haroon, M. A., 238, 251
Harrick, N. J., 231, 251
Hass, M., 26, 41, 320, 330
Hasselman, D. P. H., 314, 315
Haul, R., 135, 145, 221, 227
Haven, Y., 153, 175
Heckman, R. W., 175, 232, 250
Helfferich, F., 165, 175, 253, 276
Helberg, J. N., 187, 189
Henderson, H., 170, 176
Hendren, J. K., 31, 42
Hensley, A. L., 216, 227
Herd, L. R., 28, 37, 41, 42
Herman, H., 68, 73
Herndon, J. R., 36, 42
Heroux, L., 193, 208
Herrell, D. J., 13, 21
Hertl, W., 216, 222, 223, 227
Hesse, W., 111, 114
Heston, W. M., 243, 252
Hetherington, G., 28, 41, 105, 106, 113, 168, 176, 230, 250, 320, 330

Heynes, M. S. R., 109, 113
Hiatt, G. D., 38, 43
Hibbins-Butler, D. C., 320, 330
Hicks, J. F. G., 28, 41
Hill, C. F., 39, 43
Hillig, W. B., 21, 84, 93, 95, 96, 97, 281, 283, 285, 294, 298, 300, 309
Hilsch, R., 13, 21
Hindly, N. K., 185, 186, 189
Hirsch, J., 178, 188
Hlavac, J., 240, 251
Hobson, J. P., 220, 227
Hochstrasser, G., 223, 227, 228
Hockey, J. A., 214, 216, 222, 226, 227
Hodgson, B. P., 311, 314
Hodkin, F. W., 237, 251
Hofmaier, G., 15, 22, 105, 106, 113
Hogfeldt, E., 256, 276
Hollabaugh, C. M., 168, 175
Holland, L., 213, 226, 243, 252
Holloway, D. G., 312, 315
Hood, H. P., 44, 72
Hopkins, J. L., 198, 208
Hopkins, T. E., 36, 42
Hopper, R. W., 72, 73
Horiuchi, S., 170, 176
Hosegawa, H., 284, 295
Hubbard, D., 226, 228
Hughes, K., 147, 174
Hull, A. W., 242, 252
Hu-Ming, Cheng, 203, 209
Hummel, F. A., 82, 96
Hunt, J. D., 93, 97
Hutchins, J. R., 1, 6

Ignatowicz, S., 156, 175
Iizima, S., 226, 228
Iler, R. K., 243, 252
Imaoka, M., 284, 295
Ing, S. W., 182, 188
Inglis, C. E., 286, 295
Inglis, G. B., 188
Irmann, F., 233, 250
Irwin, G. R., 291, 295
Isard, J. O., 147, 162, 166, 174, 175, 193, 194, 195, 208, 236, 251, 313, 315
Ivanov, A. O., 35, 42
Ivanov, E. K., 328, 331
Ivkin, E. B., 182, 188
Izumitani, J., 78, 96

Jack, K. H., 28, 41, 105, 106, 113, 168, 176, 230, 250, 320, 330
Jackson, K. A., 93, 95, 97
Jaeckel, R., 125, 137, 144
Jagdt, R., 201, 209
Jantsch, O., 22
James, P. F., 50, 72
Jellyman, P. E., 325, 331
Jenckel, E., 203, 209
Jensen, A. T., 222, 227
Joffe, V. A., 179, 188
Johansson, G., 272, 277
Johari, G. P., 196, 208
Johnson, H., 290, 295
Johnson, J., 133, 145
Johnson, J. R., 153, 175
Johnston, J. W., 283, 289, 294, 295, 299, 309
Johnston, W. D., 233, 250
Jones, G. O., 1, 6, 11, 21
Jonscher, A. K., 182, 188
Jorgensen, P. J., 124, 144
Jorgensen, S. S., 222, 227, 243, 252
Judge, J. S., 248, 252

Kano, R., 113, 114
Kanta, D. L., 219, 227
Karle, J., 27, 41
Karpechenko, V. G., 179, 188
Karreman, G., 271, 276
Karsch, K. H., 203, 209
Kats, A., 135, 145, 320, 330
Kauzmann, W., 114, 120
Kay, L. A., 87, 96
Keating, D. T., 37, 42
Kelen, J., 235, 251
Kellar, A., 95, 96, 97
Keller, H., 324, 331
Kennan, A. G., 272, 277
Kennedy, G. C., 244, 252
Kennedy, J. C., 105, 106, 113
Kennicott, P. R., 170, 176
Kepak, F., 224, 228
Kerkhof, F., 293, 295, 303, 309
Khar'yuzov, V. A., 147, 174
Kieth, H. D., 87, 96
Kindl, B., 221, 227
King, C. B., 168, 175
King, S. V., 26, 41
Kingery, W. D., 67, 73, 199, 209, 240, 251

Kinton, G. L., 219, 227
Kirby, E. M., 323, 331
Kirby, P. L., 200, 209
Kirchener, J. A., 156, 175
Kiselev, A. V., 214, 219, 226, 227
Kistler, S. S., 312, 315
Kitano, J., 140, 145, 237, 251
Kitchener, J. A., 109, 113
Klemm, W., 303, 309
Kleppa, O. J., 39, 43
Kolbeck, A. G., 115, 120
Kolomiets, B. T., 181, 182, 185, 188, 189
Konnert, J. H., 14, 27
Kooi, E., 196, 208
Kordes, E., 321, 323
Kornfeld, G., 256, 276
Kouznetzov, A Ya., 225, 228
Krause, J. T., 196, 203, 208, 209
Krebs, H., 36, 42
Kreider, K. G., 240, 251
Kriedl, N. J., 6, 7, 31, 42, 67, 73, 319, 330
Kristionsen, L. A., 324, 331
Kritz, H. M., 33, 42
Kroger, C., 235, 251
Krogh-Moe, J., 31, 34, 42, 80, 96, 324, 331
Krohn, O. A., 313, 315
Krutter, K., 24, 41
Kuchida, T., 34, 42
Kugler, G. C., 272, 277
Kuhl, C., 231, 251
Kumar, S., 327, 331
Kurkjian, C. R., 31, 41, 105, 106, 108, 113, 117, 120, 196, 198, 203, 208, 209, 327, 331

Laber, M. M., 185, 189
Laberge, N., 107, 110, 112, 114
Lacharme, J. P., 162, 175
LaCourse, W. C., 181, 188
Laing, K. M., 199, 209
Laird, J. A., 15, 22, 91, 97
Landel, R. F., 111, 112, 114
Lange, F. F., 314, 315
Langmuir, I., 220, 227
Laska, H. M., 124, 144
Laws, F. A., 191, 208
Lawson, A. W., 178, 188
Lay, L. A., 82, 87, 96
Leadbetter, A. J., 26, 35, 39, 41, 42, 43
Leamy, H. J., 95, 97
Lebedev, E. A., 182, 185, 188, 189
Lebland, M., 174, 176
Lecron, J. A., 67, 73, 199, 200, 209
Lee, H. A., 237, 251, 320, 330
Lee, R. W., 121, 144, 230, 251
Leeds, R. E., 232, 250
Lengyel, B., 162, 175
Leontewa, A., 15, 22, 91, 96
Levanthal, M., 31, 33, 41, 42
Levin, E. M., 47, 56, 72
Lewchuk, R. R., 44, 72
Li, P., 15, 22
Liebau, F., 29, 41
Lifshitz, E. M., 66, 73
Lifshitz, J. M., 189
Lillie, H. R., 104, 107, 108, 113
Ling, A. C., 109, 110, 113
Lippencott, E. R., 323, 331
Litovitz, T. A., 113, 114
Little, L. H., 214, 222, 226, 227
Littleton, J. T., 156, 175
Loh, E., 322, 330
Loopstra, L. H., 40, 43
Lotkova, E. N., 324, 331
Low, M. J. D., 217, 218, 222, 227
Lowengrub, M., 287, 295
Lummerzheim, D., 235, 251
Lumsden, J., 60, 73
Lundberg, M. H., 80, 96
Lyklema, J., 224, 228, 275, 277
Lyle, A. K., 239, 251
Lynch, G. T., 180, 188
Lyng, S., 80, 96

MacCrone, R. K., 68, 73
MacDowell, J. F., 69, 73
Macedo, P. B., 51, 53, 72, 107, 110, 112, 113, 114, 115, 120
MacGillavry, C. H., 40, 43
Mackenzie, J. D., 6, 7, 15, 22, 30, 42, 91, 96, 105, 106, 109, 113, 135, 145, 147, 174, 177, 181, 188
MacMillan, N. H., 281, 294
Mader, S. R., 13, 21, 40, 43
Maghrabi, C., 37, 42
Magill, J. D., 18, 22, 91, 97
Mahle, S. H., 196, 208
Maider, A., 15, 22, 91, 97
Majer, V., 224, 228
Maklad, M. S., 67, 73

Markov, Y .F., 328, 331
Male, J. C., 182, 188
Mandelkern, L., 95, 97
Mansingh, A., 195, 208
Marboe, E. C., 5, 7
Markali, J., 80, 95
Marsh, D. M., 281, 288, 294, 295
Martin, E. H., 178, 188
Martlew, D., 240, 251
Marx, J. W., 197, 208
Materova, E. A., 260, 276
Mattox, D. M., 241, 252
Matzke, H., 133, 145
Maurer, R. D., 77, 90, 96, 141, 145, 173, 176
Mautmun, R. W., 224, 228
Mazurin, O. V., 51, 72, 162, 175, 198, 207, 208, 209
McAfee, K. B., 124, 144
McClanahan, J. L., 224, 228
McCurrie, R. A., 67, 73
McDonald, R. S., 214, 218, 226, 227
McDuffie, G. E., 113, 114
McMillan, P. W., 50, 72, 75, 84, 96, 178, 188, 311, 314
McMurdie, H. F., 47, 72
McSwain, B. D., 323, 331
McVay, G. L., 150, 162, 175, 201, 209
Mears, P., 38, 43
Medona, R., 222, 227
Meier Zu Kocker, H., 236, 251
Meiling, G. S., 15, 22, 91, 96, 108, 112, 113
Meitzler, A. H., 198, 208
Mercer, R. A., 82, 87, 96
Merker, L., 115, 120
Metcalf, A. G., 300, 309, 313, 315
Meyer, B., 36, 42
Meyer, H., 235, 251
Mezard, R., 28, 41
Michal, E. J., 233, 250
Michener, J. W., 28, 41
Micus, G., 91, 97
Mie, G., 330, 332
Milberg, M. E., 29, 34, 41, 42, 323, 331
Miller, A. A., 109, 113
Miller, A. R., 325, 331
Miller, C. E., 186, 189
Miller, D., 125, 137, 144
Miller, R. A., 203, 209
Miller, R. P., 82, 87, 96
Milnes, G. C., 147, 174
Mishra, U. D., 49, 72
Mitchell, S. A., 216, 227
Mochel, E. L., 313, 315
Mohyuddin, J., 199, 208
Moore, H., 327, 331
Moorjami, K., 13, 21, 185, 189
Morain, M., 141, 145, 232, 250
More, D. G., 241, 252
Morelock, C. R., 94, 97
Morey, G. W., 2, 6, 104, 108, 113, 118, 120, 146, 174, 195, 208, 213, 226, 243, 252, 283, 294, 306, 309, 317
Morley, J. G., 91, 96
Morgan, A. M., 218, 227
Morgan, S. O., 226, 228
Moriya, Y., 50, 72
Morningstar, O., 24, 41
Morrow, B. A., 222, 227
Morterra, C., 222, 227
Mott, N. F., 183, 184, 188, 293, 295
Mould, R. E., 281, 285, 288, 294, 295, 296, 309, 311, 314
Moulson, A. K., 133, 145, 230, 250
Moynihan, C. T., 115, 120
Mozzi, R. L., 24, 31, 41, 42
Mukherje, S. P., 80, 96
Mulfinger, H. O., 125, 144, 235, 251
Munakata, M., 179, 188
Murthy, M. K., 34, 42, 323, 331
Muschelischivili, N. J., 287, 295
Musselin, M. J., 196, 208
Mydlar, M. F., 31, 42
Myerson, R. L., 203, 209

Nademlynska, H., 240, 251
Nagel, S. R., 91, 97
Nagels, P., 182, 188
Nakagawa, K., 78, 96
Nakoneczney, N., 118, 120
Nannoni, R., 196, 208
Napolitano, A., 53, 72, 112, 114, 118, 120
Narten, A. H., 27, 41
Nath, P., 327, 331
Naudin, F., 67, 68, 69, 73
Navez, M., 217, 227
Navias, L., 195, 208
Nazarova, T. F., 182, 188
Negri, E., 221, 227
Neilson, G. F., 68, 73

Nemer, L. N., 141, 145
Neumann, K., 15, 22, 91, 97
Newhart, J. H., 182, 188
Nicolsky, B. P., 260, 276
Niekerk, M., 257, 276
Nishihama, K., 284, 295
Nordberg, M. E., 44, 72, 219, 227, 247, 252, 313, 315
North, P., 233, 250
Northover, W. R., 185, 189
Norton, F. J., 121, 138, 144, 145
Notz, N., 272, 277
Nowich, A. S., 13, 21, 40, 43

Oakley, D. R., 200, 209
Oberlies, F., 26, 41
Obukhov-Denisov, V. V., 323, 324, 331
O'Connor, T. L., 243, 252
Oel, H. J., 141, 145, 147, 174
Ohlberg, S. M., 29, 41, 44, 72
O'Keefe, J. G., 33, 42
Olcott, J. S., 313, 315
Ordway, F., 26, 41
Otto, K., 34, 42
Ovshinsky, L. R., 37, 42, 185, 186, 187, 189
Owen, A. E., 146, 147, 158, 174, 181, 188, 192, 196, 208
Owen, G. D. T., 36, 42

Padden, F. J., 87, 96
Papanikolau, E., 173, 176
Paris, M., 170, 176
Park, P .J., 33, 42
Parke, S., 327, 331
Parker, A. J., 223, 227
Parks, G. A., 105, 106, 113
Parks, G. S., 117, 120
Parratt, M. E., 288, 295
Parsons, J. L., 323, 331
Parsons, J. M., 29, 41
Pask, J. A., 15, 22, 91, 96, 170, 176, 240, 252
Pasternak, R. A., 36, 42, 222, 227
Patridge, J. H., 240, 252
Patrina, J. B., 179, 188
Paul, A., 233, 250, 327, 331
Pauling, S., 20
Pearce, M. L., 235, 236, 251
Pearson, A. D., 185, 186, 188, 189
Peberovskaya, J. L., 179, 188
Pechenkina, R. S., 325, 331
Peck, W. F., 185, 189
Peck, W. Y., 182, 188
Peglar, R. J., 222, 227
Peri, J. B., 216, 222, 227
Perkins, W. G., 121, 144
Petch, N. J., 293, 295
Peter, K. W., 281, 294
Peters, C. R., 29, 41
Pethica, B. A., 216, 227
Petit, G. D., 37, 42
Pwtrovykh, N. V., 179, 188
Philipp, H. R., 322, 330
Phillips, A. V., 84, 96
Phillips, C. J., 281, 294
Pike, R. G., 226, 228
Pincus, A. G., 34, 42, 55, 72
Platts, D. R., 239, 251
Plazek, D. J., 18, 91, 97
Plesset, M. S., 165, 175
Plieth, K., 35, 42
Plumat, E., 53, 72
Plummer, W. A., 15, 22, 106, 113
Polk, D. E., 40, 43
Pollak, M., 182, 188
Pomeranz, D. J., 242, 252
Porai-Koshitz, E., 27, 36, 41, 42, 50, 73, 90, 96
Potts, J. C., 240, 251
Powell, R. E., 109, 113
Pozza, F., 29, 41
Prade, J. A., 11, 21
Prasad, S. N., 330, 332
Prebus, A. F., 28, 41
Preston, E., 236, 251
Preston, F. W., 296, 309
Prins, J. A., 1, 6
Priqueler, M., 230, 250
Proctor, B. A., 283, 289, 294, 295, 299, 309, 312, 314
Proctor, J. P., 325, 331

Quets, J. M., 222, 227

Ram, A., 330, 332
Ramasubramanin, N., 217, 227
Ramsay, M. W., 168, 176, 320, 330
Rana, M. A., 166, 175, 244, 252
Randall, J. F., 27, 41

Randall, J. T., 232, 250
Rawson, H., 5, 7, 12, 20, 21, 22, 32, 34, 39, 42, 43, 56, 72, 109, 113, 178, 179, 188
Ray, N. H., 248, 252, 313, 314
Rechnitz, G. A., 272, 277
Redfearn, M. W., 325, 331
Reilly, M. H., 323, 331
Reuber, E., 35, 42
Reugh, J. D., 117, 120
Reyes, J. M., 195, 208
Reynolds, M. B., 140, 145
Ricca, F., 222, 227
Rice, M. J., 172, 176
Richardson, F. D., 236, 251
Richter, H., 293, 295, 303, 309
Riebling, E. F., 35, 42
Rindone, G. E., 76, 96, 199, 200, 202, 203, 209, 317
Ritland, H. N., 118, 120
Ritter, J. E., 297, 309
Robbins, C. R., 47, 72
Roberts, G. J., 134, 145, 230, 250
Roberts, J. P., 133, 134, 145, 230, 250
Robinson, P. L., 12, 21, 27, 41
Rogers, W. A., 124, 144
Rogers, P. S., 80, 84, 90, 96
Roiler, M., 182, 188
Rooksby, H. R., 27, 41
Rosenblatt, G., 180, 188
Rotger, H., 199, 208, 209
Roth, L. M., 323, 331
Roth, W. L., 172, 176
Rothermel, D. L., 313, 315
Rothermund, V., 256, 276
Rowe, J. J., 244, 252
Rowell, M. H., 258, 276
Roy, D. M., 12, 20, 21, 22, 231, 250
Roy, R., 48, 72, 314, 315
Rudow, H., 237, 251
Ruffa, A. R., 323, 330
Rukhlyadev, Y. V., 181, 188
Ryden, W. D., 178, 188
Ryder, R. J., 199, 209, 239, 251

Sager, M., 195, 208
Sah, C. T., 169, 176
Sakaino, T., 51, 72
Sakka, S., 311, 314
Samaddar, B. S., 240, 251
Samis, C. S., 233, 250
Sarjeant, P. T., 20, 22
Sarkar, A., 51, 72
Sartain, C. C., 178, 188
Sawai, I., 80, 96
Savin, G. N., 287, 295
Sayre, E. J., 12, 21, 40, 43
Schaeffer, H. A., 147, 174
Schardin, H., 292, 295
Scherer, G., 92, 97
Schiller, H., 173, 176
Schmid, A. P., 180, 188
Schmidlin, F., 182, 188
Schmitz, G. K., 300, 309, 313, 315
Scholze, H., 5, 7, 125, 144, 195, 208, 213, 226, 317
Scholze, N., 230, 250
Schonert, K., 303, 309
Schuhmann, R., 233, 250
Schultz, M. M., 260, 276
Schultz, P. C., 5, 7
Schroder, N., 311, 314
Schwabe, K., 272, 277
Schwartz, M., 147, 174
Schwenker, R. O., 170, 176
Schwerdtfeger, K., 240, 251
Scott, W. D., 12, 15, 21, 22, 91, 96
Scroggie, B., 34, 42
Sears, G. W., 93, 95, 97
Sedden, E., 153, 175, 207, 209
Segel, L. L., 180, 188
Sella, C., 217, 227
Sello, H., 169, 176
Setkina, O. N., 325, 331
Seward, T. P., 69, 73
Sewell, P. A., 218, 227
Shackelford, J. F., 121, 127, 144
Shand, E. B., 5, 7, 281, 294
Shartsis, L., 53, 72
Shaw, R. R., 50, 72, 326, 331
Shelby, J. E., 121, 137, 144, 145, 201, 203, 209
Sherburne, C. L., 297, 309
Shermer, H. F., 53, 72
Shilo, V. P., 181, 188
Shimiza, M., 222, 227
Shnaus, V. E., 115, 120
Shore, A. C., 15, 22
Shteinberg, Y. G., 325, 331
Shultze, G., 3, 6, 149, 174, 254, 276

Siderov, J. A., 323, 331
Sidgwick, N. V., 247, 252
Sigel, G. H., 322, 330
Sigety, E. A., 327, 331
Silver, A. H., 31, 42
Simmons, J. H., 53, 72
Simon, J., 319, 330
Sivertsen, J. M., 197, 208
Skatulla, W., 326, 331
Slavyanskii, V. T., 238, 251
Slyozov, V. V., 66, 73
Smekal, A., 292, 295
Smetana, J., 224, 228
Smith, A. W., 222, 227
Smith, C. A., 34, 42
Smith, P. L., 134, 145
Sneddon, J. N., 287, 295
Sosman, R. B., 5, 7
Southwick, R. D., 288, 295, 296, 309
Sovolev, N. N., 323, 324, 331
Soyer, M., 180, 188
Spaght, M. E., 105, 106, 113
Spinner, S. S., 118, 120
Srinivasan, G., 51, 72
Stacy, M. H., 311, 314
Stanworth, J. E., 20, 22, 148, 174, 178, 179, 188
Stavely, L. A. K., 18, 21
Stearne, P. E., 223, 227
Steinkamp, W. E., 200, 209
Stern, K., 258, 259, 276
Stevels, J. M., 30, 39, 42, 43, 148, 174, 178, 188, 192, 208, 320, 330
Stober, W., 243, 252
Stock, A., 12, 21
Stocker, H. J., 186, 189
Stolen, R. H., 320, 325, 330, 331
Stone, F. G. A., 38, 43
Stookey, S. D., 74, 77, 80, 90, 96, 241, 252, 313, 315, 329, 332
Stranski, J. N., 35, 42
Streltsina, M. V., 51, 72
Strnad, Z., 236, 251
Stroud, J. S., 77, 96
Stuart, D. A., 142, 145, 300, 309
Studt, P. L., 121, 127, 144
Stuke, J., 180, 188
Su, G. J., 15, 22, 323, 325, 331
Sucov, E. W., 135, 145
Sugarman, B., 312, 315
Sun, K. H., 19, 20, 21, 22
Suschke, H. D., 272, 277
Swarts, E. L., 235, 251
Swets, D. E., 121, 144
Swift, H. R., 87, 91, 92, 96
Symmers, C., 312, 315
Szymanski, A., 185, 189

Tadros, T. F., 224, 228, 257, 275, 277
Takayangi, M., 15, 22
Tamann, G., 3, 6, 12, 21, 111, 114
Tanford, C., 38, 43
Taylor, H. E., 192, 208
Taylor, J. A. G., 214, 226
Taylor, J. D., 202, 209
Taylor, N. W., 134, 145
Terai, R., 153, 162, 175
Tegetmeier, F., 3, 6
Thege, W. G., 36, 42
Thiel, K., 12, 21
Thilo, E., 12, 21, 39, 43
Thomas, D. G., 18, 21
Thornton, H. K., 241, 252
Tien, H. T., 224, 228
Tien, T. Y., 82, 96
Tilton, L. W., 4, 27
Tippet, E. J., 153, 175, 207, 209
Title, R. S., 37, 42, 180, 188
Tobolsky, A. V., 36, 42
Tochon, J., 166, 175
Tolmacheva, T. A., 260, 276
Tomlinson, J. W., 109, 113, 156, 175
Tomozawa, M., 52, 68, 72, 73, 78, 96
Tool, A. Q., 118, 120
Tooley, F. V., 5, 7
Toop, G. E., 233, 250
Topping, J. A., 194, 195, 208
Totesh, A. S., 51, 72
Towers, H., 170, 176
Trap, H. J. L., 30, 42, 178, 188
Tremere, D. A., 169, 176
Tress, H. J., 234, 251
Truchlarova, M., 240, 251
Tsekhomskii, V. A., 207, 209
Tsien, L. C., 288, 295
Tsuei, C. C., 12, 21
Turkalo, A. M., 28, 41, 44, 45, 46, 69, 72, 78, 84, 96, 285, 295, 298, 309
Turkdogan, E. T., 236, 251
Turnbull, D., 13, 17, 21, 22, 40, 43, 58, 69,

73, 91. 94, 96, 97, 105, 106, 111, 113, 114
Turner, J. A., 327, 331
Turner, N. H., 215, 227
Turner, W. E. S., 3, 44, 72, 153, 175, 207, 209, 236, 237, 251
Turner, W. H., 320, 327, 330, 331
Tuttle, O. F., 12, 21
Twaddell, V. A., 181, 188
Tweer, H., 51, 72, 107, 110, 112, 114
Tyler, A. J., 214, 216, 226, 227

Uhlmann, D. R., 18, 22, 50, 69, 72, 73, 91, 92, 93, 94, 96, 97, 108, 112, 113, 115, 120, 161, 175, 193, 208, 326, 331
Umhauer, N., 303, 309
Urbain, G., 15, 22, 105, 106, 113
Urnes, S., 28, 39, 41, 43
Utsumi, Y., 311, 314

Viapolin, A. A., 36, 42
Valkenburg, A. V., 323, 331
Valenkov, N., 27, 41
Van Amerongen, C., 192, 208
Van Hook, A., 15, 22
Van Loan, P. R., 178, 188
Van Reenan, T. J., 257, 276
Van Wazer, J. R., 35, 42
Vargeime, V. V., 327, 331
Varner, J. R., 306, 309
Varshneya, A. K., 170, 176
Vaughan, D. E. W., 124, 144
Vaugin, L., 200, 209
Veasey, T. J., 90, 96
Veltri, R. D., 158, 175
Venderovitch, A. M., 173, 176
Vergano, P. J., 15, 22, 91, 96
Verkerk, B., 153, 175
Vermeer, J., 173, 176
Verprek, O., 240, 251
Vleeskens, J. M., 216, 227
Vogel, H., 111, 113
Vogel, W., 53, 72, 90, 96, 326, 331
Volf, W. B., 5, 7
Volger, J., 192, 208
Volmer, M., 15, 22, 58, 73, 91, 97
Vostrov, G. A., 138, 145
Vuilland, G. E., 12, 21
Vukcevich, M. R., 28, 41

Wagner, C., 66, 73, 135, 145
Wagstaff, F. E., 13, 14, 21, 22, 50, 51, 72, 91, 96
Wallis, G., 242, 252
Wallner, H., 292, 295
Walters, H., 243, 252
Walters, L. C., 123, 144
Walton, H. F., 256, 276
Warbury, E., 3, 6, 146, 174
Ward, J. B., 312, 315
Warren, B. E., 3, 7, 24, 28, 39, 41, 42, 43, 55, 72
Warrington, D., 50, 72
Warshaw, J., 48, 72
Weaver, E. A., 232, 250
Weber, A. H., 217, 227
Weber, N., 160, 175, 205, 209
Weinberg, T. J., 327, 331
Weir, C. E., 323, 331
Weiser, K., 37, 42
Weiss, A., 12, 21
Wessel, H., 326, 331
Westergaard, H. M., 287, 295
Westman, A. E. R., 35, 42
Weyl, W. A., 5, 7, 219, 237, 251, 317, 330
White, J. T., 109, 113
Whitney, J., 283, 289, 294, 295, 299, 309
Wiecker, C., 12, 21, 39, 43
Wiecker, W., 12, 21, 39, 43
Wiederhorn, S. M., 281, 289, 294, 295, 300, 303, 306, 309
Wikby, A., 272, 277
Wilcox, F. L., 272, 277
Willard, J. E., 109, 110, 113
Williams, E. L., 135, 145, 175, 232, 250
Williams, F., 188
Williams, G. A., 121, 144
Williams, M. L., 111, 112, 114
Williamson, J., 80, 94, 96, 97
Wilmot, G., 325, 331
Wilson, C. G., 154
Wilson, H. A., 93, 97
Winchell, P., 148, 170, 174, 176, 235, 251
Winding, C. C., 38, 43
Winklemann, H., 327, 331
Winks, F., 44, 72
Winter, A., 12, 21
Wollast, B., 247, 252
Wong, J., 170, 176, 319, 330

Woods, K. N., 121, 144
Worster, E., 323, 331
Wright, A. C., 35, 39, 42, 43
Wüstner, H. W., 121, 144

Yager, W. A., 226, 228
Yamaguchi, J., 170, 176
Yamane, M., 51, 72
Yang, L., 170, 176
Yannas, J., 11, 21
Young, D. M., 220, 227
Young, G. J., 214, 226
Young, L., 13, 21, 36, 42
Young, P. A., 328, 331

Zachariasen, W. H., 3, 7, 18, 19, 21, 22, 24, 26, 28, 41
Zagar, L., 173, 176
Zaman, M. S., 239, 251
Zarson, D. C., 185, 189
Zarzycki, J., 28, 35, 39, 41, 42, 43, 67, 68, 73, 69, 147, 174, 255, 252
Zeise, H., 220, 227
Zellen, R., 328, 331
Zener, C., 64, 73, 143, 145, 204, 209
Zhuravlev, L. T., 214, 226
Zlatkin, L. B., 328, 331

Subject Index

Acidic or basic oxides, 233
Acids, dissolution in, 247
Adherence oxides, 241
Adhesion, 213, 240
Adsorption, chemical, 222
  physical, 218
  and potentials of glass electrodes, 274
  from solution, 224
Agglomeration, in phase separation, 68
Aluminum oxide and, anionic size, 266
  chemical durability, 245
  hydroxyl groups, 231
  ion exchange selectivity, 267
  optical absorption, 320
  static fatigue, 305
  water solubility, 230
Alkali silicates, compositions, 102
  dielectric loss, 194
  dissolution in water, 245
  ion exchange selectivity, 258, 273
  ionic diffusion in, 161, 163
  mechanical loss, 201, 207
  optical absorption, 325
  phase separation, 49
  properties, 102
  structure, 30
  surface structure, 216
  viscosities, 103, 108
Annealing, 117
Annealing point, 101

Bases, dissolution in, 246
Bonding, 23
  in dissolution in water, 243
  glass strength, 283
  in oxidation, 233
Books, 5
Borate glasses, density, 115
  heat capacity, 116
  molecular diffusion in, 137
  optical absorption, 323, 326
  phase separation, 50
  structure, 31
  viscosity, 107
Boron oxide anomaly, 33
Borosilicates, chemical durability, 51, 245
  compositions, 102
  crack propagation, 305
  diffusion in, 138, 168
  ion exchange selectivity, 258, 274
  leaching, 217
  mechanical loss, 207
  molecular solubility, 124
  phase separation, 51, 217
  properties, 102
  static fatigue, 298
  surface adsorption, 221
  viscosity, 103
Boroxyl groups, 31, 32
Bubbles, diffusion from, 141
  in fining, 237
  in strength studies, 289

Carbon dioxide, in fining, 238
  reaction with, glass, 235
    glass surface, 223
Chain structure, 35
Chalcogenide glasses, electronic conduction
    in, 178, 180, 186
  optical absorption, 328
  structure, 35
Chemical durability, 242
  crack propagation, 306
Chemical reactions, on glass surface, 222

with water, 134, 229
Clearing temperature, 50
Coarsening of a phase, 66, 68
Color and sulfates, 239
metallic particles, 329
transition metal ions, 328
Compositions, commercial, 4, 102, 321
Conductivity, electrical, 146
and diffusion, 152, 157
in high fields, 173
measurement, 148
surface, 225
water and, 149, 150
Coordination numbers, 24, 29, 33
of transition metal ions, 328
Consolute temperature, 47, 50, 56
Correlation in diffusion, 153
Cracks, branching, 291, 294
in fracture, 286
propagation of, 291, 301, 303
revealed with sodium vapor, 288
stress at, 287
velocity of, 293, 305
Crystal morphology, 87
Crystallite structure, 27
Crystallization, 13, 74
and heat flow, 90
and mechanical loss, 203
nucleation of, 75
and strengthening, 313
surface, 74, 313
theories of, 93
velocity, 15, 90, 92
Crystal structure, 30

Definitions, 1
Delayed failure, *see* Static fatigue
Density, 102, 115
Dielectric constant, 195
Dielectric loss, 117, 190
effect of frequency, 192
measurement, 190
relation with electrical conductivity, 194
theories, 204
Diffusion, and chemical reaction, 134
and electrical conductivity, 152
of gases, 128, 131
in ion exchange, 244
inter ionic, 164
molecular, 131
molecular, 131
in phase growth, 64
theories, 142, 170
Dissolution, in melts, 240
in water, 242
Distribution coefficients, in ion exchange, 256, 262
between two phases, 260, 274

E Glass composition, 289, 299
static fatigue, 299
strength, 289
Einstein Equation, 153
Elastic modulus, and crack propagation, 293
and molecular diffusion, 142
and strength, 283
Electronic conduction, 177
theories, 183

Fictive temperature, 118
Fining, 237
Fire-polishing, 311
Flaws, surface, in fracture, 284
Fracture, 281
and crack length, 288
flaw theories, 284
Free volume, 111, 125
Fused salts, ion exchange with, 258

Gases, reaction with glass, 229
solubility in glass, 121
Germanate glasses, infrared absorption, 323
mechanical loss, 203
molecular diffusion in, 137
structure, 35
viscosity, 105
Glass-ceramics, 74
Gold particles, diffusion to, 141
optical absorption, 329

Growth rates of, crystals, 15, 90, 92
phases, 64

Hackle, 291, 294
Hall mobility, 182
Halogens, on glass surface, 222
Heat capacity, 116
History, 2
Hopping conduction, 183
Hydrated layer, 167, 244

Hydrofluoric acid, dissolution of glass in, 247
reaction with glass surface, 222
Hydrogen, diffusion, 133, 141
reaction with glass, 141, 231
Hydrogen ions, 147
as an hydroxyl group, 232
in interionic diffusion, 166, 168
and mechanical loss, 200
from sulfur dioxide, 236
Hydroxyl groups, acid ionization constants, 223
in diffusion, 134
on glass surface, 214
from hydrogen, 231
infrared, spectra, 214
from water, 230

Immiscibility, *see* Phase separation
Infrared spectra, 214, 218, 320
Internal friction, *see* Mechanical loss
Ion exchange, 253
in dissolution, 245
equilibrium, 256
at glass surface, 217
and mechanical loss, 202
selectivity, 256, 262
and strengthening, 312
with two phases, 260, 274
with water, 244
Ionic distribution and electric conductivity, 172
in alkali silicate glasses, 29
between melts and glass, 253, 256
in vitreous silica, 28
Ionic salts, 38
viscosity, 110

Light scattering, 317
and phase separation, 50
Luminescence, 317

Mechanical loss, alkali ion peak, 198
and hydrogen ions, 200
effect of, composition, 201
frequency and temperature, 197
intermediate peak, 199
measurement, 196
theories, 204
Melting rate, 90, 240
Memory state in electronic conduction, 186
Metallic glasses, 40
Metallic particles, in nucleation of crystals, 75
optical absorption, 329
Mixed-Alkali effect, in electrical conductivity, 162
in mechanical loss, 201
Morphology, of crystals, 87
of separated phases, 67
Mossbauer spectroscopy, 31

Nernst-Einstein equation, *see* Einstein equation
Nernst-Planck equations, 164, 271
Nitrogen, reaction with glass, 235
Nuclear magnetic resonance, 31, 33
Nucleating agents, in crystallization, 80
impurity particles, 77
mechanisms, 81
metal particles, 76
oxides, 80
in phase separation, 51
Nucleation in glass formation, 16
in crystallization, 75
by phase separation, 82
mechanisms, 84
in phase separation, 60
theory, 57
transient, 62

Optical spectra, 30, 214, 319
Organic liquids, 18, 40
viscosity, 110
Organic polymers, 38
crystallization of, 95
Oxidation state, of glass, 232
of ions, 234
Oxygen, diffusion, 133
reaction with, glass, 135, 232
glass surface, 223
Oxygen ions, 233

Permeation of gases in glass, effect of composition, 136, 140
measurement, 128
Phase diagrams, 47
Phase separation, 44
and chemical durability, 245
free energy, 54

and ion exchange, 260
nucleation of, 60
relation to valence, 48
theories, 53
Phosphate glasses, electronic conduction in, 179
mechanical loss, 203
structure, 34
ultra-violet absorption, 323
viscosity, 105
Potential of glass electrode, and ion exchange equilibrium, 255
measurement, 268, 272
and surface adsorption, 274
theories, 269, 272
Proof test in static fatigue, 307
Pyrex glass, *see* Borosilicates
composition, 102

Random network, 24
Refractive index, 317
Regular solution model, 54
and ion exchange selectivity, 264

Sealing, 240
Selectivity, and anionic size, 266
in ion exchange, 256, 262
sequences, 263
theories, 264
Selenium, structure of, 36
Semiconduction, *see* Electronic conduction
Silica, vitreous, adsorption from solution, 224
bond angles in, 26, 27
chemical adsorption, 222
compositions, 321
crack propagation, 305
crystallization, 14
dielectric loss, 196
dissolution in, melts, 240
water, 242
electrical conductivity, 158
entropy, 28
gas solubility in, 122
ion exchange selectivity, 258, 273, 276
ionic diffusion in, 157, 168, 170
manufacturers, 321
mechanical loss, 203, 207
molecular diffusion in, 131, 133
optical adsorption, 320, 322
physical adsorption, 218
reaction with hydrogen, 231
solubility of water in, 230
static fatigue, 299
strength, 284, 289
structure, 24
surface structure, 213
viscosity, 105
water in, 135
Size, anionic, and exchange selectivity, 266
ionic, and diffusion, 168, 172
molecular, and diffusion, 131, 143
Soda-Lime glass, composition, 102
crack propagation, 293, 304
density, 115
dielectric loss, 194
electrical conductivity, 155
ion exchange selectivity, 258
ionic diffusion in, 159, 163, 165
mechanical loss, 207
phase separation in, 50, 60
properties, 102
static fatigue, 297
strength, 288
viscosity, 103, 107
Softening point, 101
Solubility, carbon dioxide, 235
of gas molecules, 121
of glass in water, 242
hydrogen, 231
theories, 125
water, 230
Spherulites, 87
Spinodal decomposition, 67
Static fatigue, and crack propagation, 303
experimental results, 297, 303
proof tests, 307
and surface treatment, 298
theories, 300
Stokes-Einstein equation, 67, 156
Strength, 281
distribution of, 285
theoretical, 281
Strengthening, 310
Stress, accelerated reactions, 300
at crack tips, 287, 301
and ionic diffusion, 149, 159
and molecular diffusion, 134
relaxation, 117, 119
and sealing, 241
Stress-intensity factors, 291, 306

Sulfur, structure of, 36
Sulfur dioxide, in fining, 239
  reaction with glass, 236
Surface conductivity, 225
Surface energy, 213
  and strength, 290
Surface structure, 213
Switching in electronic conductors, 185

Thermal expansion, 101, 102, 115
  and sealing, 241
Thirsty glass, 44, 217
Transition, glass, 99
  and electrical conductivity, 155, 161
Transition metal ions, coordination numbers, 328
  glass-forming tendency, 19
  optical absorption, 327
  oxidation state, 234

Uses of glass, 4, 102

Viscosity, 101
  definition, 101
  measurement, 103
  theories, 110
Vycor glass, 44, 217

Wallner lines, 292
Water, diffusion, 135
  effect on electrical conductivity, 149, 150
  reaction with glass, 134, 229, 242
  solubility in glass, 121, 134
Weathering, 167, 248
Working point, 101

X-Ray diffraction, 24, 25, 29, 31, 36, 37
  small-angle, 68

Zachariasen's Rules for glass formation, 19